Springer Series on Atomic, Optical, and Plasma Physics

Volume 80

AF181671

For further volumes:
http://www.springer.com/series/411

The Springer Series on Atomic, Optical, and Plasma Physics covers in a comprehensive manner theory and experiment in the entire field of atoms and molecules and their interaction with electromagnetic radiation. Books in the series provide a rich source of new ideas and techniques with wide applications in fields such as chemistry, materials science, astrophysics, surface science, plasma technology, advanced optics, aeronomy, and engineering. Laser physics is a particular connecting theme that has provided much of the continuing impetus for new developments in the field, such as quantum computation and Bose–Einstein condensation. The purpose of the series is to cover the gap between standard undergraduate textbooks and the research literature with emphasis on the fundamental ideas, methods, techniques, and results in the field.

Andrey V. Korol · Andrey V. Solov'yov

Polarization Bremsstrahlung

Springer

Andrey V. Korol
Andrey V. Solov'yov
Frankfurt Institute for Advanced Studies
Goethe Universität Frankfurt
Frankfurt
Germany

ISSN 1615-5653 ISSN 2197-6791 (electronic)
ISBN 978-3-662-50695-0 ISBN 978-3-642-45224-6 (eBook)
DOI 10.1007/978-3-642-45224-6
Springer Heidelberg New York Dordrecht London

Printed on acid-free paper

Springer is part of Springer Science+Business Media (www.springer.com)

Preface

In this book we present the recent advances in the theory of Polarizational Bremsstrahlung (PBrS) of structural particles (atoms, ions, clusters, etc.) as well as of the related phenomena occurring in collisional processes. The first theoretical predictions and experimental evidence of the PBrS phenomenon were made in the 1970–1980s. More reliable quantitative description of this type of radiation has become possible in the last two decades due to the rapid development of computational facilities. Within this period a number of theoretical, numerical, and experimental studies have been performed aimed at better understanding of the process. In this book we review the results of this research as well as of the earlier works in which the key ideas were formulated.

The subject is highly relevant since the phenomena considered in the book have a number of possible applications. Thus, the study of various radiative mechanisms can lead to creation of novel light sources operating in various parts of the electromagnetic spectrum. Analyzing spectral and angular distribution of the photon emission, one can extract information on the interaction of the colliding particles, and on their structure and dynamic internal properties. Accurate quantitative description of the radiative mechanisms allows one to determine radiative energy losses of particles in various media, which is needed, for example, for the plasma diagnostics, astrophysical applications, and medical applications which involve radiation therapy, etc.

The phenomena considered are of fundamental character and of general physical interest.

The material is arranged as follows. In the introductory chapter we describe qualitatively the typical features of PBrS, introduce main concepts and terminology, and present a review of the theoretical and experimental activity in the field. In Chap. 2 we outline the general quantum formalism which is used for theoretical quantitative analysis of the BrS radiation emitted in non-relativistic collisions of charged projectiles with many-electron targets. Specific features of PBrS in collisions with atomic hydrogen (or a hydrogen-like ion) are discussed in Chap. 3. In Chap. 4 we present the results of numerical calculation of spectral and angular distribution of BrS in electron collisions with many-electron atoms. The comparison is carried out on various approximations used for the calculations. BrS emitted in non-relativistic collisions of structural particles (atoms, ions) is discussed in detail in Chap. 5. The peculiarities in the BrS process due to relativistic motion of the projectile particle and/or atomic electrons are analyzed in Chap. 6. Chapter 7 is devoted to PBrS and the related phenomena in electron collisions with metallic clusters and fullerenes.

The atomic system of units, Inline Equation $\hbar = m_e = e = 1$, is used through the book except for Sects. 6.3 and 6.5, where the relativistic system $\hbar = c = 1$ is applied.

We express our gratitude to M. Ya. Amusia, N. B. Avdonina, A. S. Baltenkov, L. V. Chernysheva, J.-P. Connerade, F. J. Currell, L. G. Gerchikov, Walter Greiner, G. F. Gribakin, A. N. Ipatov, M. Yu. Kuchiev, A. G. Lyalin, O. I. Obolensky, R. H. Pratt, S. A. Sheinerman, A. S. Shulakov, and I. A. Solov'yov, together with whom many of the presented theoretical and numerical results were obtained.

We are grateful to V. M. Buimistrov, D. A. Varshalovich, E. T. Verkhovtseva, Yu. N. Demkov, V. K. Ivanov, C. A. Quarles, M. G. Kozlov, O. V. Konstantinov, N. L. Manakov, V. M. Mikushkin, V. N. Ostrovsky, N. A. Cherepkov, V. R. Shaginyan, D. G. Yakovlev, V. L. Yakhontov for the interest in our research and for stimulating discussions.

Financial support from the Russian Academy of Science, Russian Foundation for Basic Research, the International Science and Technology Center, the International Science Foundation, the Alexander von Humboldt Foundation, the Royal Society of London, the Royal Society of Sweden, Volkswagen Foundation, INTAS, European Commission, granted to us at various stages of the research, is gratefully acknowledged.

Frankfurt, September 2013

Andrey V. Korol
Andrey V. Solov'yov

Contents

Acronyms

AC	Atomic Cluster
BA	Born Approximation
BrS	Bremsstrahlung
DPWA	Distorted Partial Wave Approximation
DR	Dielectronic Recombination
HF	Hartree–Fock
OBrS	Ordinary Bremsstrahlung
PBrS	Polarizational Bremsstrahlung
PR	Polarizational Recombination
PRA	Plasmon Resonance Approximation
QM	Quasi-Molecular
RPA	Random Phase Approximation
RPAE	Random Phase Approximation with Exchange
RR	Radiative Recombination

Chapter 1
Introduction

In this monograph we consider a broad range of phenomena which appear in the process of bremsstrahlung (BrS) due to many-particle internal dynamic of the colliding particles. According to modern concepts, there are only few "elementary" particles which are regarded as structureless and point-like ones, while most of the objects in Nature have an internal structure. Clusters, molecules, atoms, atomic nuclei, and even nucleons are the examples of structural particles. It is hardly possible to describe the dynamic properties of diverse range of particles within the framework of a single approach due to essentially different nature of structural objects. Therefore, in this book we discuss in detail only a limited class of structural objects.

We focus on the collective effects appearing in the BrS process with atoms, ions and atomic clusters.

A distinctive feature of such systems is the existence of collective dynamic modes which involve all (or most of) particles of the system into a correlated motion. Such collective excited states were initially predicted for protons and neutrons in nuclei [164, 318] before their experimental observation [77], and later were called giant dipole resonances (see, e.g., [127]). Other examples of collective modes in finite systems include plasmon-type oscillations of electrons in many-electron systems: atoms, clusters, and fullerenes.[1] The collective excitations appear, in particular, as maxima in the photoabsorption cross sections for nuclei [77, 127], atoms [8], metallic clusters [93], and fullerenes [83].

The presence of the collective modes in a many-body systems often leads to the fact that the properties of the system can not be properly described within the framework of single-particle models. An adequate description of the behavior of many-body systems can only be achieved by means of consistent many-body approaches which take into account the correlation interaction between the particles in the system.

Another important feature of a many-body system is in the possibility of a coherent interaction with an external field. It appears when the characteristic length of the interaction exceeds the average distance between particles in the system

[1] See [222] for a comparative analysis of similarities and differences between collective excitations in various systems, as well as for the historical survey.

A. V. Korol and A. V. Solov'yov, *Polarization Bremsstrahlung*,
Springer Series on Atomic, Optical, and Plasma Physics 80,
DOI: 10.1007/978-3-642-45224-6_1, © Springer-Verlag Berlin Heidelberg 2014

[81, 279, 320]. This feature appears in many radiative and scattering processes involving many-particle systems.

In collisions involving structured particles, the conditions often occur for various interference and diffraction phenomena. These manifests themselves in the form of various peculiarities (minima, maxima, oscillations, cusp etc.) in the cross sections. These features, which can be detected experimentally, contain information about the characteristics of the system and the process.

The aforementioned properties of many-body systems are very general and fundamental. However, for a number of systems and processes, these properties have been understood and described relatively recently. An accurate quantitative analysis of complex many-body processes became possible only in recent years due to a rapid development of computer technology and experimental hardware. Description of recent achievements in one of the directions in the field of physics of many-body systems is the aim this monograph.

The book discusses the radiative processes occurring in collisions involving structural particles. A variety of the collision processes can be mentioned as the examples. These include binary collisions of atoms, ions, clusters, as well as collisions of "elementary" particles, electrons, positrons, mesons, protons, etc. with atomic, ionic and clusters targets. In the latter case, the structure of a charged projectile, even if there is one, is irrelevant in the collisional and radiative processes which occur on the atomic scale. The problem of radiative collisions is rather common and diverse. The variety of types of the colliding particles (structural and structureless, charged and neutral), of the interactions between them (static Coulomb, retarded electromagnetic, nuclear, weak), as well as the different kinematic conditions for the photon emission, all this makes the problem of radiative collisions quite general and diverse. Studying the radiation process one can extract important information on the interaction of the colliding particles, their structure, intrinsic properties. Quantitative analysis and comparison of various radiative channels allows one to determine the radiative losses of particles, which are needed, for example, in plasma diagnostics, in astrophysical applications. Therefore, the study of different mechanisms of radiation is of current importance.

Bremsstrahlung (BrS) is, perhaps, most known radiative collisional process, in which a charged projectile, moving in an external static field of a target (nucleus, ion, atom, etc.), emits the radiation of the continuous spectrum. The basic description of BrS, both classical and quantum, can be found in textbooks (see, e.g., [81, 192, 214, 278]). For the first time, the emission process accompanying the deceleration of the charged projectiles was observed by Röntgen in 1895. The term "bremsstrahlung" was proposed later by Sommerfeld, who explained the observed X-ray spectrum as the electromagnetic emission by stopping of cathode-ray electrons in the anode medium [396].

In the X-ray range, the BrS appears in collisions of charged particles with atoms. Initially, it was commonly acknowledged that the problem of the BrS emission in a charged-projectile–atom collision can be solved in a static approximation, when the atom is treated as a source of a screened Coulomb field acting on the projectile. Below in the book this process is called "ordinary BrS" (OBrS).

Omitting a discussion of the OBrS process within the framework of classical electrodynamics (see, e.g., [214]), we note that first quantum-mechanical description of OBrS was given in the papers by Bethe and Heitler [84], Sommerfeld [396, 396], and Sauter [371, 372, 374] (a thorough historical survey one finds in [347–350]).

From the first observation till now the OBrS process has been studied, theoretically and experimentally, in wide ranges of the incident energies of various projectiles, of the emitted photons, for different emission geometries, and for a number of atomic and ionic targets. Review of experimental and theoretical results obtained in the field one finds in [7, 9, 40, 189, 233–235, 288, 327, 347–351, 358, 404, 414]. The spectral [352] and spectral-angular distributions [230] of OBrS are tabulated over wide ranges of energies of projectile electron and emitted photon and for a number of atomic targets.

Not pretending to overview in full all the variety of theoretical approaches which have been used to describe the OBrS process, we mention that these range from the first order plane-wave Born approximation (both non-relativistic and relativistic, with and without the screening effect taken into account, see e.g. [81]) to more sophisticated ones. The latter include the corrections due to the Elwert factor and its modifications [70, 73, 131, 281], the use of Sommerfeld-Maue functions [81, 131, 189, 281], the approaches based on the use of classical [227, 334] and semi–classical [234] theory. Perhaps the best available theory for the OBrS cross sections (these include the spectral distribution, the spectral-angular distribution and the triply differential cross section which accounts for the angular distribution of the scattered particle) is based on the use of the distorted partial-wave expansion (DPWA) of the projectile wavefunction. This scheme has been applied to study the OBrS process of non-relativistic projectiles in the dipole-photon approximation [405, 406, 445, 446]. The most adequate description of the process has been obtained by applying the relativistic DPWA and using the multipole expansion for the projectile-photon interaction operator [133, 223, 281, 380, 381, 407–412, 443]. Despite the difficulties of numerical nature (see the discussion in [223, 381]), the relativistic DPWA and multipole series method produces best results for the single, double and triply differential OBrS cross sections over broad range of incident energies (hundreds eV—tens MeV for a projectile electron/positron), and for all atomic targets [380].

The concept of BrS as of the process of a photon emission by a charge braking in a fixed static field (electric, magnetic, nuclear) has become firmly established in the textbooks (see, e.g., [76, 81, 174, 176, 192]). However, in [14, 16, 99–101, 166, 343, 432, 454, 455] it was demonstrated, that along with the OBrS process, there is another mechanism of the photon emission, which is as follows. A charged projectile particle, interacting with the target electrons, induced a dipole moment on the target atom. The magnitude and the direction of the moment alter in the course of the collision leading thus to the radiation emission. Following [414] we call the radiation generated via this mechanism "polarizational" BrS (PBrS).

The OBrS and PBrS mechanisms are illustrated by Fig. 1.1. We note that this simplified schematic picture adequately illustrates the processes in the case of fast non-relativistic collisions. The peculiar features of slow and relativistic collisions will be discussed below.

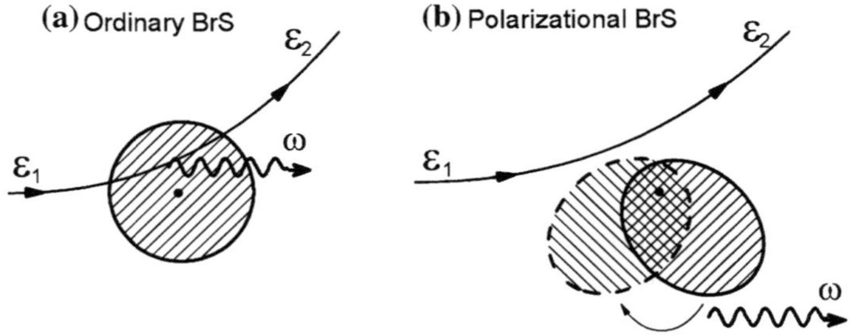

Fig. 1.1 Schematic representation of the ordinary and polarizational BrS processes. Ordinary BrS is the photon emission of a charged projectile accelerated in the static field of the target. Polarizational BrS mechanism considers the photon emission of the target electrons, virtually excited by the projectile. Virtual excitation of the electrons is equivalent to polarization of the target

The magnitude of the moment induced on the target atom during the collision, and, consequently, the intensity of PBrS, are determined by intensity of the electric field of a projectile particle, and virtually are not dependent on its mass [6, 33, 34, 98]. As a result, in collisions, for example, of an electron and a proton (of the same velocity) with an atom the intensities of PBrS are comparable in a broad range of photon energies ω [98], whereas the OBrS intensity of a proton is $(m_p/m_e)^2 \sim 10^6$ times less than that of an electron [81]. Another important feature to be mentioned, is that it is exactly the polarizational mechanism which gives rise to the BrS radiation in collision of neutral particles, for example, atoms [33, 34]. In this case, the qualitative description of the PBrS process is as follows. The colliding structural particles experience reciprocal polarization, so that each acquires the induced dipole momentum. The total dipole moment of the system, equal to the sum of the two induced momenta, alters in the course of the collision resulting in the photon emission. We stress, that in the system of two (or more) structural particles, the PBrS mechanism dominates in the radiation spectrum in a broad range of ω [33, 34].

Below in the book we discuss in detail the qualitative and quantitative characteristics of PBrS for various types of colliding particles and the different collision velocities, describe the methods for analytical and numerical calculation of the radiation spectrum, and in some cases carry out the comparison with the available experimental data. In the introductory section we will only mention the most prominent properties and feature of PBrS.

Percival and Seaton [341] were, apparently, the first who demonstrated the important role of electronic degrees of freedom in forming the spectrum of radiation in electron–atom collisions. They showed, that for the photon energies ω close to the transition energies $\omega_{n0} = \varepsilon_n - \varepsilon_0$ in the target atom, the BrS process acquires a resonant character: the incident electron excites the atom into the discrete state $\varepsilon_n = \varepsilon_0 + \omega$ which, then, relaxes via the photon emission. Thus, the total BrS spectrum will contain a series of narrow resonances over smooth OBrS background.

Some time later, Hubbard and Rose [197] considered similar process but with respect to the resonance transition in the atomic nucleus. They termed this mechanism of the photon emission as "nuclear" BrS and showed, that if the "nuclear" BrS is sufficiently large compared to the OBrS, then "...one may expect to find peaks, due to nuclear bremsstrahlung and to the interference effects, superimposed on the background of electron bremsstrahlung" ([197], p. 340).

Of course, due to the difference in the electron and nucleon masses, the intensity of the BrS radiation arising from atomic resonances will exceed that from nuclear resonances by orders of magnitude. Nevertheless, both of the mentioned radiative channels clearly indicate that, in some cases, the internal dynamic structure of the target must be taken into account when constructing the BrS spectrum.

In the 1970s in the papers [14, 16, 98–101, 166, 343, 432, 454, 455] an important step forward was made. It was shown, that the dynamic polarization of the target atom influences the continuous BrS spectrum formed in non-relativistic collisions in a broad range of photon energies rather than in the vicinities of the discrete atomic excitations. Later, the theory of PBrS was extended to the domain of relativistic collisional velocities [35, 57]. It was established theoretically that two mechanisms, the ordinary and the polarizational ones, shape the unified BrS process, so that the total BrS amplitude is written as a sum of the ordinary and polarizational terms. The waves, emitted via the two possible channels, interfere. Therefore, the total cross section of BrS contains the interference term in addition to the OBrS and PBrS contributions. In the case of the constructive interference, the PBrS amplitude compensates the part of OBrS amplitude which is due to the screening of the nucleus by the cloud of atomic electrons. In electron collisions with many-electron target, such a compensation occurs shell by shell with the increase of the emitted photon energy [12, 13, 69]. In the limit $\omega \gg I_{1s}$ (with I_{1s} being the 1s ionization potential of a target atom) the PBrS leads to a complete compensation of the screening, so that the BrS is virtually emitted due to the incident electron acceleration in the field of the bare nucleus [100, 236]. The phenomenon of partial (or complete) de-screening of the atomic nucleus was called "stripping effect" [12]. A "stripping" approximation, based on this effect, is efficient for obtaining BrS spectra for photon energies greater than the ionization thresholds of atomic (ionic) subshells [13, 69, 73, 252, 346, 356, 357]. Let us note that in atomic collisions of light positively charged projectiles,—positrons, the interference becomes destructive, and the effect of the atom "stripping" is absent [98, 237].

PBrS is a universal radiative mechanism dependent on the internal dynamic structure of colliding particles. As a result, PBrS appears not only in collisions with atomic targets, but also with more complex structural particles, for example, fullerenes [31, 154, 177], metallic clusters [114, 150, 154], rare-gas clusters [162, 163] etc. In fact, the PBrS mechanism of radiation appears in those cases where at least one of the colliding particles has an internal structure.

For several cases which allow for the exact analytical treatment of the PBrS process the corresponding formulae were derived, and the essential role of PBrS in the full spectrum of radiation has been ascertained. These include the PBrS formed in fast electron–hydrogen atom collision [124, 125], in electron/positron collisions with muonic hydrogen [22, 41], and in the collision of a charged particle with positronium [32]. A universal character of the PBrS mechanism, i.e. its (qualitative) independence on the type of the interaction between a projectile and a target was demonstrated in [15, 17, 23] for the processes of neutron– and neutrino–atom scattering (see also the earlier publications [418, 419]). The PBrS manifests itself in nuclear collisions [42], and in nuclear reactions such as α, β, γ decays or nuclear fission process [46, 387].

The theory of PBrS in collisions of heavy charged particles with atoms/ions was developed. In [98] analytical treatment of the exactly solvable problem of the PBrS arising in a positron and a proton collision with a hydrogen atom was given. It was shown that over wide range of the photon energies the BrS intensities for both projectiles are of the same order of magnitude. The PBrS formalism for a bare-ion—atom collision [167, 168, 204–207] was developed and the performed calculations allowed to explain the earlier measured experimental data on the BrS spectrum of protons [208, 339] (see also the review papers [202, 203, 215]). A more general treatment of the BrS arisen in the collision of two complex colliders (atoms, ions), which accounts also for the recoil effect of the nuclei, was carried out in [33, 34].

The important feature of the PBrS mechanism is that it leads to the emission in collisions of two electrically neutral objects possessing an internal structure. In [33, 34] for the first time it was demonstrated that the intensive BrS emission can appear in atom–atom collision. In this case the radiation is due to the mutual virtual polarization of the colliders. The total induced dipole moment of the system alters during the collision, and this results in the photon emission. The general formalism developed in the cited papers allowed to express the BrS amplitude via the generalized atomic polarizabilities of the colliders. It was demonstrated that no dipole photon emission appear in symmetric collisions. The specific features (due to the Coulomb repulsion) of the PBrS formed in ion–ion collision were studied in [44]. The numerical calculations carried out in [33] have shown that the radiation intensity in the collision of two neutral (but different) atoms is comparable to that formed in the collision where one of the colliders is substituted with a charged heavy particle or an electron of the same velocity. Later the formalism, initially developed for non-relativistic atomic collisions, was extended to the relativistic domain [38, 39, 43]. The theory of PBrS in slow atomic collisions was also developed [388].

For the same collision velocity, the intensity of radiation in collision of two fast structural particles due to the PBrS mechanism is equal, at least in the order of magnitude, to that formed in electron–atom collision [33, 34]. Surprising is the fact that this result remains valid even in collisions of neutral atoms.

In general case, it is not an easy task to calculate the parameters of PBrS formed in a collision of two atomic particles. This is due to a number of particles involved into the scattering and radiation processes: the nuclei and the bound electrons. However, for fast collisions (including the relativistic ones) the problem is simplified, allowing one to express the PBrS cross section in terms of well-known atomic characteristics,—the form-factors and dynamic polarizabilities. This result reflects the physical mechanism of the formation of PBrS in fast collisions. As mentioned above, the radiation emitted by each colliding atom appears due to its polarization by the static field of the partner atom [33, 34]. The intensity of PBrS is enhanced for the atoms with larger polarizabilities and for the partner-atoms with stronger static field. In the collision of two neutral atoms the OBrS is virtually absent whereas the PBrS can be generated quite intensively. To be features is the PBrS arisen in symmetric atomic collisions (i.e., those which involve two identical atomic particles). In this case, the dipole PBrS is strongly suppressed as compared to that formed in asymmetric collisions. The physical reason for this effect is that the mutually induced dipole momenta compensate each other leading to the absence of the total dipole momentum in a symmetric system [33, 34]. Therefore, the PBrS emission in symmetric collisions can appear either into higher multipoles (e.g., the quadrupole radiation) or due to the relativistic effects (the Doppler effect and the aberration of light) [43].

The PBrS mechanism of radiation can be studied experimentally provided the final state of the target (or a projectile) is detected in coincidence with the photon detection. However, such experiments are quite challenging (although not at all inaccessible, see, for example, [3, 196, 220]), therefore, the total radiative spectrum is also of interest. In addition to the BrS emitted in the process occurring without excitation of the colliding particles (the so-called, "elastic" BrS), the total spectrum includes the contribution of all "inelastic" BrS channels [414], which accounts for all the variety of the BrS processes accompanied by the excitation or ionization of of the colliders. The analysis carried out in [36, 37, 98, 414] has demonstrated, that in all types of atomic collisions occurring either with non-relativistic or relativistic there exists a broad range of photon energies within which the "elastic" BrS channel dominates the total spectrum of radiation. The physical reason for this effect is as follows [36, 37]. In the "elastic" BrS process the waves emitted by each electron are coherent, as in the process of the Rayleigh scattering of light. On the contrary, the contribution of electrons to the "inelastic" part of the spectrum is incoherent, similar to what happens in the Compton scattering process. Thus, the contribution of "elastic" channel to the total spectrum exceeds that of "inelastic" ones by a factor proportional to the number of emitting electrons, which can be large in the case of many-electron colliding systems. The aforementioned arguments are quite universal. They can be applied to both non-relativistic and relativistic collisions, they are sensitive neither to the mass of the colliding particles nor to the type of interaction between them.

Peculiar features in spectral distribution of BrS (known in literature as "giant resonances" in connection with the photoionization process, see, e.g., [8, 112]) appear in the photon energy range in the vicinity of the 4d-subshell ionization potentials in Xe, Ba and rare-earth elements [382, 426, 427, 447, 451–453]. The magnitude of the dipole polarizability of these atoms is quite large in the the photon energy range $100 \dots 200$ eV, therefore, the BrS spectrum is mainly formed via the polarizational mechanism. The giant resonances in the spectra for Ba and La are due to the radiative decay of the continuous spectrum excited states of the type $4d^9 4f^n \varepsilon f$. The resonances in La and the subsequent rare-earth elements are due to the decay of the discrete spectrum autoionizing excited states $4d^9 4f^{n+1}$ [447].

A peculiar feature of the electron structure of rare earth atoms is in the spatial localization of the f-electrons which appear as a specific behaviour of the 4d subshell photoabsorption cross section the atoms in the solid state. Namely, the 4d-subshell cross section of solid state is more atom-like and is almost independent of the details of the condensed matter electron structure [448, 449, 452]. This fact allowed one to use the isolated atom data on the photoionization [364, 437] and the dynamic polarizability to calculated the PBrS part of the spectrum [242–248].

The possibility of powerful peaks in spectral dependence of BrS in the photon energy range above or in the vicinity of the ionization potential of an easily polarized, many-electron atomic subshell was predicted theoretically in [14]. In this spectral range most of the radiation is emitted via the polarizational mechanism. Later, the relationship between these peaks and the giant resonances in the photoabsorption spectra was established [45]. In Chap. 4 we discuss theoretical models and the obtained results for various electron–atom collisions and in a broad range of photon energies.

Let us briefly discuss the methods used to calculate the BrS spectra with account for the PBrS mechanism. We will focus on the two regions of the photon energies: in the vicinity of the giant resonances and in the vicinity of the inner-shell thresholds. These regions are of a particular importance for the experimental probing of PBrS [157–162, 190, 333, 345, 346, 354, 356, 357, 363, 401, 426–430, 436].

The first calculations of the PBrS spectra within aforementioned photon energy ranges were performed for Ar, Xe and La atoms within the framework of non-relativistic plane-wave Born approximation (BA) [13, 45, 69, 428] and for Xe within the non-relativistic DPWA scheme [20, 28].

The main difficulty in numerical computation of the PBrS amplitude is in accurate calculation of the dynamic response of the target, which in many cases is governed by many-electron correlations, thus implying the use of methods of many-body theory. Within the framework of the BA the PBrS amplitude is expressed in terms of the generalized dynamic polarizability $\alpha(\omega, q)$ of the atom, where ω is the emitted photon energy and q stands for the momentum transferred to the atom in the course of the collision [414]. This quantity contains all information on the dynamic response of the target to the field of the projectile particle and of the emitted photon. For many-electron targets, where the electron correlation effects play important role, the calculation of $\alpha(\omega, q)$ can be carried out within the RPAE framework [8, 18]. In connection with the PBrS problem the RPAE scheme was first used in [13, 20, 69]. However, the RPAE scheme, including only part of the correlation corrections, does not provide good description of the photoionization of Ba and La atoms in the vicinity of the giant resonances (see, e.g., [8]). In this case, the theoretical results can be improved by applying the Generalized RPAE (GRPAE) framework [8, 24] which accounts for the relaxation of the atomic core, or using the many-body perturbation theory [276]. However, even these more complicated schemes do not guarantee high accuracy of the calculated values of the photoabsorption cross section and of the dynamic polarizabilities of many-electron atomic targets.

The PBrS arisen in fast electron (positron) collisions with atomic and ionic many-electron targets was considered in [59–63, 66, 365] within the framework of statistical atomic model [292]. An extended review of the methods and approximations which were elaborated and applied to describe the dynamic atomic response within the framework of the Thomas-Fermi model one finds in the review papers [65, 66].

The static atomic potential distorts the wavefucntion of the scattering electron in the domain of intermediate incident energies, $\varepsilon_1 = 10^{-1} \ldots 10^1$ keV. To take this into account one can consider the scattering process within the DPWA framework. This scheme was used in [20] to construct the PBrS amplitude and to calculate the total BrS in collision of a 600 eV electron on a Xe atom in th vicinity of the 4d giant resonance. The obtained result was compared to the the available experimental data [426, 427] revealing reasonable agreement between the experimental and theoretical data (see also [30]). The DPWA scheme utilized in [20] implied that the dynamic atomic response (considered within the RPAE) was calculated for each partial wave separately, which was quite CPU time consuming. Additionally, special precaution were needed to accelerate slow convergence of the partial wave series of the PBrS amplitude [29].

More effective algorithms of the PBrS amplitude calculation in the vicinity of giant atomic resonances were elaborated in [153, 242–249, 456]. The calculation of the PBrS spectra in solid state targets with account for the photon attenuation and electron multiple scattering was carried out in [243, 246, 248]. The BrS spectra with account for the polarizational mechanism were also calculated for noble gas atoms in broad range of photon energies [249, 250, 253].

In collisions with heavy atoms (or ions) polarization of the inner atomic shells contributes to the high-frequency domain of the PBrS spectrum. The dynamics of these electrons must be described in terms of the relativistic Dirac equation. The general formalism for the PBrS cross section calculations was developed and the representative numerical results were presented in [254, 255, 258, 259]. Efficient method for the PBrS amplitude calculation in the high-frequency limit was proposed in [256]. The methods exploits the formalism of the Coulomb Green's function and the scaling properties of the generalized atomic polarizability.

The PBrS radiation in relativistic collisions of complex atomic particles acquires specific features as compared to the non-relativistic collisions. As the collision velocity increases, $v_1 \to c$, the spectrum and the angular distribution of PBrS emitted by a relativistic complex projectile is strongly influenced by the Doppler shift and the aberration of light [39]. In the cited paper it was demonstrated, that in the limit of high-photon energies, $\omega > I_{1s}$, where the atomic polarizability behaves as $\propto 1/\omega^2$, the radiation of the projectile-atom is concentrated in the narrow cone along its velocity with the opening angle $\theta_{\mathbf{k}} \sim 1/\gamma$, where $\gamma = \varepsilon_1/mc^2$ is the relativistic Lorentz factor. On the contrary, in the domain of small photon energies, $\omega \ll I$ (here I stands for the first ionization potential), when the polarizability reduces to its static value, the radiation is emitted into the wide cone $\theta_{\mathbf{k}} \gg 1/\gamma$, within which it is weakly anisotropic. Therefore, as it was pointed out in [39, 43] for sufficiently high velocities of the collision there is a principal possibility to separate the radiation emitted by each of the colliding complexes. Another important feature of the BrS process at relativistic velocities, which is also due to the Doppler and the aberration effects, is that the intensity of the dipole radiation formed in symmetric collisions does not vanish.

The PBrS of the target becomes modified due to the retardation effect in the interaction with the projectile particle. The most important feature which appears in the BrS process in relativistic atomic collisions of is the logarithmic increase of the PBrS cross section with the incident energy ε_1. This effect was noted for the collisions of structureless charged particles with atoms [35, 57, 255, 258] as well as for relativistic ion-atom and ion-ion collisions [39]. Qualitatively, the logarithmic growth is due to the increase in the effective distances between the projectile and the target at which the photon emission occur. Indeed, a relativistic one interacts with the target not only by its Coulomb field but also (in the ultra-relativistic case, predominantly) by the field of transverse virtual photons (see, e.g., [2]. The effective radius of this field, $R_{\mathrm{eff}} \propto \gamma \propto \varepsilon_1$, increases infinitely as $v_1 \to c$ [35]. As a result, the range of distances at which a projectile effectively polarizes a target increases leading the logarithmic growth of cross section. Relativistic effects lead to a peculiar features in the effect of coherence radiation of atomic electrons in atom-atom and atom-ion (ion-ion) collisions [43]. Chapter 6 of the book is devoted to the discussion of relativistic effects in PBrS.

Emission of the continuous spectrum radiation in slow atomic collisions was experimentally measured for the first time in [369]. Electron states in a system of two slowly moving atoms have a pronounced molecular character. In such a system the radiation can be emitted following the transition between the different terms of the quasi-molecule (QM) [369] as well as due to the alternation of the dipole moment of the system during its motion corresponding to a given term [96].

The radiation accompanying the electronic transition between different QM levels is called molecular orbital radiation. It is formed in a broad range of photon frequencies [173]. Theoretical [95, 178, 284, 305] and experimental [316, 317] considerations, which followed the first experimental detection, demonstrated that molecular-orbital radiation (MOR) can be interpreted as the radiative decay of the vacancy in a quasi-molecule. The width of the line is determined mainly by the collisional mechanism. The radiative transitions corresponding to small inter-nuclear distances can be interpreted in terms of the united atom with the nuclear charge equal to the sum of nuclear charges of the colliding atoms [139, 316, 317]. The molecular orbital radiation was investigated in collisions of light [96, 284, 305] and heavy [229] quasi-molecular systems.

The PBrS process was considered in [388] in collisions of slow atomic particles. The relative contributions of PBrS and MOR into the total radiative spectrum were analyzed. It was demonstrated that the PBrS dominates in the domain of large impact parameters and low photon energies ω. This dominance is a result of mutual polarization of the colliding particles, which is mostly effective at large distances. It was noted, that for very small values of ω, the intensity of OBrS due to the nucleus acceleration being proportional to $1/\omega$, becomes comparable to that of PBrS which decreases as ω^3. In slow ionic collisions the radiation can arise due to the capture of an electron from one colliding particle by the field of the partner. In the cited paper it was shown that the intensity of the radiation during the electron capture process can be calculated as a product of the PBrS intensity and the probability of the capture.

In adiabatically slow atomic collisions the photon can be emitted only at the instants when the splitting ω_{fi} between the energy levels ε_f and ε_i of the quasi-molecule is equal to the photon energy ω. The matching condition $\omega = \omega_{fi}$ can be met more than once in the course of the collision. For instance, the photon of energy ω can be emitted at the incoming and outgoing branches of the colliding ions (atoms) trajectory. As a result, the waves, emitted from spatially different parts of the trajectory, interfere giving rise to the oscillations in the BrS cross section as a function of ω and the impact parameter. The interference phenomenon was recognized theoretically [96, 284, 305] (see also [173, 388]) and later was observed at the experiments [315, 377, 413].

In [270, 271] it was demonstrated that the dynamic response function of any quantum adiabatic system as a function of ω has a peculiarity in the vicinity of the frequency of the external field applied to the system. This peculiarity is similar to the one which appears in the photon distribution in the case of Fresnel diffraction of light at a screen edge. The established Frensel peculiarity is of a universal nature. It appears in many physical processes in which the polarizability of the system plays an essential role.

The mechanism of the radiation emission in the collision of a slow charged particle with the excited hydrogen is somewhat different, compared with PBrS and MO mechanisms. More accurately, in addition to these types of radiation there is a peculiar source of low-frequency photon emission [240]. In this case the radiation is generated by the rotating dipole moment of the hydrogen during the collision. The specific feature of the hydrogen atom is in the existence of the linear Stark effect (see, e.g., [279]). The electric field of the projectile splits the initially degenerated levels of the excited hydrogen. Atomic states with a given principal number form a Stark multiplet. The components of the multiplet already possess a dipole moment. The vector of this dipole moment rotates following the movement of the projectile, and the radiation appears as a result of this rotation. In addition to this mechanism, which in [240] was termed "rotational" BrS, there is another mechanism, called "transitional" BrS, which is due radiative transitions between the components of the Stark multiplet. The latter is an analogue of the MO radiation. In the cited paper the quantitative analysis was carried out of the contributions of the "rotational" and "transitional" BrS to the emission spectrum.

The "rotational" BrS mechanism is intrinsic to the systems with the linear Stark effect. For systems possessing a quadratic Stark effect, PBrS appears due to rotation of the dipole moment *induced* during the collision. This latter mechanism is called PBrS [7, 9, 414]. Of course, PBrS occurs in collisions of a charged particle with the hydrogen atom in the ground [99, 124, 125] and in the excited state [253]. However, there is a distinction between PBrS and "rotational" BrS [240]. In the latter case the radiation is emitted as a result of re-orientation of already existing dipole moment while to generate the PBrS it is necessary to polarize the atom first. This makes "rotational" BrS more intense tan PBrS in a broad range of photon energies.

Many-body effects play an important role not only in the processes of scattering and radiation emission in atomic collisions but also in collisions involving more complex structural objects such as metal clusters and fullerenes.

Prior to discussing these phenomena, let us include a brief review of theoretical models used to describe the structure and dynamic properties of metallic clusters and fullerenes.

The valence electrons of the atoms which form a metallic cluster are fully delocalized. To some extent, this feature is also valid for fullerenes, where the delocalization of electrons takes place in the surface layer in the vicinity of the fullerene's cage. Quite often, it is the delocalized electrons that play an important role in the formation of the cross section of various collision processes [179, 268].

With the discovery of electronic shell structure in free metal clusters [128, 129, 232] the essential role of the quantized motion of delocalized valence electrons in the mean field created by ions in a cluster has been understood. The results of experiments, carried out under different conditions, suggested that the properties of alkali and other simple metal clusters are not too much sensitive to the details of the ionic core structure (see, e.g., [92, 94, 191, 311] for a review). This feature validates the applicability of the jellium model the framework of which implies a uniform distribution of positively charged ionic background over the clusters volume. The description of the delocalized electrons is given in terms of single particle wavefunctions obtained within a self-consistent quantum-mechanical mechanical approach (for example, the Hartree-Fock (HF) approximation, the Local Density Approximation (LDA)) using the Hamiltonian which includes the interaction with the electrostatic potential of the ionic background. We stress that this framework is valid for clusters with delocalized valence electrons only. For the sake of clarity, below in this section we use the term "metal cluster" when referring to such systems.

The jellium model has been successfully applied to simple alkali metals (Na, K) clusters, where one valence electron per atom is delocalized (see, e.g., [297, 298, 311] and references therein). The jellium model electronic shell closures for alkali-metal clusters define the magic numbers $N = 8, 20, 34, 40, 58, 92$ that are in a good agreement with the experimental data [179]. The jellium model provides a very useful basis for studying various collision processes, such as photoabsorption [209], photoionization [83, 212], elastic [146, 147] and inelastic scattering [147, 149–151], electron attachment [113, 200], photon emission [148, 154], atomic cluster fission process [296, 301, 302] and others, involving metal clusters. On the basis of the jellium model one can develop *ab initio* many-body theories, such as the Random Phase Approximation with Exchange (RPAE) [8, 18] and effectively solve many-electron correlation problem even for relatively large cluster systems. Note that the jellium model can be generalized by accounting for the collective ion background vibration dynamics [152, 155]. Review of these methods in their application to study collision processes involving atomic clusters one can find in [390, 391].

In spite of its great success for the simple alkali metal clusters, the jellium model can not be used for all types of atomic clusters. Thus, clusters of divalent metals, such as Be, Mg, and Ca, differ from the jellium model predictions at least at small cluster sizes. In this case, bonding between atoms is expected to have some features of the van der Waals type of bonding, because the electronic shells in the divalent atoms are filled. Clusters of divalent metals are very appropriate to study non-metal to metal transition, testing different theoretical methodologies and conceptual developments of atomic cluster physics. However, relatively little work was done so far on the exploration of the alkali-earth metal clusters in comparison with that for the alkali-metal clusters (see, e.g., [130, 303, 393] and references therein).

When jellium model fails to describe the electronic cluster structure it is necessary to employ direct *ab initio* methods for calculation of the electronic and ionic cluster structure. These methods are based on the molecular dynamics simulations where all or at least valence electrons in the system are treated accurately and can employ the time-dependent density functional theory (TDDFT), configuration interaction (CI) methods and random-phase approximation (RPA) (for reviews see, e.g., [130, 182, 394] and references therein).

Recently, significant attention was paid to the magnesium clusters. Various properties of the clusters, such as their structure, the binding energy, optical response, ionization potentials, HOMO-LUMO gap, average distances and their evolution with the cluster size have been investigated theoretically (see [303, 394] and references therein). Theoretical methods used in the cited papers are based on the density functional theory and the many-body perturbation theory.

Metal clusters and fullerenes in general are highly polarizable systems. This is borne out by the extremely large photoionization cross sections of the giant resonances in clusters as compared with atoms (see, e.g., [94, 394] for metal clusters, [83, 195, 212, 219, 361, 378, 423] for fullerenes). Electrons of a suitable energy can thus excite a giant dipole oscillation, which will radiate, and this radiation, due to the dynamic polarizability of the target, is superimposed on the OBrS background. The significance of the dynamic polarization in the radiative processes with metal clusters has been stressed in [114, 115]. The application of this idea to the radiative electron capture and the BrS is considered in these papers. The photon emission spectrum generated in electron-fullerene collisions was described in the main logarithmic approximation in [31] and more accurately in [154].

There are two possible kinds of giant resonances, which may be excited in clusters, sèe, e.g., [94]. Firstly, one may have giant resonances of the electrons localized on individual atoms and, secondly, there are collective oscillations of delocalized electrons belonging to the whole metal cluster. These electrons can oscillate collectively giving rise to giant resonances of an entirely new kind [83, 93, 384]. The energy of the giant resonance of the first type in a cluster is comparable with the corresponding giant resonance energy of a single atom, which is about 10…100 eV, while the energy of the resonance associated with oscillations of delocalized electrons is much lower, typically being in the range 2–5 eV for metal clusters. The characteristic excitation energy of these resonances in fullerenes is 19 eV. Delocalized giant resonances are less strongly damped out than the giant resonances in atoms: several oscillations take place. In Chap. 7 we focus mainly on the manifestations of the collective motion of the delocalized electrons in the spectra of photons emitted in collisions involving clusters. In fact, almost all features of PBrS known for atomic targets can be also transferred for clusters at least qualitatively.

Let us briefly mention the methods used to calculate the cross sections in electron–cluster and electron-fullerene collisions.

The plane-wave BA is applicable to describe the scattering process when the energy ε_1 of the projectile electron surpasses the typical energy of the delocalized electrons in the target [52, 146, 147, 149, 151, 154, 169, 421, 422]. It was established in the theory of PBrS in electron-atom collisions, that in the BrS process the range of applicability of BA can be extended down to the incident energies comparable with the energy of the giant resonance [20, 246]. Therefore, in electron-fullerene collisions the BA can be used for $\varepsilon_1 \geq 50\,\text{eV}$ and for the photon energies within the range $10\ldots30\,\text{eV}$. For metallic clusters the limit of applicability of BA lowers down to $\varepsilon_1 \sim 10\,\text{eV}$. In some cases, to provide correct description of the diffraction phenomena arising in the collisions one can consider the process within the framework of the second BA [147]. For intermediate energies of the incident electron one can exploit the DPWA method [148]. In the case of slow collisions the partial wave functions of the scattering electron to calculate the partial wavefunctions of the scattering electron as solutions of the Dyson equation with the non-local polarization potential [113, 200]. This method is more accurate that that based on the use of the static local polarizational potential in the Schrödinger equation for the scattered electron [273–275].

To describe the dynamic response of the cluster or/and fullerene to the fields of the projectile and of the emitted radiation one can use the Plasmon Resonance Approximation (PRA) [146] (see also Sect. 7.2), or apply direct quantum calculations within the RPAE framework using the HF jellium model wavefunctions [89, 148, 150, 151, 180].

Excitation of plasmon oscillations in electron-cluster and electron-fullerene collisions strongly influences the cross sections of radiative processes, like bremsstrahlung and radiative capture. Plasmon excitations due to collisions of fast electrons with small metal particles have been considered by both classical [144, 145, 267] and quantum [293, 294] approaches. A purely classical description of the photon emission resulting from plasmon excitations was given in [145]. The classical framework is adequate for particles containing thousands of atoms, i.e., large clusters. For small clusters, build from hundreds atoms and less, a quantum description produces more accurate results. Quantum corrections to the classical treatment have different nature. In particular, they arise when the emitted photon energy becomes compared to that of the incident particle. Apart from this, the dynamic response of a small cluster can only be described correctly within a quantum-mechanical framework.

Collective plasmon-like excitations in metal clusters and fullerenes appear in the form of giant resonances in the PBrS spectrum. The PBrS resonances in electron-fullerene collisions were analyzed in [31, 154], the detailed quantitative and qualitative analysis for electron-cluster collisions was carried out in [114, 150, 154]. PBrS in cluster-cluster collisions was discussed in [114, 300]. The PBrS radiation in electron scattering from a metallic nanosphere embedded into a dielectric medium was investigated theoretically in [53]. In the cited papers it was shown, that in the vicinity of the plasmon resonance ω_p (which is located at 2–5 eV for metallic clusters and at ca 20 eV for C_{60}) the PBrS mechanism dominates in the emission spectrum by far exceeding the contribution of OBrS. In the regions of both small photon energies, $\omega \ll \omega_p$, and the large ones, $\omega \gg \omega_p$, the contributions of both mechanisms become comparable (see the discussion in the cited papers and [52, 273–275]).

Methods for analytical and numerical calculation of the BrS cross section in broad photon energy range were developed in [114, 115, 148, 150]. To a large extent, these methods are applicable to any polarized a system with collective electron excitations.

For the photon energies in the vicinity of the plasmon resonance, the approach for effective calculation of the PBrS cross section was proposed in [154]. It exploits the smallness of the parameter a/R, where a is the width of the surface layer in a metallic cluster (or the width of a fullerene shell) in which the collective oscillations of the electron density mainly occur, R stands for the cluster (fullerene) radius. The width a is of the same order of magnitude as the size of the cluster atom. The key element of the method (see Sect. 7.2.1 for the details) is the calculation of the dynamic response of the target. When doing this one has to account for the many-electron Coulomb interaction in a cluster (fullerene) [83, 366, 367]. The method proposed allowed one to evaluate the dependence of the generalized dynamic polarizability of a metallic cluster and the fullerene C_{60} on the transferred momentum q in the domain $q \ll 1/a$ and on the photon energy in the vicinity of $\omega_{\rm p}$. It was demonstrated, that the polarizability is determined by the following three parameters: the radius R of the system, the plasmon frequency $\omega_{\rm p}$, and on the width Γ of the plasmon resonance.

The method allowed one to derive the expression for the PBrS cross section beyond the logarithmic approximation used in [31], and to establish the dependence of the PBrS intensity on the velocity v_1 of the incoming electron [154]. In particular, it was demonstrated that for $\omega \approx \omega_{\rm p}$ the maximum of the intensity is achieved for $v_1 \sim \omega_{\rm p} R$. The spectral and spectra-angular distribution of the PBrS radiation were analyzed in a broad range of photon energies and various emission geometries. The diffraction oscillations in the PBrS cross section were revealed, which turned out to be of the same physical origin as those in the cross section of elastic electron scattering from clusters.

In [148, 150]. the calculations of the PBrS cross sections were performed within the framework of the many-body theory, and the comparison was carried out with the results followed from the PRA. On the basis of this analysis, it was concluded that the PRA reproduces all main features of the PBrS cross sections quite well.

Alternative simple method to calculate the generalized polarizability of a cluster target in the vicinity of the plasmon resonance is based on the use of the Kramers-Kronig dispersion transformation and the relation between the imaginary part of the dipole polarizability and the photoabsorption cross section. This method was widely used to calculate the PBrS part of the spectrum in electron-atom collisions [10, 45, 260]. For electron-fullerene collision it was adapted in [31], and for metallic clusters it was used in [260].

The PBrS process, as well as the OBrS one, corresponds to the radiative free-free transition with respect to the projectile electron. However, the PBrS mechanism appears in the free-bound electron transitions as well, when the initially free electron becomes bound to a target forming negatively, neutral or positively charged complex depending on the initial charge state of the target. In [113, 115, 200] the radiative capture was studied in electron-cluster collisions. It was shown, that the polarization mechanism of radiative capture is dominant in a wide range of incident electron energy equal and above the energy of the plasmon resonance in the metallic cluster target. This result was first obtained within the PRA [115], and then it was validated further within the framework of the quantum-mechanical many-body theory applied to the radiative capture to small sodium and potassium clusters [113, 200] (see Sect. 7.4).

In connection with electron–atomic-ion collisions the polarizational channel of the radiative recombination was considered in [62, 64, 65, 102, 103, 238, 239]. The details one finds in Sect. 4.8.

The PBrS mechanism appears in the radiation spectrum of charged particles in dense medium both amorphous [50, 87, 221, 328, 329] and (poly-)crystalline [54, 55, 58, 68, 88, 170, 299, 329, 332]. The PBrS has much in common with the transitional radiation [414] which appears due to the scattering of the electric field of a projectile on the macroscopic inhomogeneities in the medium [156].

In an amorphous medium, a projectile experiences uncorrelated collisions with the constituent atoms. As a result, the total intensity of BrS is proportional to the number of atoms n_{a} (per unit volume), i.e., it is equal to the incoherent sum of the intensities formed in collisions with individual atoms.

Passing through a crystalline media along a line close to any crystallographic direction, a projectile "feels" the periodicity of the potential. The periodicity strongly influences the emission process when a coherence length $L_{\mathrm{c}} \approx 2\varepsilon_1^2/\omega m^2 c^3$ [81, 400] exceeds the period l of the potential [340, 400, 416]. In this case, the amplitudes of the waves emitted in collisions of the projectile with each of these $N_{\mathrm{c}} \sim L_{\mathrm{c}}/l$ may add coherently, so that the resulting emitted intensity will be enhanced by a factor $\sim N_{\mathrm{c}}$. For each emission angle $\theta_{\mathbf{k}}$ the coherence effect will be mostly pronounced if θ and ω satisfy the condition $l/(L_{\mathrm{c}}\theta) = 2\pi n$, with n being an integer [340, 400, 416]. This process is commonly termed as *coherent* BrS. thus distinguishing it from *incoherent* BrS occurring in an amorphous medium.[2]

With the exception for the consideration presented in Sect. 4.4, in our book we do not consider the influence of the density effect on the PBrS for energetic projectiles penetrating through dense (i.e., non-gaseous) medium. For further information, the interested reader is referred to the papers cited above. Another issue which is left beyond the scope of the book is the multiphoton PBrS process [67, 165, 265, 266].

[2] It should be noted that the position disorder of the particles in medium is not a sufficient condition for the absence of coherent radiation. For example, a charged particle propagation in a dielectric amorphous medium with the speed exceeding the phase velocity of light in the medium gives rise to the coherent Vavilov-Cherenkov radiation (see [90] for a historical survey and the bibliography).

Chapter 2
Bremsstrahlung in Collisions of Structureless Charged Particles with Atoms and Ions

2.1 Peculiar Features of PBrS

In this section we discuss the peculiar features of PBrS which distinguish this mechanism from OBrS and which strongly influence the total emission spectrum. Focusing on the physical nature of these phenomena and to avoid unnecessary complexities, we consider the BrS process of a non-relativistic charged structureless particle with mass m and charge Z_0 on a spherically-symmetric many-electron target (called an atom, for brevity) within the framework of the simplest approximations, which are (a) the plane-wave first Born approximation for the scattering process, and (b) the dipole-photon scheme for the description of the photon-atom and photon-projectile coupling. Such treatment, although quite often being insufficient for the quantitative description of the process, allows one to carry out a simple qualitative analysis and to explain all specific features of PBrS.

Let $\mathbf{p}_j$ and $\varepsilon_j = p_j^2/2m$ denote the momenta and energy of initial ($j = 1$) and final ($j = 2$) states of the projectile. The initial and the final states of the target atom are assumed to be the ground one. To simplify the formalism we consider the BrS process in the collision with a spherically symmetric neutral atom. This restriction is rather of a formal but not of a principal nature, since the effects described below also occur in the BrS process on a target with a ground state of an arbitrary symmetry.

In the course of the collision, a photon can be emitted via one of the two mechanisms, which are schematically described by Fig. 1.1. The diagrammatic representation of the BrS amplitude is presented in Fig. 2.1 [14]. The thin solid line stands for the projectile, the wave functions of which are described by plane waves $\exp\left(\mathrm{i}\mathbf{p}_j \cdot \mathbf{r}\right)$. The long-dashed line corresponds to the emitted photon which is characterized by the energy ω and the unit vector of polarization $\mathbf{e}$. In the upper two diagrams, which illustrate the OBrS process, the short-dashed vertical line depicts the interaction of the projectile with the atomic static field. The lower diagrams describe the PBrS process in which the Coulomb interaction between the projectile and atomic electrons (vertical dashed line) leads to the virtual excitation $0 \rightarrow n$ of the atom.

A. V. Korol and A. V. Solov'yov, *Polarization Bremsstrahlung*,
Springer Series on Atomic, Optical, and Plasma Physics 80,
DOI: 10.1007/978-3-642-45224-6_2, © Springer-Verlag Berlin Heidelberg 2014

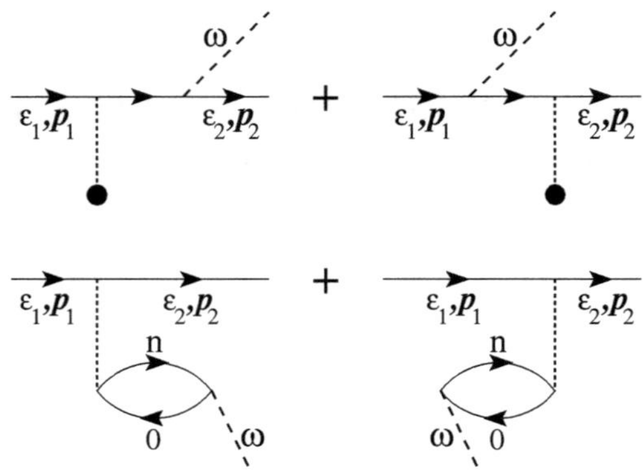

Fig. 2.1 Diagrammatic representation of the amplitude of BrS of a fast electron on an atom in the Born approximation. *Two upper* diagrams describe ordinary BrS, *two lower* ones—polarizational BrS

Considering the two mechanisms of the photon emission one derives the following expression for the total amplitude of the process [14, 100, 454] (see also [414] and references therein):

$$f_{\text{tot}} = f_{\text{ord}} + f_{\text{pol}} = \frac{4\pi(\mathbf{e}\cdot\mathbf{q})}{q^2}\left[\frac{Z_0^2}{m}\,\frac{Z - F(q)}{\omega} + Z_0\,\omega\,\alpha(\omega, q)\right], \qquad (2.1)$$

where $\mathbf{e}$ and ω are the photon polarization vector and energy, $\mathbf{q} = \mathbf{p}_1 - \mathbf{p}_2$ is the momentum transfer.

The first term in (2.1) describes the OBrS part of the amplitude. It is proportional to $Z - F(q)$, where Z is the charge of the nucleus and $F(q) = \int d\mathbf{r}\rho(r)\exp(i\mathbf{q}\cdot\mathbf{r})$ is the form-factor of atomic electrons ($\rho(r)$ stands for an isotropic distribution of the electron charge). Hence, f_{ord} is dependent on the static distribution of the charge in the atom. The PBrS amplitude f_{pol} is expressed via a generalized atomic dynamic polarizability $\alpha(\omega, q)$ which appears as a result of the action of two field on the atom: the field of the projectile and the electromagnetic field of the emitted photon.[1]

The first feature which clearly distinguishes between the two mechanisms follows immediately from (2.1). The OBrS amplitude is inversely proportional to the mass of projectile, while the polarizational part is independent of it. The explanation for this fact follows from the basic principles of electrodynamics (see, e.g., [2]). In the OBrS process it is the projectile that emits the photon. The intensity of this radiation is proportional to the square of the projectile acceleration in the external field of a target. In turn, the acceleration is proportional to $1/m$ and this dependence appears in f_{ord}. In contrast, during the PBrS process the projectile serves as a source of the external field acting on the atomic electrons, and thus the amplitude of this process is almost insensitive to the variations of m [414]. Moreover, the intensity of PBrS for a heavy projectile is comparable and can be even higher than that of an electron of the same velocity [33, 98].

[1] Explicit formulae for $\alpha(\omega, q)$ is presented in Sect. 2.2.1, (2.14).

Other qualitative differences between the two mechanisms of the photon emission one can trace by comparing the dependencies of $f_{\rm ord}$ and $f_{\rm pol}$ on the photon energy and on the momentum transfer.

The OBrS amplitude is a smooth function of ω. The only peculiarity appears in the soft-photon region $\omega \to 0$ where a simple perturbative approach, giving an infinite magnitude of $f_{\rm ord}$, fails to describe the process. This phenomenon, known as the "infrared catastrophe", had been recognized and understood long ago (e.g. [2]). The q dependence of $f_{\rm ord}$ is concentrated in the factor $Z - F(q)$. The atomic form-factor $F(q)$ is the Fourier image of the electron charge distribution and is a monotonically decreasing function of q. Qualitatively, the value $F(q)$ defines the number of atomic electrons inside the sphere of a radius $r \sim 1/q$. Hence, this function reaches its maximum value at $q = 0$ where $F(0) = Z$ and decreases monotonically with the increase of q. In the case of large q, $\lim_{q\to\infty} F(q) = 0$. The natural scale to measure the magnitude of q is the inverse radius of the target, $R_{\rm at}^{-1}$. Thus, the amplitude of OBrS is large for $q > R_{\rm at}^{-1}$ while in the region $q \ll R_{\rm at}^{-1}$ it becomes negligibly small. Such behaviour has a clear explanation [348]. To radiate a photon via the ordinary mechanism a projectile must penetrate inside the atom, at a distance $r < R_{\rm at}$, where a strong nuclear potential $-Z/r$ is less screened by the electron cloud. In the opposite limit, when $r \gg R_{\rm at}$, the nucleus is fully screened by the electrons (in the case of a neutral target) and the probability for a projectile to get the acceleration and to radiate vanishes.

The PBrS appears as a result of the alteration of the atomic dipole moment induced during the collision.There are two external fields—the field the photon and the Coulomb field of the projectile—which act on the atom in this process. The dynamic response of the target depends, therefore, on the parameters of both fields. Formally, it is reflected in the dependence of the generalized dynamic polarizability $\alpha(\omega, q)$ on two variables. We use the term 'generalized' when addressing to $\alpha(\omega, q)$ in order to stress the dependence on q, and, thus, to distinguish this quantity from the dipole dynamic polarizability, $\alpha_{\rm d}(\omega)$, to which $\alpha(\omega, q)$ reduces in the limit of small transferred momenta:

$$\lim_{q\to 0} \alpha(\omega, q) = \alpha(\omega)\,. \tag{2.2}$$

The dependence on q appears because of the action of the external Coulomb field of the projectile. This field distorts the electrons' orbits and induces a dipole moment of the atomic system. The dipole polarization of the electron cloud is most pronounced if the Coulomb field of the projectile is uniform on the scale of $R_{\rm at}$, i.e. when the projectile is outside the target, $r \gg R_{\rm at}$. These distances correspond to small values of the transferred momentum $q \ll R_{\rm at}^{-1}$ where, in accordance with (2.2), the PBrS amplitude, as well as the cross section, can be expressed through $\alpha_{\rm d}(\omega)$. For small distances, $r \ll R_{\rm at}$, the field of the projectile is almost spherically symmetric, so that it induces small dipole moment on the target. Hence, in contrast to the OBrS process, it is the large distances between the projectile and the target which are of the most importance for the PBrS mechanism [45, 454].

Therefore, although in the case of light projectiles (electron, positron) two mechanisms of BrS formation must be treated simultaneously, it is possible, in principle, to distinguish the photons emitted via the polarizational mechanism from those which are formed through the ordinary one. To do this it is necessary to detect the emitted photon in coincidence with the scattered particle. Then, the "polarizational" photons will be observed together with the projectile scattered at small angles (corresponding to the transferred momenta $q < R_{\rm at}^{-1}$), while large scattering angles correspond to the emission of the "ordinary" photon. This consideration is valid for neutral atoms but does not hold for ions, since in the latter case a long-range Coulomb potential leads to the emission of the photon via the ordinary mechanism for small-angle scattering of the projectile.

The dependence of $\alpha(\omega, q)$ on ω reflects the ability of the electron cloud to be dynamically polarized by an external electromagnetic field of a given frequency. In a many-electron atom, the electrons are distributed among the atomic subshells. Each subshell is characterized by an ionization potential I. In terms of classical mechanics, this corresponds to the frequency of the rotation of electrons of a given subshell around the nucleus. Using this analogy, one may say that the dynamic response of the electron cloud to the external field increases for those ω which are close to the ionization thresholds of the target subshells. Therefore, in the region

$$I_1 < \omega < I_{1\rm s}, \tag{2.3}$$

where I_1 and $I_{1\rm s}$ stand, respectively, for the ionization potentials of the outermost shell and of the $1s$ shell, the function $\alpha(\omega, q)$ is non-monotonic with extrema in vicinities of the ionization potentials [6, 7, 414, 454] .

This happens, in particular, when the photon energy lies within the region of a giant dipole resonance of the photoionization cross section of a many-electron atomic subshell [8]. As far as we are aware, for the first time wide maxima in the emission spectra were observed experimentally in electron scattering from solid-state Ba, La and Ce [285, 286]. Later the theoretical explanation was given [432] which related the maxima to the virtual excitations of the $3d$-subshell electrons. In [451] a powerful maximum was observed in the emission spectrum in electron-La collision. In the subsequent paper [45] for the first time an important conclusion was drawn about the common nature of the giant resonances in photoionization and those in PBrS spectra. To reveal this similarity one recalls that at large distances between the projectile and the atom the amplitude of the PBrS process is proportional to the dipole dynamic polarizability, $\alpha_{\rm d}(\omega)$. The imaginary part of this quantity is related to the photoionization cross section $\sigma_\gamma(\omega)$ through (e.g. [81]): $\mathrm{Im}\, \alpha_{\rm d}(\omega) = c/4\pi\omega\ \sigma_\gamma(\omega)$, where $c \approx 137$ is the speed of light. Since the OBrS amplitude is real, the modulus square of the imaginary part of $f_{\rm pol}$ enters the total BrS cross section as an additive term. Therefore, a maximum in $\sigma_\gamma(\omega)$ manifests itself in the BrS spectrum as well reflecting the collective nature of the dynamic response of atomic electrons.

Although based on the assumption that main contribution to $f_{\rm pol}$ comes from the region of large distances $r \gg R_{\rm at}$, which does not always lead to a correct quantitative result, the qualitative arguments of [45] provided a clear physical explanation of the nature of powerful maxima in emission spectra. The experiments, carried out later, supported the theoretical prediction. The maxima in the BrS spectra were measured for Ba and several rare-earth elements [382, 430, 450, 453], for La and for atoms from the lanthanum group [452], for Xe [160–163, 401, 426–428, 430] and for Ba [430] (see also the review paper [429] in [358]). In all these experiments, performed for various energies of the incoming electron (ranging from several hundreds of eV up to several keV), the powerful BrS maxima were observed for photon energies within the ranges of the giant resonances in the photoionization cross section of the $4d$ subshells.

In general, in the whole ω-region defined by (2.3), a highly non-monotonic behaviour of the generalized polarizability of a many-electron atom results in a series of peculiarities (maxima, minima, cusps) in the total BrS spectrum. The important role of the polarizational mechanism in forming the total BrS spectra in atomic collisions over a wide range of photon energies was analyzed theoretically (see the reviews [7, 9, 65, 66, 260, 261, 335, 351, 358, 414, 435] and experimentally [157–159, 161, 203, 345, 346, 356, 357, 435].

For the photon energies noticeably higher than the $1s$ ionization potential, $\omega \gg I_{1s}$ (but still in the dipole-photon domain, $\lambda \gg R_{at}$, with $\lambda = 2\pi c/\omega$ being the radiation wavelength) the atomic electrons can be effectively treated as (quasi-)free ones. Consequently, the ω dependence of $\alpha(\omega, q)$ is much like that for the cloud of free electrons. The leading term in the expansion of $\alpha(\omega, q)$ in powers of ω/I_{1s} reads [12,100]:

$$\alpha(\omega, q) \approx -\frac{F(q)}{\omega^2}\,. \tag{2.4}$$

As a result, for a fast electron ($m = 1$, $Z_0 = -1$) the total amplitude (2.1) reduces to the BrS amplitude on a bare nucleus [100]:

$$f_{tot}\Big|_{\omega \gg I_{1s}} \approx \frac{4\pi(\mathbf{e}\cdot\mathbf{q})}{q^2}\,\frac{Z}{\omega}\,. \tag{2.5}$$

This shows that for large ω the atomic electrons do not participate in the screening of the nucleus and do not contribute to the BrS cross section. Thus, in the region $\omega \gg I_{1s}$ the polarizational channel results in a (dynamic) de-screening of the nucleus.

The physical reason for this effect (following [12] we use the term 'stripping' effect) is that, for $\omega \gg I_{1s}$, the electrons of all atomic subshells may be treated as free ones [100]. If the incident electron is also free (the Born approximation), then there is no dipole radiation by a system of free electrons [278].

These arguments were exploited in [12] to construct an approximate expression for the total BrS amplitude for photon energies lower than the $1s$-shell ionization threshold. To this end, the target electrons are divided into two groups, the 'inner' and the 'outer' electrons. The 'inner' electrons have binding energies $I_{in} > \omega$, and, therefore, their orbits are not distorted noticeably by an external electromagnetic field of a frequency ω. As a result, the contribution of the inner electrons to the PBrS amplitude is ignored. The 'outer' electrons have binding energies I_{out} less than ω, so that they behave as quasi-free particles under the action of the field, and contribute to $f_{\rm pol}$ in accordance with (2.4) where $F(q)$ must be substituted with the form-factor of the outer electrons, $F_{\rm out}(q)$. As a result the total BrS amplitude acquires the form

$$f_{\rm tot}\Big|_{I_{\rm out}<\omega<I_{\rm in}} \approx \frac{4\pi(\mathbf{e}\cdot\mathbf{q})}{q^2}\,\frac{Z - F_{\rm in}(q)}{\omega}\,, \tag{2.6}$$

where $F_{\rm in}(q)$ stands for the form-factor of the inner electrons. This expression demonstrates that the outer electrons do not participate in the screening of the nucleus (or, in other words, the nucleus is 'stripped' by a total number $N_{\rm out}$ of the outer electrons) [12]. The physical reason for this partial 'stripping' is as formulated above: for $\omega \gg I_{\rm out}$ the outer electrons are quasi-free, and, thus, no dipole radiation is emitted by the system 'projectile electron + the outer electrons'.

Although the 'stripping' approximation does not account for specific details of the PBrS amplitude in the vicinity of each threshold, it allows one to estimate the behaviour the BrS cross section below and above the threshold. In particular, the asymmetry of the giant resonances in the experimentally measured emission spectra [382, 426, 452] was explained. The experiments indicated that the magnitude of total BrS cross section far above the 4d-thresholds exceeds that below the threshold. The explanation is as follows. Equation (2.6) suggests, that the difference between the amplitude $f_{\rm tot}$ at $\omega \gg I_j$ (I_j is the ionization potential of atomic jth subshell) exceeds $f_{\rm tot}$ at $\omega \ll I_j$ by a factor proportional to $F_j(q)$. The latter can be estimates as $F_j(q) \approx N_j$, where N_j is the number of the electrons in the subshell. Therefore, the difference between the two amplitudes is proportional to N_j, and the increase in the cross section is $\sigma(\omega \gg I_j) - \sigma(\omega \ll I_j) \propto N_j^2$ [12].

This qualitative explanation was confirmed by numerical calculations [13] carried out within the framework of the non-relativistic BA. Later, the 'stripping' approximation was extended beyond the BA [73, 236, 251, 252].

In the range of photo energies below the first ionization threshold, I_1, of the target,one can expect the decrease in the contribution of the PBrS channel to the total spectrum. It follows from (2.1), that this limit $f_{\rm pol} \propto \omega\alpha(0, q) \sim \omega\alpha_{\rm d}$, where $\alpha_{\rm d}$ is a static dipole polarizability of the target. Thus, $f_{\rm pol} \to 0$ as $\omega \to 0$ in contrast to the OBrS term which behaves as $\propto 1/\omega$. Thus, the ratio $f_{\rm pol}/f_{\rm ord} \sim m\omega^2\alpha_{\rm d}/e$ vanishes as ω goes to zero. For low but non-zero values of ω the contribution of the PBrS channel strongly depends on the magnitude of the static polarizability. The higher the magnitude of $\alpha_{\rm d}$ is, the wider is the ω interval where the contribution of $f_{\rm pol}$ might be noticeable.

These arguments are valid also beyond the range of validity of the first BA, on the basis of which (2.1) was derived. In the low-frequency limit, the OBrS amplitude is expressed in terms of the elastic scattering amplitude, $f_{\rm ord} \propto f_{\rm el}/\omega$ (e.g., [81]). On the other hand, the PBrS process occurs most effectively at large distances between the projectile and the target, where the wavefunction of the projectile is not distorted strongly by the potential, so that one can extend the range of applicability of the BA for the PBrS channel even to the domain of low projectile velocities. Therefore, the ratio of the amplitudes still is proportional to $\omega^2\alpha_{\rm d}$. In [455], the role of the polarizational channel was studied in the inverse BrS process (i.e. BrS absorption) for slow electrons scattered from atoms. It was demonstrated that for an $e^{-}-$Ar scattering the polarizational mechanism changes noticeably the absorption coefficients, whereas for an $e^{-}-$Ne scattering its influence is much less. This quantitative effect is due to large difference in the static polarizabilities: $\alpha_{\rm d}^{\rm Ar} = 11.10$ a.u. and $\alpha_{\rm d}^{\rm Ne} = 2.66$ a.u. [360]. Rather strong effect of the target polarization on the BrS spectra was reported in [142, 186–188] for low-energy ($\varepsilon_1 = 0.4 \dots 3.5\,$eV) electron–rare-gas atom collisions.

In the collisions of slow heavy particles with atoms both the OBrS and the PBrS mechanisms fail, as a rule, to describe adequately the radiation spectrum. In such processes another channel, known as molecular orbital radiation, becomes important (see, e.g., [173]). Nevertheless, the PBrS is important in asymmetric slow collisions of atoms and ions in the region of large impact parameters [388]. The intensity of the OBrS is negligibly small due the large masses of the colliders.

In the collision of a slow charged particle with an excited hydrogen atom, H*, there appears additional and quite peculiar source of low-frequency photon emission [240]. The specific feature of the hydrogen atom introduces the linear Stark effect (see, e.g., [279]). The electric field of the projectile splits initially degenerated levels of H*. Atomic states with a given principal number form a Stark multiplet. The components of the multiplet already possess a dipole moment. The vector of this dipole moment rotates following the movement of the projectile, and the radiation appears as a result of this rotation. We stress that this mechanism of BrS is intrinsic for systems with a linear Stark effect. This is also a distinguishing feature from the 'real' PBrS which also appears because of the alteration of the target's dipole moment. In the latter case, it is really the induced dipole moment intrinsic for systems with a quadratic Stark effect. As a result the character of these spectra at low frequencies are quite different [240].

Once the internal dynamic structure of a target is taken into account, the next logical step is to consider the radiative processes which are accompanied by the excitation or ionization of the target. Following [39, 56] we call BrS processes of this type 'inelastic' BrS contrary to the 'elastic' one, when the target does not change its state after the collision.

It is important to establish the contribution of inelastic channels to the total emission spectrum. This is not purely of theoretical interest since experimentally it is quite difficult to separate elastic and inelastic channels. To do this it is necessary to observe the final state of the target with simultaneous detection of the photon. The comparison of the contributions of the 'elastic' and 'inelastic' channels is discussed in more detail in Sects. 5.3 (non-relativistic collisions) and 6.4 (relativistic collisions). Here we only mention the main results obtained.

It has been demonstrated that over a wide region of the photon frequencies, the elastic channel dominates over the inelastic one in the total BrS spectrum for both heavy [37, 56, 98, 414] and light [36, 56, 98, 414] projectiles scattered on a many-electron atom. Semi-quantitatively, the cross sections of the elastic BrS, of both the ordinary and the polarizational nature, exceed those of the inelastic by a factor Z. The explanation is as follows [36, 37]. During the elastic BrS the contributions of each atomic electron to the polarizational part of the total amplitude (see (2.1) are coherent, as in the process of Rayleigh scattering of light. Considering the case of a neutral atom and having in mind that the ordinary part of the elastic BrS spectrum is approximately proportional to the nuclear charge squared, one finds that the total elastic cross section is proportional to Z^2. In contrast, during the inelastic BrS the contributions of each electrons must be summed in the cross section rather than in the amplitude. Hence, the inelastic BrS cross section is proportional to Z and is parametrically small in the case of a many-electron target, when $Z \gg 1$.

The region of the photon frequencies, in which the coherence effect plays an essential role, is estimated as [36, 37, 56]

$$I_1 < \omega < \frac{v_1}{R_{\rm at}} \tag{2.7}$$

where v_1 is the initial velocity of projectile. Beyond the region of coherence inelastic BrS becomes more important. An exception of this rule occurs in collisions of fast heavy charged particles with atoms/ions collisions (as well as in atom-atom, ion-atom and ion-ion) in the region of high photon frequencies. In this region the process of inelastic BrS has a threshold, which is equal to $\omega_{\rm max} \approx v_1^2/2$. However, the elastic BrS takes place at higher energies, up to $\omega \geq 2v_1^2$, dominating in this region in the total photon emission spectrum. Therefore, the photon energy range $v_1^2/2 \leq \omega \leq 2v_1^2$ is convenient for the observation of the elastic PBrS. We note that in this ω region there is a peculiar feature in the spectrum of PBrS [241] similar to that which occurs in inelastic scattering, where it is known as the Bethe ridge [279]. In more detail we discuss this phenomenon in Sect. 4.7.

Numerical comparison of the relative role of elastic and inelastic channels in proton-atom collisions was performed in [59, 237, 241].

In electron/positron–many-electron atom scattering elastic BrS dominates parametrically over the inelastic one in the region (2.7). However, if one is interested in accurate data on the total BrS cross section it is necessary to include the inelastic channels into the computational scheme. Up to now, mostly due to technical difficulties, numerical investigations of the role of inelastic BrS have not been as extensive as in the elastic case. The achievements in this field include the model theoretical study carried out in [428, 456] in connection with the experimental data on the intensity of the BrS spectrum in e^-+Xe collision as a function of the incoming electron energy ε_1 [401, 428] (see also [157, 158] for the experimental data in non-relativistic e^-+Ar collision and the calculations made in [51] for the relativistic e^-+Ar).

For low-Z targets the absolute magnitudes of both terms from (2.1), $f^{\rm ord}$ and $f^{\rm pol}$, and the terms $f^{(m)}_{\rm ord}$ and $f^{(m)}_{\rm pol}$ from the inelastic BrS amplitude (see (5.22) in Sect. 5.3) are all of the same order. In this case, it is the charge of the projectile which introduces peculiarities in the total radiative spectrum. It can be shown that, both for high frequencies of the photon, $\omega \gg v_1/R_{\rm at}$ [37, 98] and for the low ones, $\omega \ll 1$ [37, 237] the role of inelastic channels is negligibly small compared with the elastic BrS in the case of electron scattering, while for the positron-atom collision the situation is the opposite. It occurs mainly because of the difference in the behaviour of the interference between the ordinary and polarizational amplitudes in inelastic BrS. The interference is negative in the case of the electron projectile and positive in the positron case. The simplest way to trace this effect is to consider the inelastic BrS amplitude (5.22) in the limit of high photon frequencies, as was done above for elastic BrS. Having done this, one notices that for a projectile electron the ordinary and polarizational terms cancel each other out [37, 100, 414] while for a positron the effect is opposite [98, 414].

To conclude this section we mention theoretical approaches used to describe the BrS process in non-relativistic collisions. These can be subdivided into three parts: (a) methods applied to describe the scattering process, (b) models used for the interaction with the photon, and (c) approximations used to describe the dynamic atomic response.

The theoretical approaches to the scattering process range from the plane-wave BA to more sophisticated ones. Since the OBrS phenomenon has much longer history, these methods were first tested in application to this process and later on were applied to the PBrS problem. In application to the BrS problem in electron–atom scattering the models beyond the plane-wave BA used in both the non-relativistic and the relativistic domains include the corrections due to the Elwert factor and its modifications [70, 73, 131, 281],the use of Sommerfeld-Maue functions [81, 131, 281], the approaches based on the classical [135, 227]and semi-classical [49, 65, 234] scattering theories. The best available results have been obtained using the distorted partial-wave expansion of the projectile wavefunction. This scheme has been applied to study the ordinary BrS process of non-relativistic projectiles in the dipole-photon approximation [405, 406, 445, 446]. The most adequate description of the process has been obtained by applying the (relativistic) DPWA accompanied by the multipole series for the projectile-photon interaction operator, see, for example, the review articles [189, 327, 347, 348, 350] and research papers [133, 223, 281, 380, 381, 407–412, 443]. and references therein.

In many papers on the PBrS problem the non-relativistic Born approximation was used for both light (a positron, an electron) and heavy (a proton, an ion) projectiles. Although the range of validity of the Born approximation for PBrS is larger than for OBrS, to obtain more accurate data on the total BrS cross sections of a light projectile it is necessary to go beyond this scheme. Therefore, the non-relativistic DPWA formalism was developed [20, 27–29, 244] and applied to calculate the cross sections $\mathrm{d}\sigma$ and $\mathrm{d}^2\sigma$ over a broad spectral range [246, 248–250, 253, 257, 365] for non-relativistic electrons scattered on many-electron atoms. The partial-wave approach was also used to study the PBrS process of slow electrons [166, 273–275, 455]. To describe low-energy ($\varepsilon_1 \sim 10^0 \dots 10^1$ eV) electron scattering from highly polarizable targets (metallic clusters, fullerenes) it is necessary to go beyond the frozen-core static potential (even supplemented with the static long-range polarizational potential $V_{\rm pol} = -\alpha_{\rm d}(0)/2r^4$). To this end, the approach was developed [113, 200] to calculate the partial wavefunctions of the scattering electron as solutions of the Dyson equation with the non-local polarization potential [19, 107, 175].

In most of the papers the dynamic atomic response to the joint actions of the field of the projectile and of the radiation field was treated within the frame of the non-relativistic dipole-photon theory (with exception for several papers mentioned below). Even within this framework the accurate calculation of the generalized polarizability $\alpha(\omega, q)$ is not a simple task. Apart from the case of a hydrogen atom (or hydrogen-like ion) where the analytical evaluation is possible [99, 125, 257] one has to use more sophisticated approaches to calculate this quantity. The methods known to us include the HF-based calculations with the inclusion of many-body corrections, [12, 13, 20, 69, 244–248, 253], the statistical model of the atom [59–66] based on the theory of the inhomogeneous electron gas [292]. Another semi-empirical approach for effective and quite accurate calculation of $\alpha(\omega, q)$ in the vicinity of giant resonances was proposed initially for many-electron targets, [242, 243, 245], and later on was applied to fullerene and clusters [114, 115, 200, 300, 391]. The approach proposed in [256] is based on the use of the non-relativistic Coulomb Green function, and is valid for the calculation of $\alpha(\omega, q)$ in the vicinities of K- and L-shells.

Beyond the dipole-photon approximation the PBrS was considered in collisions of a non-relativistic heavy projectile with many-electron atom [38, 39, 56, 57, 168, 389]. In these papers the corrections of the order $kR_{at} \ll 1$ were considered and it was demonstrated that they lead to the additional modification of the angular distribution of the radiation. More systematic analysis of the non-dipole corrections has become available recently within the framework of the full relativistic description of the PBrS process, see Chap. 5.

2.2 Non-Relativistic Distorted Partial Waves Approximation

In this section we consider the BrS emission formed in electron–atom collisions in the domain of intermediate energies ε of the incoming electron. The term 'intermediate energy' implies that the initial, ε_1, and final, ε_2, energies are, on the one hand, not too high to justify the application of the Born approximation, but on the other hand, not too low to implicate the consideration of the exchange effects between the projectile and the target electrons. An adequate description of the collision process at intermediate energies is obtained by using the distorted partial-wave approximation (DPWA) (see, for example, [330]). It is used below to construct the BrS amplitude and evaluate the partial-wave series of the cross sections [20, 29].

2.2.1 DPWA Series for BrS Amplitude

The total BrS amplitude

$$f_{tot} = f_{ord} + f_{pol} \tag{2.8}$$

treated in the first order of the non-relativistic perturbation theory in the electron–dipole-photon interaction, can be presented by the Feynman diagrams presented in Fig. 2.2. The first diagram stands for the OBrS amplitude, f_{ord}, while the last two describe the term f_{pol}. The solid straight lines in the diagrams describe the initial

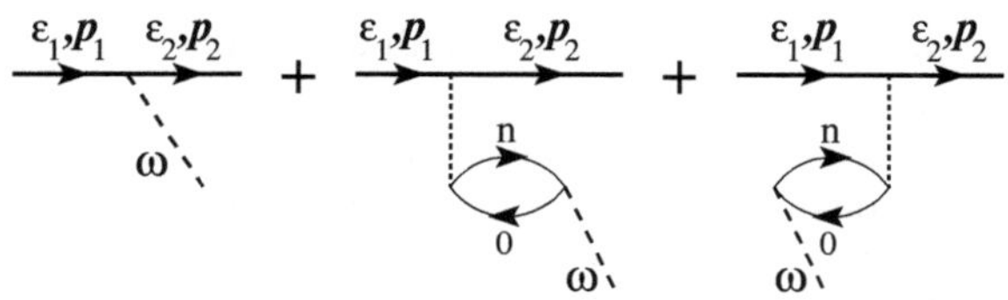

Fig. 2.2 Diagrammatic representation of the amplitude of BrS in the lowest order of the non-relativistic perturbation theory in electron–dipole-photon photon interaction. The *first diagram* describes the OBrS process, the *second* and the *third* one stand for PBrS

and the final states of the incoming particle (for clarity, we will term the projectile as 'electron') in the static field of the target atom (or ion). Other notations are as in Fig. 2.1.

As drawn, the BrS amplitude does not account for the exchange between the projectile and the target electrons. We also assume that the electron energy before, ε_1, and after, ε_2, the collision are high enough in comparison with the ionization potentials of the atomic subshells which provide the main contribution to the target polarization by the electromagnetic field of the frequency ω. The latter satisfies the strong inequality $\omega R_{\rm at}/c \ll 1$ which implies the validity of the dipole approximation, when the emission wavelength $\lambda = 2\pi c/\omega$ exceeds greatly the atomic size $R_{\rm at}$.

General analytical expressions for the terms $f_{\rm ord}$ and $f_{\rm pol}$, corresponding to the presented diagrams, are as follows:

$$f_{\rm ord} = \left\langle \mathbf{p}_2^{(-)} \left| \mathbf{e} \cdot \mathbf{r} \right| \mathbf{p}_1^{(+)} \right\rangle, \tag{2.9}$$

$$f_{\rm pol} = -\sum_n \left[\frac{\langle 0 \left| \mathbf{e} \cdot \mathbf{D} \right| n \rangle \left\langle \mathbf{p}_2^{(-)} n \left| \hat{V} \right| \mathbf{p}_1^{(+)} 0 \right\rangle}{\omega_{n0} - \omega - \mathrm{i}0} + \frac{\left\langle \mathbf{p}_2^{(-)} 0 \left| \hat{V} \right| \mathbf{p}_1^{(+)} n \right\rangle \langle n \left| \mathbf{e} \cdot \mathbf{D} \right| 0 \rangle}{\omega_{n0} + \omega} \right]. \tag{2.10}$$

Here $|\mathbf{p}_1^{(+)}\rangle$ and $|\mathbf{p}_2^{(-)}\rangle$ denote, correspondingly, the initial and the final scattering states of the electron with the asymptotic momenta $\mathbf{p}_1$ and $\mathbf{p}_2$. The '$\pm$' superscripts correspond to the outgoing ('$+$') and to the incoming ('$-$') spherical waves in the asymptotic form of the electron wavefunction. The general form of the DPWA series for these states are:

$$\left| \mathbf{p}^{(\pm)} \right\rangle = 4\pi \sqrt{\frac{\pi}{p}} \sum_{lm} \mathrm{i}^l \exp\left(\pm \mathrm{i}\delta_l(p)\right) \frac{P_\nu(r)}{r} Y^*_{lm}(\mathbf{n}_{\mathbf{p}}) Y_{lm}(\mathbf{n}_{\mathbf{r}}) \,. \tag{2.11}$$

Here δ_ν are the phaseshifts, the notation ν stands for a set of quantum numbers (p, l). The radial wave functions $P_\nu(r)$ satisfy the Schrödinger equation in the static potential created by the target atom in its ground state (the so-called 'frozen core' approximation). The factor in front of the sum implies that the radial wave functions are normalized to the delta function of energy measured in Rydberg. The notation

$Y_{lm}(\mathbf{n_a})$ stands for the spherical harmonics [420] dependent on the spherical angles of the unit vector $\mathbf{n_a}$ along the vector $\mathbf{a}$.

Vector $\mathbf{D}$ in (2.10) is the operator of the dipole interaction of the atomic electrons with the electromagnetic field, $\omega_{n0} = E_n - E_0$ is the energy of the atom's transition from the ground state 0 to the virtually excited state n (including excitations into the continuum).

The PBrS amplitude (2.10) can be expressed in terms of the generalized dynamic polarizability $\alpha(\omega, q)$ [244]. To facilitate this, one can make use of the Fourier transformation of the operator $\hat{V}$, which stands for the Coulomb interaction between the projectile ($\mathbf{r}$) and the atomic electrons ($\mathbf{r}_\mathrm{a}$):

$$\hat{V} = \sum_a \frac{1}{|\mathbf{r} - \mathbf{r}_\mathrm{a}|} = \frac{1}{2\pi^2} \sum_a \int \frac{\mathrm{d}^3\mathbf{Q}}{Q^2} \exp(-\mathrm{i}\mathbf{Q} \cdot (\mathbf{r} - \mathbf{r}_\mathrm{a})). \tag{2.12}$$

Here the sum is carried out over all electrons.

Introducing (2.12) into (2.10) one derives:

$$\begin{aligned} f_\mathrm{pol} &= -\frac{1}{2\pi^2} \int \frac{\mathrm{d}^3\mathbf{Q}}{Q^2} \left\langle \mathbf{p}_2^{(-)} \left| \mathrm{e}^{-\mathrm{i}\mathbf{Q}\cdot\mathbf{r}} \right| \mathbf{p}_1^{(+)} \right\rangle \\ &\times \sum_n \left[\frac{\langle 0\,|\mathbf{e}\cdot\mathbf{D}|\,n\rangle \langle n\,|\sum_a \mathrm{e}^{\mathrm{i}\mathbf{Q}\cdot\mathbf{r}_\mathrm{a}}|\,0\rangle}{\omega_{n0} - \omega - \mathrm{i}0} + \frac{\langle 0\,|\sum_a \mathrm{e}^{\mathrm{i}\mathbf{Q}\cdot\mathbf{r}_\mathrm{a}}|\,n\rangle \langle n\,|\mathbf{e}\cdot\mathbf{D}|\,0\rangle}{\omega_{n0} + \omega} \right]. \end{aligned} \tag{2.13}$$

For a spherically symmetric target the $\sum_n[\ldots]$ is related to the generalized dynamic polarizability of the atom, $\alpha(\omega, Q)$:

$$\mathrm{i}(\mathbf{e}\cdot\mathbf{Q})\alpha(\omega, Q) = \sum_n \sum_a \left[\frac{\langle 0\,|\mathbf{e}\cdot\mathbf{D}|\,n\rangle \langle n\,|\,\mathrm{e}^{\mathrm{i}\mathbf{Q}\cdot\mathbf{r}_\mathrm{a}}|\,0\rangle}{\omega_{n0} - \omega - \mathrm{i}0} + \frac{\langle 0\,|\,\mathrm{e}^{\mathrm{i}\mathbf{Q}\cdot\mathbf{r}_\mathrm{a}}|\,n\rangle \langle n\,|\mathbf{e}\cdot\mathbf{D}|\,0\rangle}{\omega_{n0} + \omega} \right]. \tag{2.14}$$

In the limit of small Q the polarizability $\alpha(\omega, Q)$ reduces to the dipole dynamic polarizability, $\alpha(\omega)$:

$$\lim_{Q\to 0} \alpha(\omega, Q) = \alpha(\omega) = \frac{2}{3} \sum_n \frac{\omega_{n0}\ \langle 0\,|\mathbf{D}|\,n\rangle \langle n\,|\sum_a \mathbf{r}_\mathrm{a}|\,0\rangle}{\omega_{n0}^2 - \omega^2 - \mathrm{i}0}. \tag{2.15}$$

Thus, for f_pol one can write

$$f_\mathrm{pol} = -\frac{\mathrm{i}}{2\pi^2} \int \mathrm{d}\mathbf{Q}\, \frac{\mathbf{e}\cdot\mathbf{Q}}{Q^2} \left\langle \mathbf{p}_2^{(-)} \left| e^{-\mathrm{i}\mathbf{Q}\cdot\mathbf{r}} \right| \mathbf{p}_1^{(+)} \right\rangle \alpha(\omega, Q). \tag{2.16}$$

Expression (2.16) is obtained within the frame of the DPWA. In connection with the PBrS process was applied for the first time in [20] (see, also, [28, 29]). Expression (2.16), however, differs from the formulae for f_pol presented in the cited works in that that it explicitly establishes a relationship between the amplitude of PBrS and the generalized polarizability of the target. This form of representation of $f_{\rm pol}$ has several advantages.

In many cases the dynamic response of an atom to the field of a projectile and that of a photon, which leads to the PBrS emission, bears essentially many-electron features. Therefore, its correct description implies the use of the methods developed within the framework of many-body perturbation theory. The fact, that $f_{\rm pol}$ can be expressed only via $\alpha(\omega, Q)$, which contains all the necessary information on the atomic dynamic response, can simplify theoretical and numerical analysis of the PBrS process (see, Chap. 4 for the examples).

The representation of $f_{\rm pol}$ in the form (2.16) allows one to easily carry out the Born limit of the PBrS amplitude. For doing this, one substituting the distorted waves $\left|\mathbf{p}_{1,2}^{(\pm)}\right\rangle$ with the wavefunctions of a free movement, $\left|\tilde{\mathbf{p}}_{1,2}^{(\pm)}\right\rangle = \exp(\mathrm{i}\mathbf{p}_{1,2}\cdot\mathbf{r})$, and accounting for the relation

$$\langle \tilde{\mathbf{p}}_2^{(-)} |\exp(-\mathrm{i}\mathbf{Q}\cdot\mathbf{r})| \tilde{\mathbf{p}}_1^{(+)}\rangle = (2\pi)^3\delta(\mathbf{q}-\mathbf{Q}),$$

one obtains an expression for $f_{\rm pol}$ within the frame of the plane-wave Born approximation (cf. (2.1)):

$$f_{\rm pol}^{\rm B} = -4\pi\mathrm{i}\frac{\mathbf{e}\cdot\mathbf{q}}{q^2}\,\alpha(\omega, q). \tag{2.17}$$

This expression for the PBrS amplitude as well as the corresponding formulae for the spectral and angular distribution of the radiation were presented for the first time in [14].

Using (2.11) in (2.9) and (2.16) one obtains, after some angular algebra, the expansion of $f_{\rm ord}$ and $f_{\rm pol}$ in partial series. The structure of both series is similar and is given by the following general expression:

$$f = \frac{16\pi^3}{\sqrt{p_1p_2}}\sqrt{\frac{4\pi}{3}}\sum_{l_1l_2}\mathrm{i}^{l_1-l_2}\mathrm{e}^{\mathrm{i}\left(\delta_{l_1}(p_1)+\delta_{l_2}(p_2)\right)}(-1)^{l_>}\sqrt{l_>}\,T_{l_2l_1}\left(\mathbf{n_e},\mathbf{n}_{\mathbf{p}_1},\mathbf{n}_{\mathbf{p}_2}\right)R_{l_2l_1}. \tag{2.18}$$

Here $l_2 = l_1 \pm 1$ in accordance with the dipole selection rules, $l_> = \max\{l_1, l_2\}$. The factor $T_{l_2l_1}\left(\mathbf{n_e},\mathbf{n}_{\mathbf{p}_1},\mathbf{n}_{\mathbf{p}_2}\right)$, dependent on the angular variables of the vectors $\mathbf{e}$, $\mathbf{p}_1$ and $\mathbf{p}_2$, is equal to

$$T_{l_2l_1}\left(\mathbf{n_e},\mathbf{n}_{\mathbf{p}_1},\mathbf{n}_{\mathbf{p}_2}\right) = \sum_{\substack{m_1m_2\\ m}}(-1)^{m_2}\begin{pmatrix} l_2 & 1 & l_1\\ -m_2 & m & m_1\end{pmatrix}Y_{1m}^*(\mathbf{n_e})Y_{l_1m_1}^*(\mathbf{n}_{\mathbf{p}_1})Y_{l_2m_2}(\mathbf{n}_{\mathbf{p}_2}). \tag{2.19}$$

Here $\begin{pmatrix} l_2 & 1 & l_1\\ -m_2 & m & m_1\end{pmatrix}$ is the 3j-symbol [420].

In (2.18) the notation $R_{l_2 l_1}$ stands for

$$R_{l_2 l_1} = \begin{cases} +R^{\text{ord}}_{l_2 l_1} & \text{for OBrS} \\ \\ -R^{\text{pol}}_{l_2 l_1} & \text{for PBrS.} \end{cases}$$

The partial amplitudes of the ordinary and the polarizational BrS are expressed via the following integrals:

$$R^{\text{ord}}_{l_2 l_1} = \langle \nu_2 \| r \| \nu_1 \rangle = \int_0^\infty dr \, P_{\nu_2}(r) \, r \, P_{\nu_1}(r) \tag{2.20}$$

$$R^{\text{pol}}_{l_2 l_1} = \frac{2}{\pi} \int_0^\infty \mathrm{d}Q \, Q \, \langle \nu_2 \| j_1(Qr) \| \nu_1 \rangle \, \alpha(\omega, Q). \tag{2.21}$$

where $j_1(Qr)$ is the spherical Bessel function [1].

The ordinary BrS emission is formed mainly at the distances ρ lower and equal, in order of magnitude, to the atomic radius, $\rho \lesssim R_{\text{at}}$ (see, for example, [81, 350]. In contrast, in the PBrS process the large distances $\rho \sim p_1/\omega$ play the most important role [45, 455]. Thus, the specific angular momenta in the ordinary BrS process are $l_{\text{ord}} \sim p_1 R_{\text{at}}$, while for the polarizational BrS, $l_{\text{pol}} \sim p_1^2/\omega$. It follows then that for sufficiently high projectile velocities $l_{\text{pol}} \gg l_{\text{ord}} > 1$. From the computational viewpoint this inequality means that the sum over l_1 and l_2 in (2.18) in the case of PBrS converges rather slowly, and one has to calculate a large number of partial terms to obtain an accurate result for f_{pol}.

To avoid the technical difficulty related to slow convergence of the partial series in f_{pol} one can take advantage of the fact that the projectile's radial wavefunctions of large orbital momenta are weakly distorted by the static atomic potential and, thus, are close to those of a free movement. Hence, it is possible to re-write the expression for f_{pol}, explicitly extracting the Born amplitude (2.17) from (2.18). Then, the remainder, $\Delta f_{\text{pol}} = f_{\text{pol}} - f^{\text{B}}_{\text{pol}}$ will be represented by a rapidly convergent partial series [244].

Having done this one obtains the following representation of f_{pol}:

$$f_{\text{pol}} = f^{\text{B}}_{\text{pol}} + \Delta f_{\text{pol}} \tag{2.22}$$

The structure of the term Δf_{pol} is given by the general expression (2.18) where the partial amplitudes $R_{l_2 l_1}$ are substituted with

$$\Delta R^{pol}_{l_2 l_1} = R^{\text{pol}}_{l_2 l_1} - \mathrm{e}^{-i\left(\delta_{l_1}(p_1) - \delta_{l_2}(p_2)\right)} \, \tilde{R}^{\text{pol}}_{l_2 l_1} \, . \tag{2.23}$$

The notation $\tilde{R}^{\text{pol}}_{l_2 l_1}$ is used for the integral

$$\tilde{R}^{\text{pol}}_{l_2 l_1} = \frac{2}{\pi} \int_0^\infty \mathrm{d}Q \, Q \, \langle \tilde{\nu}_2 \| j_1(Qr) \| \tilde{\nu}_1 \rangle \, \alpha(\omega, Q) \tag{2.24}$$

The matrix element in the integrand in (2.24) is calculated between the radial wave functions of a free particle: $\|\tilde{\nu}\rangle = (p/\pi)^{1/2} r j_l(pr)$.

2.2.2 BrS Cross Section

The two-fold differential BrS cross section, which characterizes the spectral and the angular distributions of radiation, is given by

$$\frac{d^2\sigma}{d\omega d\Omega_{\mathbf{k}}} = \frac{1}{(2\pi)^4}\frac{\omega^3}{c^3}\frac{p_2}{p_1}\int d\Omega_{\mathbf{p}_2}\sum_{\lambda}|f_{\mathrm{tot}}|^2 \tag{2.25}$$

Here $d\Omega_{\mathbf{k}}$ is the element of the solid angle of the emission along the vector $\mathbf{k}$ which is the photon momentum. The integral is carried out over the solid angle, $d\Omega_{\mathbf{p}_2}$, of the scattered electron. The sum is taken over the photon polarizations, λ. For arbitrary vectors $\mathbf{a}$ and $\mathbf{b}$ the following general formula can be applied (see, e.q., [81]):

$$\sum_{\lambda}(\mathbf{e}_\lambda \cdot \mathbf{a})(\mathbf{e}_\lambda \cdot \mathbf{b}) = \mathbf{a}\cdot\mathbf{b} - \frac{(\mathbf{a}\cdot\mathbf{k})(\mathbf{b}\cdot\mathbf{k})}{k^2}\,. \tag{2.26}$$

The cross section (2.25) defines the probability of the photon emission within the energy interval $[\omega, \omega + d\omega]$ in the solid angle $d\Omega_{\mathbf{k}}$, normalized to the flux of the incident electrons.

Integrating (2.25) over the emission angles one defines the spectral distribution of BrS, which is characterized by the differential cross section:

$$\frac{d\sigma}{d\omega} = \int d\Omega_{\mathbf{k}}\,\frac{d^2\sigma}{d\omega d\Omega_{\mathbf{k}}}\,. \tag{2.27}$$

In the dipole approximation, the double differential cross section $d^2\sigma/d\omega d\Omega_{\mathbf{k}}$ has a quite simple dependence on the emission angle $\theta_{\mathbf{k}} = \widehat{\mathbf{k},\mathbf{p}_1}$ [69]:

$$\frac{d^2\sigma}{d\omega d\Omega_{\mathbf{k}}} = \frac{1}{4\pi}\,\frac{d\sigma}{d\omega}\Big(1 - \beta(\omega)\,P_2(\cos\theta_{\mathbf{k}})\Big)\,. \tag{2.28}$$

$P_2(y) = (3y^2 - 1)/2$ is the Legendre polynomial of the second order. In analogy with the photoionization process, the quantity $\beta(\omega)$, which defines the profile of the angular distribution of the dipole radiation, is called angular anisotropy parameter [121].

Using (2.18)–(2.21) to construct the total BrS amplitude (2.8), and then substituting the result into (2.25) one obtains the following partial series for the cross section $d\sigma/d\omega$ and for the angular anisotropy parameter [29, 244]:

$$\frac{\mathrm{d}\sigma}{\mathrm{d}\omega} = \frac{32\pi^2}{3p_1^2}\frac{\omega^3}{c^3}\sum_{l=0}^{\infty}\sum_{l'=l\pm 1} l_> \left|R^{\mathrm{tot}}_{l'l}\right|^2 = \frac{\mathrm{d}\sigma_{\mathrm{ord}}}{\mathrm{d}\omega} + \frac{\mathrm{d}\sigma_{\mathrm{pol}}}{\mathrm{d}\omega} + \frac{\mathrm{d}\sigma_{\mathrm{int}}}{\mathrm{d}\omega}\,. \tag{2.29}$$

$$\beta(\omega) = \left(\frac{\mathrm{d}\sigma}{\mathrm{d}\omega}\right)^{-1}\frac{16\pi^2}{3p_1^2}\frac{\omega^3}{c^3} \times \sum_l (l+1)\left\{ l\sum_{l'=l\pm 1}\frac{\left|R^{\mathrm{tot}}_{l'l}\right|^2}{2l'+1} - 6\frac{l+2}{2l+3}\mathrm{Re}\left[\mathrm{e}^{\mathrm{i}\Delta_l}\, R^{\mathrm{tot}}_{l+1l}\left(R^{\mathrm{tot}}_{l+1l+2}\right)^*\right]\right\}, \tag{2.30}$$

where the following short-hand notations are used:

$$R^{\mathrm{tot}}_{l'l} = R^{\mathrm{ord}}_{l'l} - R^{\mathrm{pol}}_{l'l}\,, \qquad \Delta_l = \delta_l(p_1) - \delta_{l+2}(p_1)\,.$$

If one neglects the partial PBrS amplitude $R^{\mathrm{pol}}_{l_2 l_1}$ on the right-hand sides of (2.29) and (2.30), the resulting formula coincides with the known partial-wave expansion for OBrS [385, 405, 406, 446].

Presentation of the cross section as a sum of three terms, $\mathrm{d}\sigma_{\mathrm{ord}}$, $\mathrm{d}\sigma_{\mathrm{pol}}$ and $\mathrm{d}\sigma_{\mathrm{int}}$ (2.30), although being somewhat conditional, is convenient for further analysis.

In principle, it is experimentally possible to distinguish the photons emitted via the polarization mechanism from those generated in the OBrS process.

As has been already mentioned, the domains of the impact parameters in which one of the two BrS mechanisms dominates are well separated provided the condition $p_1/\omega \gg R_{\mathrm{at}}$ is fulfilled. Small impact parameters $\rho < R_{\mathrm{at}}$, which are important for OBrS, correspond to large momentum transfer, $q \sim \rho^{-1}$, or, which is equivalent, to (comparatively) large scattering angles. The polarization mechanism is of less importance, since the induced dipole moment is small. Formally, the latter statement follows from the definition of the generalized polarizability, (2.14). Indeed, for large values transferred momenta, $Q \gg 1/R_{\mathrm{at}}$ the exponent $\exp(\mathrm{i}\mathbf{Q}\cdot\mathbf{r}_a)$ rapidly oscillates in the matrix element, resulting in $\alpha(\omega, Q) \to 0$.

As the distance between the projectile and the target becomes larger than the atomic radius, the Coulomb field of the nucleus is fully screened by the electron cloud, and the OBrS radiation is suppressed. On the contrary, the contribution of PBrS for $\rho \gg R_{\mathrm{at}}$ is enhanced since the field of the projectile is nearly uniform on the scale of R_{at}, so that the polarization of the target occurs more effectively. Large distances correspond to small transferred momenta, $q \ll R_{\mathrm{at}}$ and, correspondingly, to small scattering angles.

Therefore, to distinguish between the polarizational and ordinary BrS, the experimental setup should allow one to detect the photon and the scattered electron simultaneously. Then the PBrS will be predominantly detected for small-angle scattering events, while the large-angle scattering will give rise to the photon yield through the OBrS channel.

The arguments presented above are valid for neutral atoms. For ionic targets, the long-range Coulomb field of the net charge of the ion increases the intensity of OBrS in the domain of small-angle scattering.

For the sake of completeness and for further reference let us present the formulae for $\mathrm{d}\sigma/\mathrm{d}\omega$ and $\mathrm{d}^2\sigma/\mathrm{d}\omega\mathrm{d}\Omega_{\mathbf{k}}$ written within the framework of the plane-wave Born approximation. The formulae can be obtained directly, by using the BrS amplitude (2.1) in (2.25), or as the Born limit of (2.29) and (2.30) [29]. The result can be written as follows (see also [13]):

$$\frac{\mathrm{d}^2\sigma_{\mathrm{B}}}{\mathrm{d}\omega\mathrm{d}\Omega_{\mathbf{k}}} = \frac{1}{4\pi}\frac{\mathrm{d}\sigma_{\mathrm{B}}}{\mathrm{d}\omega}\Big[1 - \beta_{\mathrm{B}}(\omega)P_2(\cos\theta_{\mathbf{k}})\Big), \tag{2.31}$$

$$\frac{\mathrm{d}\sigma_{\mathrm{B}}}{\mathrm{d}\omega} = C\int_{q_{\min}}^{q_{\max}}\frac{\mathrm{d}q}{q}\left|\frac{Z_0}{M}\Big(Z - F(q)\Big) + \omega^2\alpha(\omega,q)\right|^2, \tag{2.32}$$

$$\beta_{\mathrm{B}}(\omega) = \left(\frac{\mathrm{d}\sigma_{\mathrm{B}}}{\mathrm{d}\omega}\right)^{-1} C\int_{q_{\min}}^{q_{\max}}\frac{\mathrm{d}q}{q}\left|\frac{Z_0}{M}\Big(Z - F(q)\Big) + \omega^2\alpha(\omega,q)\right|^2 P_2(\cos\theta_{\mathbf{q}}). \tag{2.33}$$

Here $C = 16Z_0^2/3c^3p_1^2\omega$, $q_{\substack{\max\\ \min}} = p_1 \pm p_2 = p_1(1 \pm \sqrt{1-\omega/\varepsilon_1})$ are the maximum and minimum values of the transferred momentum $q = |\mathbf{p}_1 - \mathbf{p}_2|$, $\theta_{\mathbf{q}} = \mathbf{p}_1\cdot\mathbf{q}/p_1q = (q_{\max}q_{\min} + q^2)/2p_1q$.

2.3 Multipole Series for PBrS Cross Section

Let us analyze the role of quadrupole terms, as well as the higher multipoles, in the cross section of PBrS [389]. To simplify the consideration we will treat the PBrS within the BA. In Sect. 4.1 it will be demonstrated that the range of applicability of BA for the PBrS process for an electron is much broader than that for the OBrS process. As for the heavy projectiles, the BA can be applied to describe PBrS in comparatively slow collisions as well [388].

For a heavy projectile, the PBrS cross section differential in ω, the solid angle of emission and in the transferred momentum q, can be written in the form [389][2]:

$$\frac{\mathrm{d}^3\sigma_{\mathrm{pol}}}{\mathrm{d}\omega\mathrm{d}\Omega_{\mathbf{k}}\mathrm{d}q} = \frac{2Z_0^2\omega^3}{\pi c^3 v_1^2 q}\sum_{N=0}^{\infty} k^N\mathscr{A}_N(\omega,q)\,C_N(\cos\theta_{\mathbf{k}},\cos\theta_{\mathbf{q}}). \tag{2.34}$$

Here $\cos\theta_{\mathbf{q}} = \omega/v_1q + q/2p_1$, and the following notation is introduced:

[2] The formulae presented in this section can be also derived as a non-relativistic limit of general relativistic formalism for PBrS described in detail in Sects. 6.5.1 and 6.5.3.

$$\mathscr{A}_N(\omega, q) = \sum_{j=0}^{N} \alpha_j(\omega, q)\alpha^*_{N-j}(\omega, q). \tag{2.35}$$

For atomic targets with filled (or semi-filled) subshells, the expression for a multipole generalized polarizability $\alpha_j(\omega, q)$ (with $j \geq 1$)[3] reads

$$\alpha_j(\omega, q) = -4\mathrm{i} \sum_{n'l'} \sum_{l_1 l_2} \frac{\mathrm{i}^{l_1}(-1)^{l+l'}\mathscr{C}_{l\,j\,l_2 l_1\,l'}}{\omega_{n0}^2 - \omega^2 - \mathrm{i}0} \langle nl\| j_{l_1}(qr)\|n'l'\rangle\langle n'l'\|r^j \frac{\mathrm{d}}{\mathrm{d}r}\|nl\rangle. \tag{2.36}$$

Here n and l stand for the principal and orbital quantum numbers. The summations in (2.36) are carried out subject to the following conditions: (1) the integers l_2 and j are of the same parity and $l_2 \leq j$, (2) $l_1 = l_2 \pm 1$, (3) the integers l' and $l + l_1$ are of the same parity and $l + l_1 \geq l' \geq |l - l_1|$. The coefficient $\mathscr{C}_{l\,j\,l_2 l_1\,l'}$ is as follows:

$$\mathscr{C}_{l\,j\,l_2 l_1\,l'} = \frac{j!(2l+1)(2l'+1)(2l_1+1)(2l_2+1)}{(j-l_2)!!(j+l_2+1)!!} \begin{pmatrix} l & l_1 & l' \\ 0 & 0 & 0 \end{pmatrix} \begin{pmatrix} l_2 & 1 & l_1 \\ 0 & 0 & 0 \end{pmatrix}. \tag{2.37}$$

The coefficients C_N from (2.37) are expressed in terms of the Legendre polynomials as follows:

$$C_N(\cos\theta_k, \cos\theta_{\mathbf{q}}) = \sum_{L \leq N+2} K_{NL}\, P_L(\cos\theta_q) P_L(\cos\theta_{\mathbf{k}}), \tag{2.38}$$

where

$$K_{NL} = \frac{(2L+1)N!}{(N-L)!!(N+L+1)!!} \quad \text{for} \quad L \leq N\,.$$
$$K_{NN+2} = -\frac{N+2}{(2N+3)!!} \quad \text{for} \quad L = N+2.$$

In (2.38), parities of the summation index L and the integer N coincide.

Integrating (2.34) over the emission angle one arrives at

$$\frac{\mathrm{d}^2\sigma_{\mathrm{pol}}}{\mathrm{d}\omega \mathrm{d}q} = \frac{8Z_0^2\omega^3}{c^3 v_1^2 q} \sum_{N=0}^{\infty} \frac{k(2N)(2N)!}{(2N)!!(2N+1)!!} \left(1 - \frac{2N+1}{2N+3}\right) \mathscr{A}_{2N}(\omega, q). \tag{2.39}$$

This expression shows, that there is no interference of the photons of different multipolarity in the spectral distribution of the emitted radiation. The interference takes place in the spectral-angular distribution but vanishes in the cross section integrated over the emission angle.

[3] For $j = 0$ the term $\alpha_0(\omega, q)$ coincides with $\alpha(\omega, q)$ defined in (2.14).

The above-written formulae are simplified in the dipole-photon limit supplemented with the quadrupole correction. Thus, the dipole part of the double differential cross section (2.39) is as follows:

$$\left.\frac{d^2\sigma_{pol}}{d\omega dq}\right|_{dip} = \frac{16Z_0^2\omega^3}{3c^3v_1^2q}|\alpha(\omega, q)|^2. \tag{2.40}$$

The quadrupole correction to (2.40) reads:

$$\left.\frac{d\sigma}{d\omega dq}\right|_{quad} = \frac{16Z_0^2\omega^5}{15c^5v_1^2q}\mathscr{A}_2(\omega, q), \tag{2.41}$$

where

$$\mathscr{A}_2(\omega, q) = 2\mathrm{Re}\,\alpha_0(\omega, q)\alpha_2^*(\omega, q) + |\alpha_1(\omega, q)|^2.$$

The parameter of the multipole series (2.35), (2.39) is $(\omega R_{at}/c)^2$. For the photon energies comparable to the K-ionization potential of an atom with $Z \sim 10$, this parameter can be estimated as $\sim$0.1–0.2. Hence, the quadrupole term (2.3) can provide additional 10–20 % to the yield of PBrS. The quadrupole contribution increases with Z leading to a noticeable asymmetry in the angular distribution of emitted radiation with respect to the direction $\theta_{\mathbf{k}} = \pi/2$. This effect explains some discrepancy between the experimental data [205] and the results followed from the theory of dipole PBrS.

2.4 BrS Spectrum in the Tip Region

Let us estimate the relative contributions of PBrS near the high-frequency edge [27]:

$$\varepsilon_1, \omega \gg \varepsilon_2\,. \tag{2.42}$$

Thus, it is assumed that the incident electron releases nearly all of its energy via the radiative mechanisms. The target atom is supposed to be in its ground state before and after the collision.

To evaluate the amplitudes f_{ord} and f_{pol} we adopt the Born approximation for the incoming electron, $|\mathbf{p}_1\rangle = \exp(i\mathbf{p}_1 \cdot \mathbf{r})$, and consider the operator of electron–dipole-photon interaction in the form of 'velocity', $\hat{V}_\gamma = \mathbf{e} \cdot \hat{\mathbf{p}}$, rather in the form of 'length' which was used in Sect. 2.2.1 (see, for example, [385]). Then, instead of (2.9) and (2.16), one derives:

$$f_{\rm ord} = \left\langle \mathbf{p}_2^{(-)} \left| \mathbf{e} \cdot \hat{\mathbf{p}} \right| \mathbf{p}_1^{(+)} \right\rangle = (\mathbf{e} \cdot \mathbf{p}_1)\, \phi^*_{\mathbf{p}_2}(\mathbf{p}_1) \tag{2.43}$$

$$f_{\rm pol} = -\frac{\omega}{2\pi^2} \int \frac{d\mathbf{q}}{q^2}\, (\mathbf{e} \cdot \mathbf{q})\, \phi^*_{\mathbf{p}_2}(\mathbf{p}_1 - \mathbf{q})\, \alpha(\omega, q)\,. \tag{2.44}$$

Here $\phi_{\mathbf{p}_2}(\mathbf{Q})$ is the Fourier transform of the slow-electron wavefunction $\psi_{\mathbf{p}_2}(\mathbf{r})$:

$$\phi_{\mathbf{p}_2}(\mathbf{Q}) = \int d\mathbf{r}\, \psi_{\mathbf{p}_2}(\mathbf{r})\, e^{-i\mathbf{Q}\cdot\mathbf{q}}\,. \tag{2.45}$$

Let us estimate the amplitudes $f_{\rm ord}$ and $f_{\rm pol}$. In the final state the electron is slow. Therefore, in the partial-wave expansion (2.11) of its wavefunction one can retain only the s-wave.[4] Then one finds

$$\phi_{\mathbf{p}_2}(\mathbf{p}_1) = -\frac{8\pi F(p_1)}{p_1^4} \frac{\sin\delta_s(p_2)}{p_2}\,. \tag{2.46}$$

Here $F(p_1)$ is the atomic form-factor and $\delta_s(p_2)$ is the s-wave scattering phaseshift. Introducing the scattering length $L = -\sin\delta_s(p_2)/p_2$, one obtains for the OBrS amplitude:

$$f_{\rm ord} = (\mathbf{e} \cdot \mathbf{p}_1)\, \frac{8\pi F(p_1)}{p_1^4}\, L\,. \tag{2.47}$$

To evaluate the integral in (2.44) it is necessary to examine two q-regions:

$$\text{(A)} \quad R_{\rm at}^{-1} < q \ll p_1, \qquad \text{(B)} \quad |\mathbf{p}_1 - \mathbf{q}| \le R_{\rm at}^{-1}, \tag{2.48}$$

so that $f_{\rm pol} = f_{\rm pol}^{\rm A} + f_{\rm pol}^{\rm B}$. Here $R_{\rm at}$ stands for the (average) atomic radius.

In region (A) the factor $\phi^*_{\mathbf{p}_2}(\mathbf{p}_1 - \mathbf{q})$ can be estimated as follows:

$$\phi^*_{\mathbf{p}_2}(\mathbf{p}_1 - \mathbf{q}) \approx \phi^*_{\mathbf{p}_2}(\mathbf{p}_1) - \mathbf{q} \cdot \frac{\partial \phi^*_{\mathbf{p}_2}(\mathbf{p}_1)}{\partial \mathbf{p}_1} \sim \phi^*_{\mathbf{p}_2}(\mathbf{p}_1) \left(1 - \frac{4(\mathbf{q}\cdot\mathbf{p}_1)}{p_1^2}\right). \tag{2.49}$$

This leads to

$$f_{\rm pol}^{\rm A} \approx -\frac{8\omega}{3\pi p_1^2}\, f_{\rm ord} \int_{q<R_{\rm at}^{-1}} dq\, q^2\, \alpha(\omega, q) \approx -\frac{4}{9\pi R_{\rm at}^2}\, f_{\rm ord}\, \alpha(\omega)\,. \tag{2.50}$$

In the last relation in (2.50) it is taken into account that $p_1^2/2 \approx \omega$ and $\alpha(\omega, q) \approx \alpha(\omega)$ for $q \ll R_{\rm at}^{-1}$ (see (2.4)).

---

[4] We do not consider the case of resonant scattering in the final state with higher orbital momentum [432].

In region (B) only large distances $r > R_{at}$ contribute significantly to the Fourier transform $\phi_{\mathbf{p}_2}(\mathbf{p}_1 - \mathbf{q})$ as far as $p_2 \to 0$. Then

$$\phi_{\mathbf{p}_2}(\mathbf{p}_1 - \mathbf{q}) \approx (2\pi)^3\, \delta(\mathbf{p}_1 - \mathbf{p}_2 - \mathbf{q}) - \frac{4\pi L}{(\mathbf{p}_1 - \mathbf{q})^2 - p_2^2} \tag{2.51}$$

Substituting (2.51) into f_{pol}^{B} one obtains the following estimate:

$$f_{pol}^{B} \approx -f_{ord}\, \frac{\omega^2 \alpha(\omega, p_1)}{F(p_1) L} \left(1 - \beta \frac{L}{R_{at}}\right). \tag{2.52}$$

Here β is the factor of the order of 1.

Now we may estimate the ratio η of the PBrS-to-OBrS amplitudes:

$$\eta = \frac{f_{pol}}{f_{ord}} \sim -\frac{4}{9\pi R_{at}^2}\,\alpha(\omega) - \frac{\omega^2 \alpha(\omega, p_1)}{F(p_1) L} \left(1 - \beta \frac{L}{R_{at}}\right). \tag{2.53}$$

Despite the approximations made, one can deduce, basing on (2.53), for which energies $\varepsilon_1 \approx \omega$ the PBrS dominates in the tip region.

The quantities $\alpha(\omega)$ and $\alpha(\omega, p_1)$ have real and imaginary parts, and so does the ratio η. One can expect, that the inequality $|\eta| \gg 1$ might be met for photon frequencies at which $\mathrm{Im}\,\alpha(\omega)$ attains its maximum values. The latter are connected with the maxima of the photoionization cross section $\sigma_\gamma(\omega)$ since $\mathrm{Im}\,\alpha(\omega) = (c/4\pi\omega)\alpha(\omega)$.

The OBrS cross section in the tip region tends to zero as $(\varepsilon_1 - \omega)^{1/2}$ [81]. Hence, for the incident energies ε_1 close to the maxima of $\sigma_\gamma(\omega)$, the total BrS cross will also have maxima which are due to the PBrS channel.

Let us estimate numerically the magnitude of η for Xe [27]. The cross section $\sigma_\gamma(\omega)$ for Xe has a wide maximum above the threshold for the $4d$-subshell ($I_{4d} =$ 5.6 Ryd $\approx$ 76 eV). At the maximum which is located at $\omega_{max} = 7.5$ Ryd, $\sigma_\gamma(\omega)$ attains the value $\approx$30 Mb [183]. The maximum width is approximately 3 Ryd, i.e. $\omega = 6\ldots 9$ Ryd, and for the estimate of η we choose ε_1 from this energy region. Let $E_1 = 7.5$ Ryd. The atomic radius in this case is the radius of the 4d-subshell: $R_{4d} \sim (2I_{4d})^{-1/2} \approx 0.4$ a.u. The imaginary part of the dipole dynamical polarizability of the 4d-subshell is $\mathrm{Im}\,\alpha(\omega_{max}) = (c/4\pi\omega_{max})\alpha(\omega_{max})$; the scattering length for Xe is $L = -5.4$ a.u. Substituting these data into (2.53), one obtains $\eta = -4.5$.

This estimate demonstrates that for electron energies close to the maxima of $\sigma_\gamma(\omega)$ the polarizational part of the BrS amplitude can greatly exceed the ordinary one in the tip region of the spectrum.

Chapter 3
Polarizational Bremsstrahlung in Collisions with Hydrogen Atom

3.1 Generalized Dipole Dynamic Polarizability of a Hydrogen Atom

The formulae presented in Chaps. 1 and 2 indicate that to a great extent the behaviour of the total BrS cross section is determined by a generalized dynamic polarizability $\alpha(\omega, q)$ of a target atom. The generalized polarizability describes the dynamic response of the target to two external fields—the field of the photon and the Coulomb field of the projectile. For many-electron atoms such a response is determined, as a rule, by many-electron correlations which can be accounted for by applying methods of the many-body perturbation theory [8]. Therefore, in general case the calculation of $\alpha(\omega, q)$ poses quite a difficult technical problem. Therefore, in the general case the calculation of $\alpha(\omega, q)$ is a complex task that requires time-consuming numerical calculation (see, e.g., [244–248, 260]. However, for a hydrogen atom and hydrogen-like systems this problem can be solved analytically.

In this section, following [253], to calculation of the generalized dynamic polarizability of the hydrogen atom we apply the formalism of the Coulomb Green's function in the coordinate representation, $G(\mathbf{r}_1, \mathbf{r}_2; E)$. Alternatively, to evaluate $\alpha(\omega, q)$ one can use other approaches, for example, a direct analytical solution of the inhomogeneous ω-dependent Schrődinger equation [99, 397, 438, 439]. The method described below allows one to derive closed analytical expressions for the generalized polarizabilities for a hydrogenic state with arbitrary principal n and orbital quantum l numbers. As a first step, let us represent the radial wavefunctions $P_{nl}(r)$ of the hydrogen atom in terms of the derivatives of the generating function of the Laguerre polynomials [137]:

$$P_{nl}(r) = r\,\frac{2^{l+1}}{n^{l+2}\sqrt{\Gamma(n-l)\Gamma(n+l+1)}}\frac{\mathrm{d}^{n-l-1}}{\mathrm{d}t^{n-l-1}}\left[\frac{r^l \mathrm{e}^{-\lambda r}}{(1-t)^{2l+2}}\right]\Bigg|_{t=0} \tag{3.1}$$

where $\lambda = n^{-1}(1+t)/(1-t)$, and $\Gamma(.)$ stands for the Gamma-function.

A. V. Korol and A. V. Solov'yov, *Polarization Bremsstrahlung*,
Springer Series on Atomic, Optical, and Plasma Physics 80,
DOI: 10.1007/978-3-642-45224-6_3, © Springer-Verlag Berlin Heidelberg 2014

Such a representation is convenient for the calculation of matrix elements and, hence, of the polarizabilities (2.14) and (2.15) since the operation of differentiation, carried out after the evaluation of the radial integral in the matrix elements, greatly simplifies the whole calculation. Using (3.1) in the matrix elements allows one to substitute the radial wavefunctions $P_{nl}(r)$ with simple functions of the type $r^l \mathrm{e}^{-\lambda r}$ when evaluating the radial integrals. The final formula for $\alpha(\omega, q)$ for various (nl)-states is then obtained by differentiating the resulting expression with respect to λ. The latter operation can be simplified further by using the packages for symbolic calculus, e.g., *Mathematica*® developed by Wolfram Research [310].

Let us demonstrate that the sum of matrix elements, which constitute the polarizability, can be expressed in terms of the hypergeometric function ${}_2F_1(a, b; c; z)$.

Introducing Green's function one writes (2.14) as follows:

$$\alpha(\omega, q) = \mathscr{X}(E_0 + \omega, q) + \mathscr{X}(E_0 - \omega, q), \tag{3.2}$$

where the notation is introduced

$$\mathscr{X}(E, q) = -\frac{\mathrm{i}}{q} \left\langle 0 \left| \exp\left(-\mathrm{i}\mathbf{q}\cdot\mathbf{r}_1\right) G(\mathbf{r}_1, \mathbf{r}_2; E) \left(\mathbf{r}_2 \cdot \hat{\mathbf{q}}\right) \right| 0 \right\rangle. \tag{3.3}$$

with $\hat{\mathbf{q}} = \mathbf{q}/q$ being the unit vector along $\mathbf{q}$.

To proceed further let us expand Green's function in a series of the products of spherical harmonics dependent on spherical angles of the unit vectors $\hat{\mathbf{r}}_{1,2} = \mathbf{r}_{1,2}/r_{1,2}$:

$$G(\mathbf{r}_1, \mathbf{r}_2; E) = \sum_{LM} g_L(r_1, r_2; E)\, Y_{LM}(\hat{\mathbf{r}}_1) Y^*_{LM}(\hat{\mathbf{r}}_2). \tag{3.4}$$

where $g_L(r_1, r_2; E)$ is the Green function of the radial Schrödinger equation. The explicit formula for the radial Green's function in the Coulomb field $-Z/r$ reads [359]:

$$\begin{aligned} g_L(r_1, r_2; E) = {} & \frac{2\mathrm{i}\,(-1)^{L+1}}{\sqrt{r_1 r_2}} \int_1^\infty \mathrm{d}\xi \left(\frac{\xi+1}{\xi-1}\right)^{\mathrm{i}\nu} \mathrm{e}^{\mathrm{i}k\xi(r_1+r_2)} \\ & \times \frac{J_{2L+1}\left(2k\sqrt{r_1 r_2(\xi^2-1)}\right)}{\sqrt{\xi^2-1}}, \end{aligned} \tag{3.5}$$

where $k = \sqrt{2E}$, $\nu = Z/k$. This expression, being formally valid for positive values of the energy parameter E, can be analytically continued to the domain $E < 0$. One can carry out the analytic continuation in the final expression for the matrix elements from (3.2) while assuming $E > 0$ in the intermediate formulae.

Let us consider in more detail the evaluation of the generalized dynamic polarizability of the (ns) state, i.e. the one with $l = 0$.

Introducing (3.4) and (3.5) in (3.3) one integrates over the angular variables and over $r_{1,2}$. Then, carrying out the change of variable, one represents the matrix element $\mathscr{X}(E, q)$ as follows:

$$\mathscr{X}(E, q) = \frac{2^6}{n^5\, \Gamma^2(n)} \left[\frac{\mathrm{d}^{n-1}}{\mathrm{d}t_1^{n-1}} \frac{\mathrm{d}^{n-1}}{\mathrm{d}t_2^{n-1}} \frac{\mathscr{M}(t_1, t_2)}{(1-t_1)^2(1-t_2)^2} \right]_{t_1=t_2=0} . \tag{3.6}$$

The quantity $\mathscr{M}(t_1, t_2)$ stands for the integral:

$$\begin{aligned} \mathscr{M}(t_1, t_2) = \mathrm{i}\, 2^5 k^3 \int_0^1 d\tau \frac{\tau^{1-i\nu}}{\left[(y_1y_2 - x_1x_2\tau)^2 + q^2(y_2 - x_2\tau)^2\right]^3} \\ \times \left((1-\tau)(y_1y_2 - x_1x_2\tau) - k^2\tau \frac{5(y_1y_2 - x_1x_2\tau)^2 - q^2(y_2 - x_2\tau)^2}{(y_1y_2 - x_1x_2\tau)^2 + q^2(y_2 - x_2\tau)^2} \right) \end{aligned} \tag{3.7}$$

with $x_j = \lambda_j + \mathrm{i}k$, $y_j = \lambda_j - \mathrm{i}k$, $\lambda_j = n^{-1}\,(1 + t_j)/(1 - t_j)$ $(j = 1, 2)$.

The final expression for the integral $\mathscr{M}(t_1, t_2)$ can be written in terms of the sum of the hypergeometric functions of the type ${}_2F_1(m_1, m_2 - \mathrm{i}\nu, 4 - \mathrm{i}\nu, z) \equiv \mathscr{F}_{m_1m_2}(z)$:

$$\begin{aligned} \mathscr{M}(t_1, t_2) = {} & \frac{1}{y_2^5\, q^3 (3 - \mathrm{i}\nu)} \\ & \times \left\{ -\frac{\alpha_2^2}{\beta_2} \left[\frac{3\alpha_1\, \mathscr{F}_{12}(u)}{k(2 - \mathrm{i}\nu)} - \frac{6\mathrm{i}\alpha_1\, \mathscr{F}_{13}(u)}{y_2} - \frac{2\mathrm{i}(\alpha_1 + 2\mathrm{i}q)\, \mathscr{F}_{22}(u)}{\beta_2 (2 - \mathrm{i}\nu)} \right.\right. \\ & - \frac{6k(\alpha_1 + \mathrm{i}q)\, \mathscr{F}_{23}(u)}{y_2\beta_2} - \frac{8\mathrm{i}kq\, \mathscr{F}_{32}(u)}{\beta_2^2 (2 - \mathrm{i}\nu)} \\ & \left. + \frac{4\mathrm{i}k^2(\alpha_1 + 4\mathrm{i}q)\, \mathscr{F}_{33}(u)}{y_2\beta_2^2} + \frac{24\mathrm{i}k^3 q\, \mathscr{F}_{43}(u)}{y_2\beta_2^3} \right] \\ & + \frac{\alpha_1^2}{\beta_1} \left[\frac{3\alpha_2\, \mathscr{F}_{12}(v)}{k(2 - \mathrm{i}\nu)} - \frac{6\mathrm{i}\alpha_2\, \mathscr{F}_{13}(v)}{y_2} - \frac{2\mathrm{i}(\alpha_2 - 2\mathrm{i}q)\, \mathscr{F}_{22}(v)}{\beta_1 (2 - \mathrm{i}\nu)} \right. \\ & - \frac{6k(\alpha_2 - \mathrm{i}q)\, \mathscr{F}_{23}(v)}{y_2\beta_1} + \frac{8\mathrm{i}kq\, \mathscr{F}_{32}(v)}{\beta_1^2 (2 - \mathrm{i}\nu)} \\ & \left.\left. + \frac{4\mathrm{i}k^2(\alpha_2 - 4\mathrm{i}q)\, \mathscr{F}_{33}(v)}{y_2\beta_1^2} - \frac{24\mathrm{i}k^3 q\, \mathscr{F}_{43}(v)}{y_2\beta_1^3} \right] \right\} \end{aligned} \tag{3.8}$$

where the following notations have been used

$$\begin{aligned} & u = \frac{\alpha_2\, x_2}{\beta_2\, y_2}, \qquad v = \frac{\alpha_1\, x_2}{\beta_1\, y_2}, \\ & \alpha_1 = \lambda_1 + ik + iq,\ \alpha_2 = \lambda_2 + ik - iq, \\ & \beta_1 = \lambda_1 - ik + iq,\ \beta_2 = \lambda_2 - ik - iq. \end{aligned} \tag{3.9}$$

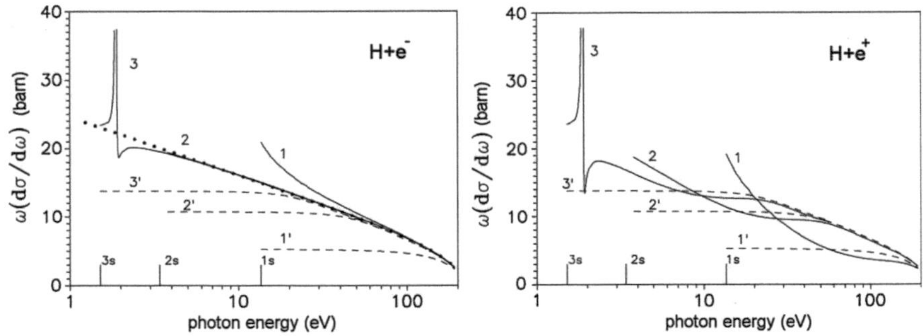

Fig. 3.1 BrS spectra, $\omega d\sigma/d\omega$, formed in the collision of 220 eV electron (*left*) and positron (*right*) with a hydrogen atom [253]. See explanation in the text for the meaning of different curves

Let us note that the right-hand side of (3.8) can be transformed to a more compact form by reducing the number of the hypergeometric functions using the recurrence relations for $\mathscr{F}_{m_1 m_2}(z)$ (see, for example, [309]). However, as it is written above, the expression is more suitable for numerical evaluation.

3.2 BrS Cross Section in Collisions with Hydrogen Atom

Two panels in Fig. 3.1 present the cross sections $\omega d\sigma/d\omega \equiv d\sigma$ of BrS in collisions of an electron (left panel) and a positron (right panel) of energy 220 eV with a hydrogen atom [253]. The calculations were performed within the plane-wave Born approximation. The solid curves 1, 2, and 3 represent the cross section of the total BrS and the dashed curves $1'$, $2'$ and $3'$—that of the OBrS for the hydrogen atom in the 1s, 2s and 3s states, respectively. The vertical lines mark the ionization potentials of these shells.

Figure 3.1 illustrates that the polarizational mechanism plays an important role in forming the total BrS spectrum for either the ground or excited atomic state. For the photon energies less than few ionization potentials the total cross section exceeds noticeably that of the ordinary BrS. As ω increases the cross section $d\sigma_{\rm tot}$ decreases monotonously being in the case of an electron larger, and in the case of a positron smaller than the OBrS cross section $d\sigma_{\rm ord}$. This feature of the total cross section (2.25) is mainly due to the constructive (for an electron) or destructive (for a positron) interference of the amplitudes of the polarizational and ordinary BrS. Similar effect was described earlier in [12, 236]. It was shown that for $\omega \gg I_{1s}$ the total BrS spectrum is effectively formed due to the projectile electron deceleration in the field of bare nucleus (so-called de-screening or 'stripping' effect. See Sect. 4.6 for more details). The dots on the left panel present the cross section of BrS of an electron in the point Coulomb field of the charge $Z = 1$. It is seen, that for $\omega \gg I_{ns}$ ($n = 1, 2, 3$) the total cross sections practically coincide with that calculated for the Coulomb field.

Comparing spectral dependencies of the emitted radiation one notices that the contribution of the polarizational BrS to the total cross section decreases with increasing principal quantum number n. This effect is associated with an increase in the radius of the electron orbit and, therefore, with decrease in the electron screening of the nucleus. Indeed, to radiate a photon via the ordinary mechanism a projectile must penetrate inside the atom, at a distance $r < R_{at}$, i.e. to move in the vicinity of the target nucleus. Therefore, the increase in R_{at} leads to that in the cross section $d\sigma_{ord}$. On the contrary, the polarizational BrS is emitted, most effectively, at large distance impact parameters where the polarization of the electron cloud is most pronounced [45, 455].

Let us note a narrow maximum, located above the 3s ionization threshold, in the spectral dependence of $d\sigma_{tot}$ for both electron and positron collision with a hydrogen atom in the 3s state. It appears as a result of the pole-like peculiarity in the dynamic polarizability of H(3s) in the photon energy range corresponding to the energy of the 3s $\rightarrow$ 2p transition.

Let us consider the features arising in the generalized dynamic polarizability of the excited states of the hydrogen atom. To do this, one can analyze the formula (2.14), calling for the sake of brevity the first term in brackets "term A", and the second term—"term B".

First we note that for the intermediate states $|n\rangle$ from the discrete part of the spectrum the energy $E_n < 0$. If E_n exceeds the energy E_0 of the initial and the final states, i.e. $\omega_{n0} > 0$, then the B-term is a monotonous function of ω whereas the A-term exhibits a resonance behaviour in the vicinity of the poles $\omega \rightarrow \omega_{n0}$. The pole $\omega = \omega_{n0}$ in the polarizability corresponds to the real two-step process, in which the atom is, at first, excited from the initial state $|\,0\rangle$ to the state $|\,n\rangle$ under the action of the Coulomb field of the projectile, and then relaxes back emitting a photon. Formally, the divergence of the polarizability can be eliminated by introducing the imaginary term $\mathrm{i}\Gamma/2$ to the energy denominator, where Γ stands for the (total) width of the state. In our consideration, however, we will consider the photon frequencies ω for which the resonance detuning exceeds the width, i.e. $\omega - \omega_{n0} \gg \Gamma$.

If the initial state is not the ground one, then for some lower-lying intermediate state $|\,n\rangle$ the inequality $0 > E_0 > E_n$, (or, $\omega_{n0} < 0$) is valid. In this case, it is the A-term which is a smooth function of ω whereas the B-term increases resonantly at $\omega \rightarrow -\omega_{n0}$. This divergence also corresponds to a real two-step process, in which the initially excited atom is de-excited via the photon emission, and then experiences the excitation $n \rightarrow 0$ interacting with the projectile. The poles $\omega = -\omega_{n0}$ of the polarizability lie above the ionization threshold $I_0 = |E_0|$ of the excited state $|\,0\rangle$.

Thus, the frequency dependence of the polarizability of the excited states of the H-atom there are poles corresponding to the real processes of excitation and de-excitation. Moreover, the poles corresponding to the de-excitation processes lie above the ionization potential of the excited states. Consequently, the PBrS spectrum, formed in the collision with an excited hydrogen atom, contains narrow lines not only below but also above the ionization potential. The only exception is the 2s-state, since its its de-excitation via the dipole transition is forbidden by the selection rules.

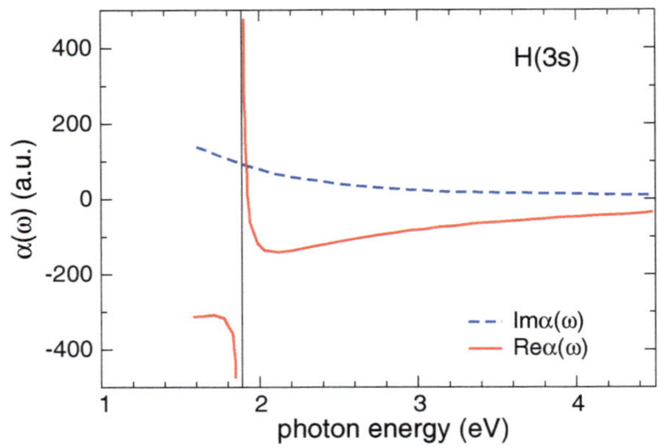

Fig. 3.2 Spectral dependence of the real and imaginary parts of the dipole dynamic polarizability $\alpha(\omega)$ for the hydrogen atom in the 3s state in the vicinity of the energy of the 3s → 2p transition. See explanation in the text

Figure 3.2 presents the spectral dependence of the dipole dynamic polarizability $\alpha(\omega)$ of the excited H(3s) atom in the vicinity of the transition energy 3s → 2p. It is seen that the real part of the polarizability has a pole-like behaviour. The imaginary part, dependent only on the A-term, decreases monotonously with ω. The pole of $\mathrm{Re}\,\alpha(\omega)$ (which is also present in the real part of the generalized polarizability $\alpha(\omega, q)$) above the 3s ionization threshold leads to the narrow line in the BrS spectrum (see Fig. 3.1).

The interference term $d\sigma_{\int}$ in the total BrS cross section is proportional to the real part of the polarizability. Hence, $d\sigma_{\int}$ changes the sign when passing the pole. In the BrS spectrum of an electron the change of the sign of $d\sigma_{\int}$ is reflected by the fact that the total BrS cross section $d\sigma_{\mathrm{tot}}$ above the pole is lower than below the pole. In the case of a positron scattering from H(3s) the behaviour of the total BrS curve is inverse: $d\sigma_{\mathrm{tot}}$ above the threshold is larger than below. greater than for the positron.

3.3 BrS in Collisions of a Slow Charged Particle with an Excited Hydrogen Atom

Peculiar features bears the low-frequency BrS formed in the collision of a slow massive charged particle with an excited hydrogen atom. It was shown in [240] that in the region of large impact parameters the BrS arises from rotation of the dipole moment of the hydrogen under the action of the electric field of the projectile, as well as due to the radiative transitions between the components of the Stark multiplet. The specific feature of the problem is in the appearance of the dipole moment of the excited hydrogen due to mixing of degenerated levels with different parity in the field of the projectile particle (linear Stark effect).

A slow charged particle is the source of the electric field **E** which splits the degenerate levels of the hydrogen atom. The atomic states with a given principal quantum number n form the Stark multiplet. The components of the multiplet are characterized by a projection of the dipole moment of the atomic state on the direction of the external electric field **E**:

$$d_n = -\frac{3n}{2}(n_1 - n_2). \tag{3.10}$$

Here $n_{1,2}$ are the parabolic quantum [279]. Equation (3.10) is valid when the Stark splitting exceeds greatly the fine-structure splitting.

An excited hydrogen atom can possess a non-zero dipole moment due to the mixing of degenerate states with different parity. This specific feature leads to the linear Stark effect in the H-atom [279]. Below we show, following [240], how this feature manifests itself in the BrS spectrum.

The electric field alters both in the absolute value and in the direction due to the motion of the projectile, i.e. $\mathbf{E} = \mathbf{E}(t)$. This results in the reorientation of the atomic dipole moment **d**. The rotation of **d** leads to a photon dipole emission, which we call 'rotational' BrS. For a slow projectile particle the reorientation occurs adiabatically without a substantial expenditure of energy. This leads to an increase in the photon yield in the low-energy part of the BrS spectrum.

The mechanism of rotational BrS is intrinsic for systems with a linear Stark effect. For systems possessing a quadratic Stark effect, the BrS radiation arises due to rotation of the dipole moment induced during the collision. Exactly this mechanism is called polarizational BrS which is discussed through the book. However, there is a distinction between PBrS and rotational BrS. Below we demonstrate that the latter is emitted as the result of reorientation of an already existing dipole moment of hydrogen while in the PBrS process it is necessary to polarize the atom first. This makes rotational BrS more intense than PBrS in a broad range of frequencies.

The projectile particle radiates itself in the field of the hydrogen via the ordinary BrS mechanism. However, for a massive projectile ($M \gg 1$) its yield is suppressed by the factor M^2 so that one can neglect the contribution of OBrS to the radiative spectrum.

3.3.1 Adiabatic Approximation

Let us consider a collision of a slow massive particle with the charge Z_0 and mass $M \gg 1$ with an excited hydrogen atom, which at the initial moment $t = -\infty$ was in the state characterized with the parabolic quantum numbers $n_1 n_2 m$ (see, for example, [279]). Assuming the impact parameter ρ to be much larger than the atomic radius one can assume the motion of the projectile along a straight line and treat its velocity **v** as a constant.

To describe the collision we introduce two coordinate systems: the rest one (XYZ) and the rotating system $(X'Y'Z')$, see Fig. 3.3. The Y-axis is directed along **v**, the projectile trajectory lies in the (YZ) plane.

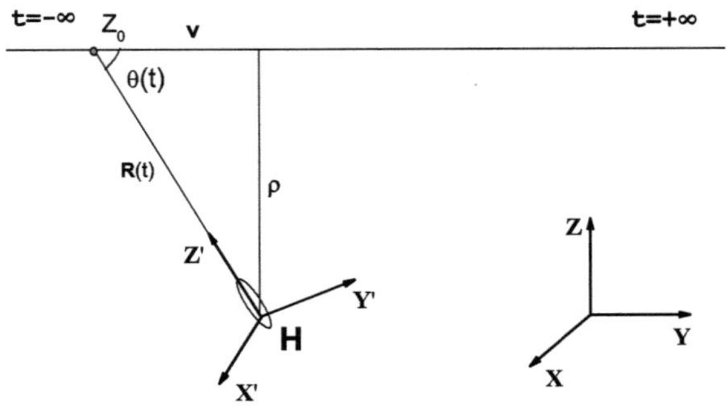

Fig. 3.3 Geometry of the collision

The origin of the rotating system is at the center-of-mass of the hydrogen, and its Z' is directed along the instantaneous position vector $\mathbf{R}(t)$ of the projectile:

$$\mathbf{R}(t) = \mathbf{v}t + \boldsymbol{\rho}. \tag{3.11}$$

The following relations are clear from Fig. 3.3 for the polar angle θ between the axes Z and Z': $\sin\theta = \rho/R(t)$, $\cos\theta = vt/R(t)$.

For an arbitrary instant t the state of the atom can be characterized by the constant quantum numbers (n_1, n_2, m) provided the adiabatic perturbation theory in the interaction of Z_0 with the atom is valid. Let us estimate the range of its applicability.

The electric field of the projectile, acting on the atom, equals to

$$\mathbf{E}(t) = -\frac{Z_0}{R^3(t)}\mathbf{R}(t). \tag{3.12}$$

The field can be treated as a perturbation with respect to the internal atomic field if the impact parameter is large:

$$\rho \gg \sqrt{Z_0}\, a_n, \tag{3.13}$$

where $a = n^2$ is the radius of the electron orbit in the state with the principal quantum number n. The condition (3.13) ensures spatial uniformity of the field $\mathbf{E}$ over the distances about the atomic size.

The Hamiltonian of the atomic electron depends on t only through the parameter $\mathbf{R}(t)$ which enters (3.12):

$$\hat{H}(\mathbf{R}) = \hat{H}_0 + \mathbf{E}(t)\cdot\mathbf{r}, \tag{3.14}$$

where $\hat{H}_0$ denotes the unperturbed atomic Hamiltonian.

The field $\mathbf{E}(t)$ splits the energy levels of the same n. The energy spacing $\Delta\mathscr{E}$ between the neighbouring levels of the Stark multiplet is equal to

$$\Delta\mathscr{E} = \frac{3n}{2}\frac{Z_0}{R^2}. \tag{3.15}$$

If variation of the Hamiltonian (3.14) with respect to the time interval $(\Delta\mathscr{E})^{-1}$ is small compared to $\Delta\mathscr{E}$, i.e.

$$\frac{\partial(\mathbf{E}\cdot\mathbf{r})}{\partial t}(\Delta\mathscr{E})^{-1} \ll \Delta\mathscr{E}$$

then the adiabatic approach describing the dynamics of the atomic electron is valid [320]. The inequality can be written as follows:

$$\omega_0 \ll \Omega_0\,, \tag{3.16}$$

where $\omega_0 = v/\rho$, and the parameter $\Omega_0 = (3nZ_0)/2\rho^2$ is the classical frequency of precession of the Coulomb orbit under the action of the external electric field $E = Z_0/\rho^2$ [181].

Combining inequalities (3.13) and (3.16) one finds the following region of the impact parameters where the adiabatic theory is applicable

$$\rho_{\min} < \rho < \rho_{\max} \tag{3.17}$$

where

$$\rho_{\min} = \sqrt{Z_0}\,a_n, \quad \rho_{\max} = \frac{3}{2}Z_0\frac{v_n}{v}\,a_n. \tag{3.18}$$

Here $v_n = n^{-1}$ is the velocity of the atomic electron in the considered state.

It follows from (3.18) that the impact parameter region is sufficiently broad if the projectile velocity is small compared to v_n:

$$v \ll \sqrt{Z_0}\,v_n. \tag{3.19}$$

The lower limit on v can be obtained assuming the photon energy to be small with respect to the kinetic energy of the projectile. Therefore

$$v \gg (\omega/M)^{1/2}. \tag{3.20}$$

Since conditions (3.16) and (3.17) are valid then the state of the atomic electron is characterized by the constant quantum numbers (n_1n_2m). The corresponding wavefunction, defined with respect to the rotating coordinate system $(X'Y'Z')$, reads:

$$\Psi(\mathbf{r},t) = \Phi_{n_1n_2m}(\xi,\eta,\varphi)\exp\left(\mathrm{i}\frac{t}{2n^2} - \mathrm{i}\int_{-\infty}^{t}\mathscr{E}(\tau)\mathrm{d}\tau\right), \tag{3.21}$$

where $\mathscr{E}(t) = E(R(t))\, d_n$ is the energy of the state $(n_1 n_2 m)$ measured from the central level of the Stark multiplet, $\Phi_{n_1 n_2, m}(\xi, \eta, \varphi)$ is the coordinate wavefunction of the state, ξ, η, φ are the parabolic coordinates. The parabolic quantum numbers n_1, n_2, m and the principal quantum number n are related via $n = n_1 + n_2 + |m|$ [279].

3.3.2 Bremsstrahlung Amplitude

The amplitude of the atomic transition $(n_1 n_2 m) \to (n'_1 n'_2 m')$ with simultaneous emission of a photon of energy ω and polarization $\mathbf{e}$ can be written as follows:

$$f = -\mathrm{i} \int\limits_{-\infty}^{+\infty} \mathrm{d}t \left\langle n'_1 n'_2 m \left| \mathbf{e} \cdot \mathbf{r} \right| n_1 n_2 m \right\rangle \exp\left(\mathrm{i}\omega t - \mathrm{i}\delta \int\limits_{-\infty}^{t} \Delta\mathscr{E}(\tau)\mathrm{d}\tau \right). \quad (3.22)$$

The parameter δ denotes the difference

$$\delta = (n_1 - n_2) - (n'_1 - n'_2). \quad (3.23)$$

Let us consider the photon energies much lower than the energy interval between the states with n and $(n + 1)$ principle quantum numbers, i.e.

$$\omega \ll n^{-3}. \quad (3.24)$$

This means that only the diagonal (with respect to n) transitions are discussed,

$$n' = n. \quad (3.25)$$

It is easier to evaluate the matrix element $\langle n'_1 n'_2 m \left| \mathbf{e} \cdot \mathbf{r} \right| n_1 n_2 m \rangle$ in the rotating coordinate system. One obtains for the amplitude (3.22):

$$\begin{aligned} \mathrm{i}f = {} & r_z \left[e_z C_z(\omega) - e_y C_y(\omega) \right]_{\delta=0} + r_y \left[e_y S_z(\omega, \nu_0) - e_z S_y(\omega, \nu_0) \right]_{\delta=1} \\ & + r_x \left[e_x S_x(\omega, \nu_0) \right]_{\delta=1}. \end{aligned} \quad (3.26)$$

Here e_x, e_y, e_z are the Cartesian components of $\mathbf{e}$, parameter $\nu_0 = \Omega_0/\omega_0 \gg 1$ according to (3.16), $r_{x,y,z}$ are the components of the matrix element $\langle n'_1 n'_2 m | \mathbf{r} | n_1 n_2 m \rangle$ (see Sect. 3.3.5), $S_{x,y,z}$ and $C_{x,y,z}$ denote the integrals over t which appears when one substitutes (3.21) into the amplitude (3.22) (they are defined and evaluated below in this section).

It is shown in Sect. 3.3.5 that the matrix element r_z is only non-zero for the diagonal radiative transition, i.e. when $n_{1,2} = n'_{1,2}$ and $m = m'$. Hence, the first term in (3.26) describes the amplitude of a photon emission due to the spatial rotation of the vector of the atomic dipole moment without changing its absolute magnitude. We call this contribution the 'rotational' BrS.

The matrix elements r_x, r_y are non-zero only for the transitions with $m' = m \pm 1$. One finds both from (3.25) and from the condition $\delta > 0$ (since the atomic electron makes a transition to a lower-lying level of a given Stark multiplet) that $\delta = +1$ in the last two terms in (3.26). These terms describe the photon emission accompanying an atomic transition to the nearest lower-lying level of the Stark multiplet. We call this process 'transitional' BrS.

When the atomic state at $t = -\infty$ is the lowest one of the multiplet ($n_2 = 0, n_1 = n - 1$) then the spontaneous transitions are forbidden, and the only mechanism is rotational BrS.

The integrals $C_{y,z}(\omega)$ in (3.26) are

$$C_y(\omega) = \int_{-\infty}^{+\infty} \mathrm{d}t \, \frac{vt}{\sqrt{(vt)^2 + \rho^2}} \exp(\mathrm{i}\omega t) = \frac{2\mathrm{i}}{\omega_0} K_1\left(\frac{\omega}{\omega_0}\right),$$
$$C_z(\omega) = \int_{-\infty}^{+\infty} \mathrm{d}t \, \frac{\rho}{\sqrt{(vt)^2 + \rho^2}} \exp(\mathrm{i}\omega t) = \frac{2}{\omega_0} K_0\left(\frac{\omega}{\omega_0}\right), \quad (3.27)$$

where $K_n(x)$ is the MacDonald function.

The integrals $S_{x,y,z}(\omega, \nu_0)$ are defined as follows:

$$S_x(\omega, \nu_0) = \int_{-\infty}^{+\infty} \mathrm{d}t \, \exp\Big(\mathrm{i}g(t)\Big)$$
$$S_y(\omega, \nu_0) = \int_{-\infty}^{+\infty} \mathrm{d}t \, \frac{vt}{R(t)} \exp\Big(\mathrm{i}g(t)\Big) \quad (3.28)$$
$$S_z(\omega, \nu_0) = \int_{-\infty}^{+\infty} \mathrm{d}t \, \frac{\rho}{R(t)} \exp\Big(\mathrm{i}g(t)\Big).$$

Here, the phase function

$$g(t) = \omega t - \int_{-\infty}^{t} \Delta\mathscr{E}(\tau)\, \mathrm{d}\tau = \omega t - \nu_0 \arctan(\omega_0 t) + \nu_0 \pi/2 \quad (3.29)$$

contains large parameter $\nu_0 = \Omega_0/\omega_0 \gg 1$.

Integrals (3.28) are well known in the theory of Stark broadening of spectral lines of hydrogen [288]. They can be expressed in terms of the confluent hypergeometric functions of the second order. However, since $\nu_0 \gg 1$ then the main contribution to the integrals can be evaluated by the stationary phase method.

The stationary phase points t_0 are found from the equation:

$$\dot{g}(t_0) = \omega - \Delta\mathscr{E}(t_0) = \omega - \frac{3nZ_0}{2R^2(t_0)} = 0, \quad (3.30)$$

which, in fact, represents the energy conservation law. The emission of a photon occurs when the energy spacing between neighbouring components of the Stark multiplet equals the photon energy ω. For a given impact parameter, ρ, (3.30) has real roots only if the condition (3.16) is fulfilled. In this case the process of transitional BrS is probable. In contrast, for $\omega \gg \Omega_0$, the roots become imaginary, and hence integrals (3.28) as well as the amplitude of transitional BrS are exponentially small.

One finds the following real roots of (3.30):

$$t_0 = \pm(\rho_0^2 - \rho^2)^{1/2} v^{-1}, \tag{3.31}$$

where

$$\rho_0^2 = R^2(t_0) = (3nZ_0)/(2\omega). \tag{3.32}$$

For a given ω the quantity ρ_0 is the maximum value of parameter ρ, providing the real solutions of the equation.

Using (3.31) one carries out the integrals (3.28):

$$\begin{aligned} S_x(\omega, v_0) &\simeq \sqrt{\frac{4\pi}{\omega v}} \frac{\rho_0}{(\rho_0^2 - \rho^2)^{1/4}} \cos\Big(g(t_0) + \pi/4\Big), \\ S_y(\omega, v_0) &\simeq \mathrm{i}\sqrt{\frac{4\pi}{\omega v}} \frac{\rho_0}{(\rho_0^2 - \rho^2)^{1/4}} \frac{y(t_0)}{R(t_0)} \sin\Big(g(t_0) + \pi/4\Big), \\ S_z(\omega, v_0) &\simeq \sqrt{\frac{4\pi}{\omega v}} \frac{\rho_0}{(\rho_0^2 - \rho^2)^{1/4}} \frac{z(t_0)}{R(t_0)} \cos\Big(g(t_0) + \pi/4\Big), \end{aligned} \tag{3.33}$$

for $\rho \le \rho_0$.

3.3.3 Cross Section

The general expression for the cross section of BrS when a target atom is initially in the state $(n_1 n_2)$ is as follows:

$$\frac{\mathrm{d}\sigma(n_1 n_2)}{\mathrm{d}\omega} = \frac{\omega^3}{2\pi c^3} \int_{\rho_{\min}}^{\rho_{\max}} \rho\, \mathrm{d}\rho \int \mathrm{d}\Omega_{\mathbf{k}} \sum_{\lambda} |f|^2 \equiv \frac{\mathrm{d}\sigma_{\mathrm{r}}(n_1 n_2)}{\mathrm{d}\omega} + \frac{\mathrm{d}\sigma_{\mathrm{t}}(n_1 n_2)}{\mathrm{d}\omega}, \tag{3.34}$$

where $\mathrm{d}\sigma_{\mathrm{r}}(n_1 n_2)/\mathrm{d}\omega$ and $\mathrm{d}\sigma_{\mathrm{t}}(n_1 n_2)/\mathrm{d}\omega$ are the cross sections of the rotational and transitional BrS, respectively. They are defined as follows:

$$\frac{d\sigma_r(n_1 n_2)}{d\omega} = \frac{4\omega^3}{3c^3} d_n^2 \int_{\rho_{min}}^{\rho_{max}} \rho \, d\rho \left(|C_y(\omega)|^2 + |C_z(\omega)|^2\right), \tag{3.35}$$

$$\frac{d\sigma_t(n_1 n_2)}{d\omega} = \frac{4\omega^3}{3c^3} |r_x|^2 \int_{\rho_{min}}^{\rho_{max}} \rho \, d\rho \left(|S_x(\omega)|^2 + |S_y(\omega)|^2 + |S_z(\omega)|^2\right), \tag{3.36}$$

In this formulae the relations $|r_x|^2 = |r_y|^2$ and $|r_z|^2 = d_n^2$ are taken into account (see Sect. 3.3.5).

Since terms with $\delta = 0$ (rotational BrS) and $\delta = 1$ (transitional BrS) in amplitude (3.26) correspond to different final atomic states there is no interference term in cross section (3.34).

3.3.3.1 Rotational BrS

Using $C_y(\omega)$ and $C_z(\omega)$ from (3.27), one obtains

$$\frac{d\sigma_r(n_1 n_2)}{d\omega} = \frac{4\omega^3 v^2}{3c^3} \frac{d_n^2}{\omega} \int_{x_{min}}^{x_{max}} x^3 \, dx \left(K_1(x)^2 + K_0(x)^2\right). \tag{3.37}$$

Here $x = \omega\rho/v$ and its minimum and maximum values are

$$x_{min} = Z_0^{1/2}(v_n/v)(\omega/\omega_n), \qquad x_{max} = (3/2)Z_0(v_n/v)^2(\omega/\omega_n). \tag{3.38}$$

Here we used the limits ρ_{min} and ρ_{max} from (3.18); $\omega_n = v_n/a_n = n^{-3}$. On the first glance the result of the calculation in (3.37) depends on the integration limits. However, if the conditions

$$x_{max} \gg 1, \qquad x_{min} \ll 1 \tag{3.39}$$

are met then the result doe not depend on x_{min} and x_{max}. In this case the adiabatic region provides the dominating contribution to the integral. Substituting ∞ and 0 in (3.37) as the upper and lower limits and performing the integration, we obtain

$$\frac{d\sigma_r(n_1 n_2)}{d\omega} = \frac{16 v^2}{3c^3 \omega} d_n^2 \tag{3.40}$$

for $\omega_{min} \ll \omega \ll \omega_{max}$. The limits ω_{min} and ω_{max} are taken from (3.39):

$$\omega_{min} = v^2/nZ_0, \qquad \omega_{max} = v/n^2\sqrt{Z_0}. \tag{3.41}$$

These limits restrict the wide region if the condition $Z_0^{-1/2} v/v_n \ll 1$ is fulfilled. This condition, in turn, is equivalent to (3.19) and valid in the adiabatic approximation.

Outside the interval defined by (3.41) the cross section $d\sigma_r(n_1 n_2)/d\omega$ tends to zero as

$$\frac{d\sigma_r(n_1n_2)}{d\omega} = \frac{8v^2}{3c^3\omega} d_n^2 \left(\frac{\omega}{\omega_{min}}\right)^2 \quad \text{for } \omega_{min} > \omega. \tag{3.42}$$

For $\omega > \omega_{max}$ the spectrum $d\sigma(n_1n_2)/d\omega$ is exponentially small, being proportional to $\exp(-2\omega/\omega_{max})$.

Let us discuss the obtained results. It is seen from (3.40) that $d\sigma_r/d\omega$ is proportional to ω^{-1}. This behaviour is similar to that of the spectrum of the OBrS in the low-frequency limit [81]. The 'infrared catastrophe' in the rotational BrS originates from the adiabatic character of the process in which the rotation of the atomic dipole moment occurs without substantial expenditure of energy. This leads to an increase of photon yield in the low-frequency part of the spectrum.

The other peculiarity of the spectrum (3.40) is its formal independence of the charge Z_0 of the projectile. Indeed, formally (3.40) is not equal to zero if $Z_0 = 0$. The settlement of this paradox is as follows. The vector of the atomic dipole moment is directed along the radius vector of the projectile only if condition (3.16) is fulfilled. If $Z_0 = 0$, then both $\mathbf{E}$ and Ω are equal to zero. Therefore there is no reorientation of the atomic dipole moment and consequently $d\sigma_r/d\omega = 0$. Formally, this result follows directly from (3.41). The lower limit tends to infinity faster than the upper limit if Z_0 goes to zero. This means the limits could not form the frequency interval where (3.40) is applicable for small enough charges.

3.3.3.2 Transitional BrS

Substituting (3.33) into (3.36) and averaging the oscillating factors, one obtains for the spectrum of transitional BrS:

$$\frac{d\sigma_t(n_1n_2)}{d\omega} = \frac{4\pi}{3c^3}\frac{\omega^2\rho_0^2}{v}|r_x|^2 \left(\rho_0^2-\rho^2\right)^{1/2}\Big|_{\rho_{min}}^{\tilde{\rho}}. \tag{3.43}$$

Here $\tilde{\rho} = \min\left\{\rho_0, \rho_{max}\right\}$ (see (3.32) and (3.17)). This restriction follows from the energy conservation law (3.30)). The region $\rho > \tilde{\rho}$ gives an exponentially small contribution to the integral. The value of $\tilde{\rho}$ is different for various ω regions:

$$\tilde{\rho} = \begin{cases} \rho_0 & \omega_{min} < \omega < n^{-3} \\ \rho_{max} & \omega < \omega_{min} \end{cases} \tag{3.44}$$

The corresponding formulae for the transitional BrS cross section are

$$\frac{d\sigma_t(n_1n_2)}{d\omega} = \frac{8\pi v^2}{3c^3\omega}|r_x|^2 \times \begin{cases} (\omega/\omega_{min})^{3/2} & \omega_{min} < \omega < n^{-3} \\ (\omega/\omega_1)^{3/2}\left(1-\sqrt{1-(\omega/\omega_1)}\right) & \omega < \omega_{min} \end{cases}, \tag{3.45}$$

where $\omega_1 = \Omega_0(\rho_{\min}) = 3/2n^3$.

Let us note that the transitional BrS is similar to the well-known molecular orbital x-ray emission (see [173]). The main difference between the two mechanisms is as follows. In molecular orbital x-ray emission the electron transitions between states with different principal quantum number are considered, while here we investigate the transitions between components of the Stark multiplet, formed during the collision.

3.3.4 Averaging Procedure

The previously derived expressions for the spectra of rotational and transitional BrS are valid when the initial and final states of the atom are totally specified, i.e. they are characterized by complete sets of the parabolic quantum numbers. However, the energy of a level of the Stark multiplet depends only on the principal quantum number n and on the difference $N = n_1 - n_2$. The degeneracy $\gamma(n, N)$ of the level with given n and N

$$\gamma(n, N) = n - |N|. \tag{3.46}$$

Therefore, (3.40), (3.42) and (3.45) must be averaged over the degenerate initial levels. Also these expressions have to be summed over all final states $(n_1' n_2' m')$ with the same $N' = n_1' - n_2'$:

$$\frac{d\sigma_{r,t}(n, N)}{d\omega} = \frac{1}{\gamma(n, N)} \sum_{n_1 n_2 m} \sum_{n_1' n_2' m'} \frac{d\sigma_{r,t}(n_1 n_2)}{d\omega} \tag{3.47}$$

The parameter δ (see (3.23)) is equal to 0 for rotational BrS and to 1 for the transitional one. Summations over $(n_1 n_2)$ and $(n_1' n_2')$ are carried out for given values of n, N and N'. Then, one obtains

$$\frac{d\sigma_r(n, N)}{d\omega} = \frac{12 n^2 N^2 v^2}{c^3 \omega} \times \begin{cases} 1 & \omega_{\min} < \omega < \omega_{\max} \\ \omega^2/2\omega_{\min}^2 & \omega < \omega_{\min} \end{cases} \tag{3.48}$$

$$\begin{aligned} \frac{d\sigma_t(n, N)}{d\omega} &= \frac{\pi n^2 (n - |N'|)(n + 2|N'| + 1)\, v^2}{c^3 \omega} \\ &\times \begin{cases} (\omega/\omega_{\min})^{3/2} & \omega_{\min} < \omega < \omega_{\max} \\ (\omega/\omega_1)^{3/2} \left(1 - \sqrt{1 - \omega/\omega_1}\right) & \omega < \omega_{\min} \end{cases} \end{aligned} \tag{3.49}$$

Here $N' = N + 1$.

The Stark splitting of the hydrogen levels arises when the quantity $\Delta\mathscr{E} = 3nZ_0/2R^2$ (see (3.15)) exceeds the fine-structure intervals. At the initial moment $t = -\infty$ the state of the hydrogen is characterized by the spherical quantum numbers (nlm). This basis is rearranged while the incident particle approaches the atom. The state in the new basis is determined by the parabolic quantum numbers $(n_1 n_2 m)$. The probability of the transition on the level $(n_1 n_2 m)$ is described by the solution of the characteristic equation in the intermediate region, where the fine-structure intervals are comparable with the Stark splitting. This problem can be solved by the simple analytical methods only for the case of small values of n [85]. The probability of populating the level with quantum numbers $(n_1 n_2 m)$ can be approximately estimated by the statistical weight of this level, which is equal to $(n - |N|)/n^2$. Then the BrS cross section of a slow particle in collision with the excited hydrogen, being at the initial state (nlm), can be derived from (3.48) and (3.49) in the following manner:

$$\frac{d\sigma_{r,t}(n)}{d\omega} = \frac{1}{n^2} \sum_{N=-n+1}^{n-1} (n - |N|) \frac{d\sigma_{r,t}(n, N)}{d\omega} \tag{3.50}$$

Carrying out the summation, one derives the expression for the cross section, which depends only on the principal quantum number of the initial state:

$$\frac{d\sigma_r(n)}{d\omega} = \frac{8n^2(n^2-1)v^2}{9c^3\omega} \times \begin{cases} 1 & \omega_{\min} < \omega < \omega_{\max} \\ \omega^2/2\omega_{\min}^2 & \omega < \omega_{\min} \end{cases} \tag{3.51}$$

$$\begin{aligned} \frac{d\sigma_t(n)}{d\omega} = & \frac{\pi v^2 n(n^2-1)(3n+2)}{6c^3\omega} \\ & \times \begin{cases} (\omega/\omega_{\min})^{3/2} & \omega_{\min} < \omega < \omega_{\max} \\ (\omega/\omega_1)^{3/2} \left(1 - \sqrt{1 - \omega/\omega_1}\right) & \omega < \omega_{\min} \end{cases} \end{aligned} \tag{3.52}$$

3.3.5 Appendix

Here we present the evaluation of the radius-vector matrix element $\langle n_1' n_2' m | \mathbf{r} | n_1 n_2 m \rangle$:

$$r_j = \frac{1}{4} \int_0^{2\pi} \int_0^{\infty} \int_0^{\infty} d\varphi\, d\xi\, d\eta\, (\xi + \eta)\, \Phi^*_{n_1' n_2' m'}(\xi, \eta, \varphi)\, x_j\, \Phi_{n_1 n_2 m}(\xi, \eta, \varphi). \tag{3.53}$$

Here $j = 1, 2, 3$ and $x_1, x_2, x_3 = x, y, z$.

The parabolic quantum numbers $(n_1 n_2 m)$ and $(n_1' n_2' m')$ satisfy the selection rules $n = n'$ and $\delta = 0, 1$ (see (3.23) and (3.25)).

Integrating over the angle ϕ, one derives

$$r_z = \frac{\delta_{mm'}}{4} \left(R^{(2)}_{n_1 n_1'}(m, m') R^{(0)}_{n_2 n_2'}(m, m') - R^{(0)}_{n_1 n_1'}(m, m') R^{(2)}_{n_2 n_2'}(m, m') \right) \tag{3.54}$$

$$\begin{pmatrix} r_x \\ r_y \end{pmatrix} = \begin{pmatrix} -i \\ 1 \end{pmatrix} \frac{\delta_{m'm+1} \mp \delta_{m'm-1}}{8}$$
$$\times \left(R^{(3/2)}_{n_1 n_1'}(m, m') R^{(1/2)}_{n_2 n_2'}(m, m') + R^{(1/2)}_{n_1 n_1'}(m, m') R^{(3/2)}_{n_2 n_2'}(m, m') \right). \qquad (3.55)$$

In these formulae integrals $R^{(\alpha)}_{n_1 n_2}(m_1 m_2)$ are defined as follows:

$$R^{(\alpha)}_{n_1 n_2}(m_1 m_2) = \int_0^\infty \mathrm{d}\rho\, \rho^\alpha f_{n_1 m_1}(\rho) f_{n_2 m_2}(\rho), \qquad (3.56)$$

where $f(\rho)$ can be expressed in terms of the Whittaker function:

$$f_{nm}(\rho) = \frac{1}{|m|!} \left(\frac{(n + |m|)!}{n!} \right)^{1/2} \rho^{1/2} M_{n+|m|/2+1/2, |m|/2}(\rho). \qquad (3.57)$$

The integrals (3.56) can be carried out using the formula [171]:

$$\int_0^\infty d\rho\, \rho^{\alpha-1} M_{k_1+\sigma_1+1/2, \sigma_1}(\rho) M_{k_2+\sigma_2+1/2, \sigma_2}(\rho)$$
$$= (-1)^{k_1+k_2} \frac{k_1!\, \Gamma(\alpha + \sigma_1 + \sigma_2 + 1)}{(2\sigma_1 + 1)_{k_1} (2\sigma_2 + 1)_{k_2}}$$
$$\times \sum_{j=1}^{k_2} \binom{k_2}{j} \binom{\alpha - \sigma_1 + \sigma_2}{k_1 - j} \frac{\Gamma(\alpha + \sigma_1 + \sigma_2 + j + 1)}{\Gamma(\alpha + \sigma_1 - \sigma_2 - k_2 + j + 1)}. \qquad (3.58)$$

Here $\Gamma(z)$ is the Gamma function and $(a)_n$ is the Pochhammer symbol.

Finally, for the dipole matrix elements we obtain

$$r_z = \delta_{mm'} \delta_{n_1 n_1'} \delta_{n_2 n_2'} d_n \qquad (3.59)$$

for $\delta = 0$ and

$$\begin{pmatrix} r_x \\ r_y \end{pmatrix} = \begin{pmatrix} -i \\ 1 \end{pmatrix} \frac{3n}{4} \left(\delta_{m'm+1} \delta_{n_2 n_2'} \delta_{n_1 n_1'-1} \sqrt{n_1 (n - n_1)} \right.$$
$$\left. \pm \delta_{mm'+1} \delta_{n_1 n_1'} \delta_{n_2 n_2'-1} \sqrt{n_2' (n - n_2')} \right) \qquad (3.60)$$

for $\delta = 1$.

3.4 BrS in Electron and Positron Collision with Positronium

Let us consider the BrS formed in the non-relativistic collision of a fast electron (or positron) with positronium (Ps), i.e., a hydrogen-like system in which a massive proton is substituted with a positron, the mass of which is equal to the electron mass. This system is unstable toward the annihilation process, $e^- + e^+ \to n\gamma$. Here the integer n depends on the total spin S. For $S = 0$ (para-positronium) $n = 2, 4, \dots$. The decay rate is the highest for $n = 2$ resulting in the lifetime $\tau \approx 1.3 \times 10^{-10}$ s. The state with $S = 1$ (orto-positronium) annihilates only into the odd number of photons with the minimum value $n = 3$. The corresponding lifetime is $\tau \approx 1.4 \times 10^{-7}$ s [2]. Both lifetimes are much larger than atomic unit of time, $\approx 2.4 \times 10^{-17}$ s, which can be used as an estimate for the collision time. Thus, Ps can be considered as a stable particle in the collisions considered below.

Although being somewhat exotic, the collision $e^{\pm}+$Ps provides a good case study of specific features of the radiative process which can be tackled analytically [32] In what follows we demonstrate that all radiation formed in the collision proceeds via deformations of the Ps, either virtual or real.[1] Specific structure of this target leads to essential difference of the BrS spectra in electron–Ps and electron–atom collisions.

Let us consider two types of the radiative collisions. First, it is the process of 'elastic' BrS in which the Ps stays in its ground state $|0\rangle = |1s\rangle$ before and after the collision. The second type includes all the processes, so-called 'inelastic' BrS, in which the photon emission is accompanied by the Ps excitation or ionization, so that the final state differs from the initial one. Hence the 'elastic' and 'inelastic' BrS are represented as follows:

$$\begin{aligned} &\text{`Elastic' BrS:} \quad e^- + \mathrm{Ps}(1s) \to e^- + \mathrm{Ps}(1s) + \gamma, \\ &\text{`Inelastic' BrS:} \quad e^- + \mathrm{Ps}(1s) \to e^- + \mathrm{Ps}^* + \gamma. \end{aligned} \tag{3.61}$$

More general description of 'inelastic' BrS in collision with many-electron atoms is presented in Sects. 5.3 (non-relativistic collisions) and 6.4 (relativistic collisions). The emphasis is made on the establishing general criteria which allow one to compare the contributions of elastic versus inelastic channels to the total radiative spectrum for various velocities of collision, photon frequencies, types of the colliding particles, etc.

The quantitative analysis of all radiative channels in $e^{\pm}$ + Ps collision deserves a special treatment since in this case all the results can be obtained analytically and have clear physical explanations.

Generally, the photon emission in either of the BrS channels, elastic or inelastic, (3.61), can occur via the ordinary and the polarizational mechanisms. Therefore, the total amplitude contains two terms:

$$f_{\mathrm{tot}}^{(m)} = f_{\mathrm{ord}}^{(m)} + f_{\mathrm{pol}}^{(m)}. \tag{3.62}$$

[1] We note that similar conclusion was drawn in [322, 323] where the general formulae describing BrS process of a positronium interacting with an external field were given.

Here, the index m indicates the final state of Ps. The elastic BrS corresponds to $m = 0$; any other index refers to the inelastic process.

Within the framework of the BA for a projectile, and in the lowest order of perturbation theory in the Coulomb interaction of the projectile with the Ps constituents, the dipole-photon amplitudes can be written in terms of the transferred momentum $\mathbf{q} = \mathbf{p}_1 - \mathbf{p}_2$ as follows (cf. (5.22)):

$$f_{\text{ord}}^{(m)} = \frac{4\pi}{q^2}\frac{F_{m0}(\mathbf{q})}{\omega}, \qquad f_{\text{pol}}^{(m)} = \frac{4\pi}{q^2} Z_0 \omega\, \mathbf{e}\cdot\mathbf{A}_{m0}(\omega,\mathbf{q})\,. \tag{3.63}$$

The charge is $Z_0 = +1$ for a projectile positron and $Z_0 = -1$ for an electron.

The matrix element $F_{m0}(\mathbf{q})$, the 'non-diagonal' form-factor, is given by the expression:

$$F_{m0}(\mathbf{q}) = \left\langle m \left| \mathrm{e}^{\mathrm{i}\mathbf{q}\mathbf{r}_+} - \mathrm{e}^{\mathrm{i}\mathbf{q}\mathbf{r}_-} \right| 0 \right\rangle, \tag{3.64}$$

where $\mathbf{r}_\pm$ the position vectors of the positron and the electron with respect to the positronium center-of-mass.

The function $\mathbf{A}_{m0}(\omega,\mathbf{q})$ describes the dynamic response of the target in terms of virtual excitations $0 \rightarrow n$ and is defined as follows:

$$\mathbf{e}\cdot\mathbf{A}_{m0}(\omega,\mathbf{q}) = -\mathrm{i}\sum_n \left\{ \frac{\omega_{nm}\,\langle m|\mathbf{e}\cdot\mathbf{D}|n\rangle}{\omega - \omega_{nm}} F_{n0}(\mathbf{q}) - F_{mn}(\mathbf{q}) \frac{\omega_{n0}\,\langle n|\mathbf{e}\cdot\mathbf{D}|0\rangle}{\omega + \omega_{n0}} \right\}. \tag{3.65}$$

Here $\mathbf{D} = \mathbf{r}_+ - \mathbf{r}_-$ is the Ps dipole moment operator. For the elastic channel, $m = 0$, the function (3.65) is proportional to the positronium generalized polarizability $\mathbf{A}_{00}(\omega,\mathbf{q}) = -\mathrm{i}\mathbf{q}\,\alpha_{\text{Ps}}(\omega,q)$, cf. (2.14).

For $m = 0$ (3.62)–(3.65) are directly connected to the elastic BrS amplitude in electron–hydrogen collision [99, 124, 125]. To see this one substitutes $\mathbf{r}_+$ with the position vector of a proton, $\mathbf{r}_{\text{p}} \approx 0$. Then, the BrS amplitude reduces to (2.1) written for $Z = 1$, $M = 1$ and $Z_0 = -1$. This immediate correspondence is due to the negligent role of the Ps recoil in the collision process. Indeed, the corrections to f_{tot} due to the Ps center-of-mass motion are of the order of magnitude v_1/c and, thus, can be omitted in the non-relativistic collision.

However, in spite of formal similarity, BrS amplitudes for electron-Ps and electron-H collisions differ essentially. On the contrary to the hydrogen atom, the Ps "diagonal" form-factor $F_{00}(\mathbf{q})$ is identically equal to zero. Therefore, the OBrS mechanism does not contribute to the "elastic" amplitude. In this connection it is worth noting, that within the BA the amplitude of elastic scattering a charged from PS is also zero being proportional to $F_{00}(\mathbf{q})$ [32, 322].

The triply differential cross section of the total BrS accompanied by the transition $0 \rightarrow m$ reads:

$$\frac{d^3\sigma^{(m)}_{\rm tot}}{d\omega dq d\Omega_{\mathbf{k}}} = \frac{\omega q}{(2\pi)^3 v_1^2 c^3}\left(\left|f^{(m)}_{\rm ord}\right|^2 + \left|f^{(m)}_{\rm pol}\right|^2\right). \tag{3.66}$$

In (3.66) the summation over the m-state fine structure has been performed. Therefore, different m are distinguished only by the principal and orbital quantum numbers. This summation resulted in nullifying the interference term in the cross section. Indeed, from (3.64) immediately follows that the OBrS process can only take place for the transitions $0 \to m$ between the states of the same parity, whereas the PBrS amplitude is non-zero if the parities are different, as it is follows from (3.65).

The absence of the interference between OBrS and PBrS is another distinguishing feature of the $e^{\pm}$+Ps process [32].

Integrating (3.66) over the emission angles, one derives the following expression for the “elastic” BrS ($m = 0$) double differential cross section:

$$\frac{d^2\sigma^{(0)}_{\rm tot}}{d\omega dq} \equiv \frac{d^2\sigma^{(0)}_{\rm pol}}{d\omega dq} = \frac{16\omega^3}{3v_1^2 c^3}\frac{|\alpha_{\rm Ps}(\omega, q)|^2}{q}. \tag{3.67}$$

In the case of a target atom possessing a heavy nucleus, the is a possibility to emit a photon via the “elastic” OBrS. Equation (3.67) suggests that it is not the case for a Ps atom, when the “elastic” BrS process is accompanied by the virtual excitation of the target.

To compare the “elastic” BrS spectra in e^-−Ps and e^-−H collisions let us consider the ratio of the cross sections:

$$\eta = \frac{d^2\sigma^{(0)}_{\rm tot,Ps}}{d^2\sigma^{(0)}_{\rm tot,H}} = \left|\frac{\alpha_{\rm Ps}(\omega, q)}{\omega^{-2}\left(1 - F_{00,\,\rm H}(q)\right) - \alpha_{\rm H}(\omega, q)}\right|^2 \tag{3.68}$$

where $F_{00,\,\rm H}(q) = 16/(4+q^2)$ is the form-factor of the hydrogen atom in the ground state.

To evaluate the right-hand side further we consider the following two limiting cases.

- In the region $q \ll 1$ one estimates the generalized polarizabilities as follows (see (2.2)): $\alpha_{\rm H}(\omega, q) \approx \alpha_{\rm H}(\omega)$, $\alpha_{\rm Ps}(\omega, q) \approx \alpha_{\rm Ps}(\omega) = 8\alpha_{\rm H}(2\omega)$. This results in

$$\eta\Big|_{q\ll 1} \approx \frac{d^2\sigma^{(0)}_{\rm tot,Ps}}{d^2\sigma^{(0)}_{\rm tot,H}} = 64\left|\frac{\alpha_{\rm H}(2\omega)}{q^2/2\omega^2 - \alpha_{\rm H}(\omega)}\right|^2. \tag{3.69}$$

  For $q/3 \ll \omega \ll I_{1s,\rm H}$ one ignores the first term in the denominator and substitutes $\alpha_{\rm H}(\omega)$ and $\alpha_{\rm H}(2\omega)$ with the static polarizability of the H-atom, $\alpha_{\rm H}(0) = 9/2$ (see, e.g., [279]).

  Then, the ratio of the cross section is $\eta = 64$.

- In the limit of large photon energies, $\omega \gg I_{1s,\mathrm{H}}$, using the relations $\alpha_{\mathrm{H}}(\omega, q) \approx -F_{00,\,\mathrm{H}}(q)/\omega^2$ and $\alpha_{\mathrm{Ps}}(\omega, q) \approx -2F_{00,\,\mathrm{H}}(q/2)/\omega^2$, which are valid for arbitrary q, one finds $\eta = 4$.
  This value can be easily deduced from the following qualitative arguments. In the limit of large ω the target (either Ps or H) can be considered as a system of (quasi-)free particles. It is well-known (see, e.g., [278]), that no dipole radiation can be emitted by a system of free non-relativistic particles with equal charge-to-mass ratio. Then, within the framework of the non-relativistic BA, the BrS of an electron on H-atom reduces to that of the incident electron on bare nucleus (the effect of the H-atom's stripping, see Sect. 4.6 for more details). For the e^-–Ps collision the radiation is emitted by the system "incoming electron + e^+". In the latter case, the reduced mass is 2 times smaller. Hence, the intensity of radiation, being inversely proportional to the reduced mass, turns out 4 times higher than in the e^-–H collision.

Integrating (3.66) over the emission angles and summing over all final states (including $m = 0$ and the integral over the continuous spectrum of the excited states) one obtains the total emission spectrum in the e^-–Ps collision:

$$\frac{\mathrm{d}^2\sigma_{\mathrm{tot,\,Ps}}}{\mathrm{d}\omega \mathrm{d}q} = \sum_m \frac{\mathrm{d}^2\,\sigma^{(m)}_{\mathrm{tot,\,Ps}}}{\mathrm{d}\omega \mathrm{d}q} = \frac{16\omega^3}{3v_1^2c^3}\frac{1}{q}\sum_m \left(|F_{m0}(\mathbf{q})|^2 + \frac{\omega^4}{q^2}\,|\mathbf{A}(\omega, q)|^2 \right). \tag{3.70}$$

Let us evaluate of the right-hand side in the limit $\omega \gg I_{1s,\mathrm{Ps}} = 6.8\,\mathrm{eV}$. In this limit the sums of first and the the second terms reduce to:

$$\sum_m |F_{m0}(\mathbf{q})|^2 = 4\left\langle 0 \left| \sin^2 (\mathbf{q}\cdot\mathbf{r}/2) \right| 0 \right\rangle .$$

$$\sum_m \frac{\omega^4}{q^2}\,|\mathbf{A}(\omega, q)|^2 = 4\left\langle 0 \left| \cos^2 (\mathbf{q}\cdot\mathbf{r}/2) \right| 0 \right\rangle .$$

Hence

$$\left.\frac{\mathrm{d}^2\,\sigma_{\mathrm{tot,\,Ps}}}{\mathrm{d}\omega \mathrm{d}q}\right|_{\omega \gg I_{1s,\mathrm{Ps}}} = \frac{64\omega^3}{3v_1^2c^3}\frac{1}{q}. \tag{3.71}$$

Integrating (3.71) over the transferred momentum from $q_{\min}p_1 - p_2 \approx \omega/v_1$ up to $q_{\max}p_1 + p_2 \approx 2\mu v_1$, where $\mu = 2/3$ is the reduced mass of the system e^-+Ps, one evaluates the spectral distribution of the emitted radiation:

$$\left.\frac{\mathrm{d}\,\sigma_{\mathrm{tot,\,Ps}}}{\mathrm{d}\omega}\right|_{\omega \gg I_{1s,\mathrm{Ps}}} = \frac{64\omega^3}{3v_1^2c^3}\ln\frac{4v^2}{3\omega}. \tag{3.72}$$

To compare the contributions of the "elastic" and "inelastic" BrS channels (3.61) to the spectrum (3.72), let us calculate the spectral distribution of the "elastic" BrS. To do this, one substitutes $\alpha_{\rm Ps}(\omega, q)$ with $-F_{00}(\mathbf{q})/\omega^2$ in the right-hand side of (3.71), and then integrates over q from $q_{\min}$ up to the cut-off value $\sim R_{\rm Ps}^{-1} \approx 1$ a.u. The result reads

$$\left.\frac{\mathrm{d}\,\sigma^{(0)}_{\text{pol, Ps}}}{\mathrm{d}\omega}\right|_{\omega \gg I_{1s,\rm Ps}} = \frac{64\omega^3}{3v_1^2c^3} \ln\frac{v}{\omega}. \tag{3.73}$$

Comparing the logarithmic factors in (3.72) and (3.73), one finds that the contribution of the "inelastic" BrS channels to the spectrum is proportional to $\ln(4v_1/3)$. This is also a specific feature of the Ps as a target. Indeed, for any other atom, either the H-atom [101] or any many-electron atom [12] the inelastic BrS processes do not contribute to the spectrum of the dipole radiation in the high-energy limit $\omega \gg I_{1s}$. As mentioned, in this case the target electrons can be treated as quasi-free ones. Therefore, the dipole photon radiation in the system "incoming electron + atom" is solely due to the alternating dipole moment of the incoming electron and the nucleus. The target electrons do not virtually participate in the process, and, therefore, they do not change their state. The situation is the opposite in the system "e^- + Ps": the alternating dipole moment is due to the incoming electron and the target's positron. The latter particle can change its state in the course of the collision, and thus, the "inelastic" BrS channels contribute to the spectrum.

Chapter 4
Cross Section of PBrS from Many-Electron Atoms and Ions

4.1 Spectral Distribution of BrS in Vicinity of Giant Resonances

It was demonstrated in Chaps. 2 and 3 that the PBrS amplitude is dependent (and in some cases, proportional to, see (2.17)) generalized atomic polarizability $\alpha(\omega, q)$. Therefore, the PBrS mechanism in forming the total radiative spectrum in electron–atom collisions will be most pronounced for those atoms and photon energy ranges where $\alpha(\omega, q)$ is especially large.

Important examples of this type are the BrS spectra in electron collisions with Xe, Ba and rare earth atoms. Due to specific features of the electron configuration, the BrS spectrum of these atoms is characterized by the presence of a wide and powerful maximum (the giant resonance) located in the vicinity of the 4d-subshell ionization subshell. In the maximum, the total BrS cross section greatly exceeds the cross section of the ordinary BrS.

The core electron configuration of the Ba atom as well as the successive atomic group of rare earth elements from La up to Lu is that of the Xe atom:

$$1s^2 2s^2 2p^6 3s^2 3p^6 3d^{10} 4s^2 4p^6 4d^{10} 5s^2 5p^6.$$

Atop of it there are two or three valence electrons ($6s^2$ or $5d^1 6s^2$), and the electrons in the partly filled 4f-subshell. The filling of the 4f-subshell starts at the Ce atom but proceeds further through the periodic table not monotonously due to the 'competition' of the 4f-, 5d- and 6s-subshells.

Occurrence of the 4f-electron in the electron configuration of the Ce atom indicates that the nucleus charge is large enough to bind the electron. It turns out that the 4f-subshell is localized being placed spatially in between the $5s^2$- and $5p^6$-subshells [140]. The effect of retracting of the 4f-subshell inside the $n = 5$ shell was predicted by Fermi in 1928 [134]. He showed, within the framework of the statistical atomic model, that with the increase in atomic number Z from 55 up to 60 the 4f-subshell 'collapses', i.e. its radius decreases sharply. Such a dependence of the $4f$-subshell

A. V. Korol and A. V. Solov'yov, *Polarization Bremsstrahlung*,
Springer Series on Atomic, Optical, and Plasma Physics 80,
DOI: 10.1007/978-3-642-45224-6_4,

on Z is due to a specific feature of the effective potential $U_l(r)$ which acts on an electron with the orbital momentum l in a many-electron atom:

$$U_l(r) = -\varphi(r) + \frac{(l+1/2)^2}{2r^2}. \tag{4.1}$$

Here $\varphi(r)$ stands for the interaction potential of the electron with other atomic constituents, and the second term describes the centrifugal potential within the quasi-classical approximation.

The centrifugal term gives rise to a potential barrier in the profile of $U_l(r)$, which separates the potential into two wells, the inner and the outer ones. As Z increases, the inner well becomes more powerful so that the l-electrons, localized primarily in the outer well, penetrate into it. The presence of the potential barrier leads to a principal deviation of the photoionization cross section $\sigma_\gamma(\omega)$ in the X-ray domain from the behaviour predicted by the hydrogenic model [120]. Instead of the monotonous decrease of $\sigma_\gamma(\omega)$ for $\omega \geq I_{nl}$ (I_{nl} stands for the ionization potential of the nl-subshell), the cross section can increase to some maximum value located above the I_{nl} threshold, and start decreasing after it. This happens, for example, in the photoionization of $4d$-subshells in Xe, Ba, La. In addition to the first maximum, in some cases the second maximum of $\sigma_\gamma(\omega)$ can be observed [132].

Qualitative explanation of the specific non-hydrogenic behaviour of $\sigma_\gamma(\omega)$ is as follows. For sufficiently high potential barrier between the inner and the outer wells, $U_l(r_0) > 0$ (here r_0 is the point of the maximum of $U_l(r)$), the wavefunction of the outgoing electron of energy $\varepsilon \ll U_l(r_0)$, being located predominantly in the region $r > r_0$, does not overlap with the atomic orbital which is bound by the inner well. As a result, the photoionization of the subshell will be characterized by small values of the cross section. However, as ω increases, the wavefunction of the photoelectron with $\varepsilon \lesssim U_l(r_0)$ may correspond to a quasi-stationary state which is bound to the inner well for some lifetime τ. This will give rise to the maximum in $\sigma_\gamma(\omega)$ the width of which Γ is proportional to τ^{-1}.

For the rare earth atoms beyond La, the wavefunctions of the $4f$ states are collapsed into the inner well. A wide maximum in the photoabsorption spectrum for these atoms, the giant resonance, is related to the discrete $4d \to 4f$ transition followed by the autoionization decay.

The giant resonances in the photoabsorption spectra for atomic Xe, Ba, La and Eu in the vicinity of the 4d-thresholds has been investigated experimentally [78, 364, 433, 448] and theoretically [8, 26, 225, 276].

In [6] it was demonstrated that resonances in the BrS spectra on many-electron atomic targets correlate in the position with those in the photoabsorption spectra. This is a consequence of the proportionality of $\sigma_\gamma(\omega)$ to the imaginary part of atomic polarizability $\operatorname{Im}\alpha(\omega)$, and of the dispersion relations for $\operatorname{Im}\alpha(\omega)$ and $\operatorname{Re}\alpha(\omega)$ (see (4.10) and (4.11)).

The full curves represent in Fig. 4.2 the dependencies of the total BrS cross section $\omega d\sigma/d\omega$ (i.e., the one which accounts for OBrS, PBrS parts as well as for the interference term, (2.29) calculated for the collision of a 250 eV electron with the Ba, La and Xe atoms for the photon energy range $\omega = 80$–200 in the vicinity of the 4d-thresholds [153, 243–247, 260]. The thresholds are marked with the vertical dashes.

The calculations were performed within the DPWA framework (see Sects. 2.2.1 and 2.2.2) to describe the scattering process of the projectile electron in the static field of the target atom.

To calculate the generalized dipole polarizability $\alpha(\omega, q)$, which enter the formula for the PBrS amplitude, (2.8) and (2.21), the following dipole transitions from the ground state 4d-, 5s-, and 5p-subshells were taken into account within the GRPAE framework:

$$4d \rightarrow (n,\varepsilon)f \text{ and } (n,\varepsilon)p; \quad 5s \rightarrow (n,\varepsilon)p; \quad 5p \rightarrow (n,\varepsilon)s \text{ and } (n,\varepsilon)d. \quad (4.2)$$

The contribution to $\alpha(\omega, q)$ due to these transitions was calculated within the Generalized RPAE (GRPAE) framework [8, 24].[1] The experimental values for the ionization 4d-thresholds of atomic Ba and La, $I_{4d}^{\mathrm{Ba}} = 99.3$ eV and $I_{4d}^{\mathrm{La}} = 113.2$ eV [364], were used in the calculations.

The contribution from the outer electrons, the $6s^2$ electrons in Ba and the $5d^1 6s^2$ ones in La, to $\alpha(\omega, q)$ was estimated using the asymptotic formula—$F_{\mathrm{out}}(q)/\omega^2$, where $F_{\mathrm{out}}(q)$ is the form-factor of the shells [12].

To describe the ground and the excited states of the Eu atom the non-relativistic Spin-Polarized HF approximation [21] was used. Within this framework, applicable to the atoms with filled or/and semi-filled subshells, the electrons of a subshell are divided into two groups, the "up"- and "down"-electrons (or, $\uparrow$ and $\downarrow$) according to the projections of their spins. Thus, each atomic subshell splits into two with different values of spin and energy.

In terms of "up"- and "down"-core's states the electron configuration of the outer shells of Eu is written as follows [26]:

$$4d^5\uparrow 4d^5\downarrow 5s^1\uparrow 5s^1\downarrow 5p^3\uparrow 5p^3\downarrow 4f^7\uparrow 6s^1\uparrow 6s^1\downarrow \ ({}^8S).$$

Energy splitting for the 4d$\uparrow$- and 4d$\downarrow$-subshell results in the following values of the ionization potentials: $I_{4d\downarrow}^{\mathrm{Eu}} = 148.7$ eV and $I_{4d\uparrow}^{\mathrm{Eu}} = 165.8$ eV.

In [244] to describe the dipole–photon interaction with Eu atom with the correlation effects taken into account, the Spin-Polarized RPAE (SPRPAE) was used [8, 21]. The inter- and intra-shell correlations were simultaneously taken into account when considering the following dipole transitions from the core's states to the excited states of the discrete and the continuous spectra:

$$\begin{aligned}&4d\downarrow\uparrow \longrightarrow (n,\varepsilon)f\downarrow\uparrow;\ 4f\uparrow \longrightarrow (n,\varepsilon)d\uparrow;\quad 4f\uparrow \longrightarrow (n,\varepsilon)g\uparrow;\\&5s\downarrow\uparrow \longrightarrow (n,\varepsilon)p\downarrow\uparrow;\ 5p\downarrow\uparrow \longrightarrow (n,\varepsilon)s\downarrow\uparrow;\ 5p\downarrow\uparrow \longrightarrow (n,\varepsilon)d\downarrow\uparrow.\end{aligned}$$

[1] Another term used for this approach is 'RPAE with relaxation', see, for example, [276].

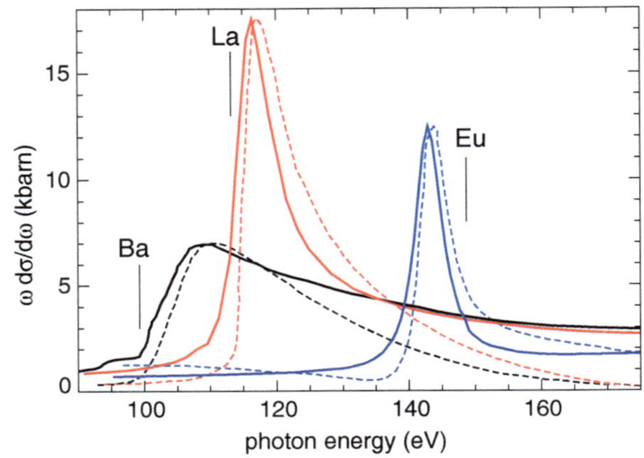

Fig. 4.1 Total BrS cross section $\omega \mathrm{d}\sigma/\mathrm{d}\omega$ (*full curves*) for $\varepsilon_1 = 250$ eV on Ba, La and Eu atoms, as indicated. *Broken curves* represent the photoabsorption spectra $\sigma_\gamma(\omega)$, which are normalized to $\omega \mathrm{d}\sigma/\mathrm{d}\omega$ in the maximum. *Vertical dashes* mark the 4d ionization thresholds

Similar to the case of Ba and La atoms, the contribution of the outer $6s \uparrow\downarrow$-subshells to $\alpha(\omega, q)$ was estimated as—$F_{6s}(q)/\omega^2$.

Figure 4.1 demonstrates that the maxima in the total BrS spectrum $\omega \mathrm{d}\sigma/\mathrm{d}\omega \equiv \mathrm{d}\sigma$ for Ba, La and Eu are located in the vicinities of the 4d-subshell ionization potentials. In the maximum, the magnitude of $\mathrm{d}\sigma$ (which is mainly due to PBrS, as discussed below in the section) exceeds that of PBrS (not shown in the figure) by factors $\approx 8, 20$ and 15 for Ba, La and Eu, respectively.

The maxima in $\mathrm{d}\sigma$ for Ba (the maximum point is $\omega_{\max} = 110.0$ eV) and La ($\omega_{\max} = 116.4$ eV) are blue-shifted with respect to the ionization potentials. In the maxima, the dominant contribution comes from the virtual excitation to and the subsequent radiative decay of the intermediate $4d^9\varepsilon f$ states of the continuous spectrum. The maximum in the BrS spectrum of the Eu atom is located at $\omega_{\max} =$ 142.9 eV which is below the $4d \uparrow$ threshold. In contrast to the Ba and La cases, the position, width and the magnitude of this maxim are related to the excitation into the discrete state which, in turn, is is strongly correlated with the $4f^6 \uparrow \varepsilon g \uparrow$ states of the continuous spectrum.

Broken curves in Fig. 4.1 represent the photoabsorption cross sections $\sigma_\gamma(\omega)$. For each atom the $\sigma_\gamma(\omega)$ to level its maximum value with that of the BrS spectrum.

Figure 4.1 illustrates close relationship between the giant resonances in $\mathrm{d}\sigma_{\mathrm{pol}}(\omega)$ and those in σ_γ [6]. We note, however, that in general case, the quantity σ_γ associated with the photoabsorption process rather than with the photoionization one alone. The former accounts not only for the ionization into the continuum but the discrete excitations as well. In those cases when excitations into the continuum dominate in the photoabsorption spectrum the maximum of σ_γ lies above the ionization threshold and so does the maximum of $\mathrm{d}\sigma(\omega)$. In connection with Fig. 4.1 this is true for Ba and La, and does not hold in the case of a Eu atom. For the latter is is known [78] that the main oscillator strength of the 4d subshell is associated with the discrete transition 4d → 4f. Therefore, the maxima of σ_γ and $\mathrm{d}\sigma(\omega)$ are located below I_{4d}.

Position of the maximum and shape of the BrS resonance curve is, generally, somewhat different from those of the photoabsorption giant resonance. This can be understood taking into account the following considerations. First, the cross section σ_γ is proportional to $\mathrm{Im}\,\alpha(\omega, q)$ whereas $\mathrm{d}\sigma(\omega)$, via its polarizational part $\mathrm{d}\sigma_{\mathrm{pol}}(\omega)$, depends on both the imaginary and real part of the polarizability. Hence, $\mathrm{d}\sigma_{\mathrm{pol}}(\omega)$ is related, but not proportional, to σ_γ. Second, as it is demonstrated below, an important role in forming the profile of total BrS spectrum is played by the interference term, $\mathrm{d}\sigma_{\mathrm{int}}(\omega)$.

For $\omega > \omega_{\mathrm{max}}$ the total BrS cross section $\mathrm{d}\sigma_{\mathrm{tot}}$ decreases with the increase of the photon energy. However, its magnitude far above the threshold exceeds that below the threshold. This feature is due to the 'de-screening' effect [12]:[2] for $\omega \gg I_{4\mathrm{d}}$ the electrons of the subshell can be considered as quasi-free ones, and thus do not possess the alternating dipole moment which leads to the photon emission. Effectively, they do not participate in the screening of the nucleus. As a result, the difference in the total BrS amplitude magnitudes above and below the threshold is a positive quantity proportional to the number of electrons, $N_{4\mathrm{d}}$ in the subshell. Correspondingly, the difference in the cross sections is proportional to $N_{4\mathrm{d}}^2$. More formally, the de-screening appears as a result of the interference of the radiation emitted via the ordinary and the polarizational mechanisms. The interference is destructive below the threshold and constructive above it. This phenomenon is illustrated in Fig. 4.2 by the broken curves which present the total cross section but without the interference term in (2.29):

$$\frac{\mathrm{d}\sigma}{\mathrm{d}\omega} \to \frac{\mathrm{d}\tilde{\sigma}}{\mathrm{d}\omega} = \frac{\mathrm{d}\sigma_{\mathrm{ord}}}{\mathrm{d}\omega} + \frac{\mathrm{d}\sigma_{\mathrm{pol}}}{\mathrm{d}\omega}. \tag{4.3}$$

It is seen, that for $\omega \gg I_{4d}$ the PBrS cross section is negligibly small in comparison with $\mathrm{d}\,\sigma_{\mathrm{ord}}$. On the other hand, the total cross section noticeably exceeds the OBrS one. This means, that the substitution (4.3) cannot be used far away from giant resonances. In other words, in a broad range of photon energies the total cross section can not reproduced as the sum of only two terms, $\mathrm{d}\,\sigma_{\mathrm{ord}}$ and $\mathrm{d}\,\sigma_{\mathrm{pol}}$. The account for the interference results also in the change of the magnitude of the asymmetry parameter of the resonance, $\xi = \alpha/\beta$, (which is most clear seen in Fig. 4.2 for Ba) and to the shift of the maximum toward larger values of ω. Therefore, for an adequate description of the total BrS spectrum all three terms on the right-hand side of (2.29) must be retained (this feature was discussed also in [28, 29]).

Let us discuss now to which extent the BrS cross section is affected by the distortion of a projectile electron's wavefunction due to the action of the static field of a target.

In Fig. 4.3 the solid curves 1 and 2 present the DPWA cross sections $\mathrm{d}\,\sigma_{\mathrm{tot}}$ and $\mathrm{d}\,\sigma_{\mathrm{ord}}$ for a 600 electron scattering on Xe atom, and for a 250 eV electron scattering on Ba and Eu atoms. The broken lines, also marked as 1 and 2, represent the same cross sections but calculated within the plane-wave BA (see (2.1)). In all cases, the

[2] The de-screening effect (another term 'stripping' effect) is discussed in Sect. 4.6.

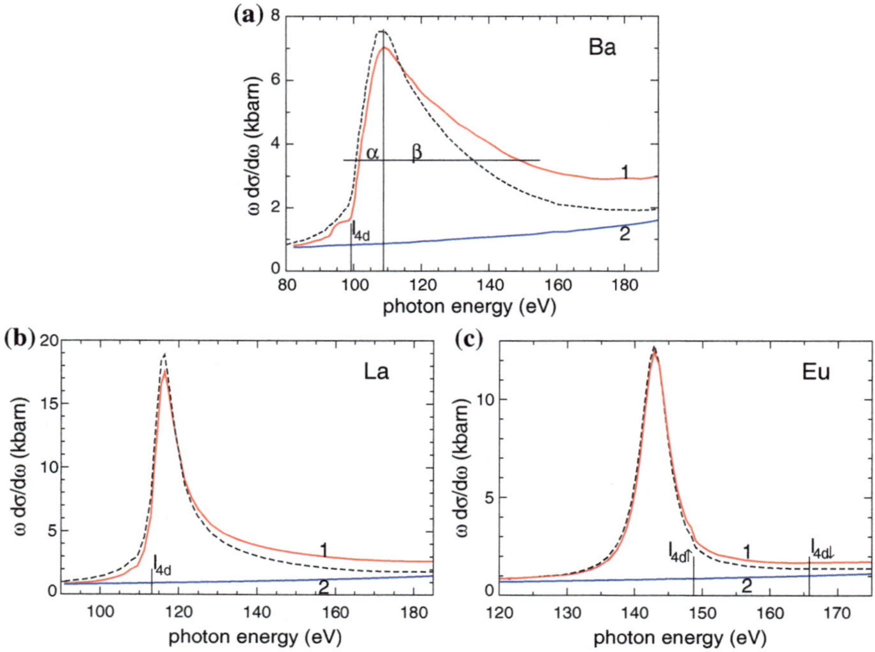

Fig. 4.2 BrS cross sections $d\sigma = \omega d\sigma/d\omega$ for $\varepsilon_1 = 250$ eV on Ba, La and Eu atoms (as indicated) calculated within the DPWA. *Curves 1* mark the total BrS, *curves 2*—the OBrS background $d\sigma_{ord}$. *Broken curves* represent the sum $d\tilde{\sigma} = d\sigma_{ord} + d\sigma_{pol}$, (4.3) [243–245]. *Vertical dashes* mark the 4d ionization thresholds

atomic response was calculated within the RPAE. It is seen that BA overestimates considerably the magnitude of the OBrS cross section and, as a consequence, that of the total BrS. The discrepancy between the solid and the broken curves 1 is not surprising, since the OBrS is formed 'inside' the atom, where the action of the atomic potential leads to a significant deviation of a projectile's wavefunction from a plane wave.

In contrast, the polarizational BrS, calculated for the same collisions and presented in Fig. 4.4 by the full (DPWA) and broken (BA) curves, is much less sensitive to the approximation used for the description of the projectile's movement. As mentioned, it happens because in this process the specific distances between the target and the projectile are greater than the atomic radius and, hence, the influence of the atom's field on the projectile is weaker.

Comparing the total BrS spectrum calculated in the DPWA and the BA (Fig. 4.3), we point out that, apart from the overestimation of the background radiation, the shape of the Born curve for $d\,\sigma_{tot}$ exhibits two other different features compared with the DPWA curve. These are: (1) the width of the maximum peak, and (2) the existence of a minimum below the 4d-threshold. These are artefacts which only demonstrate the unacceptability of the BA for the calculation of the interference term of the spectrum.

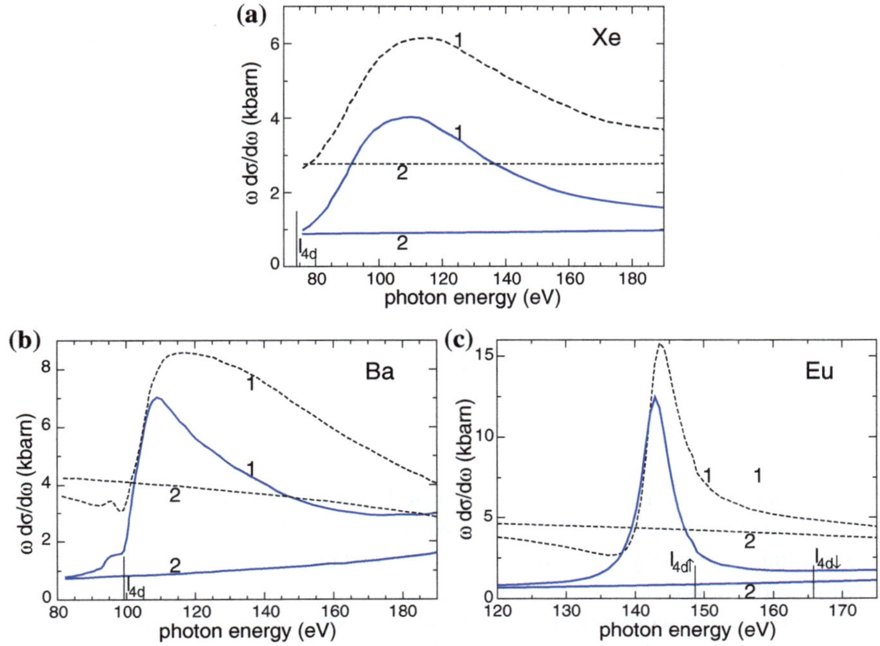

Fig. 4.3 BrS cross section $\omega \mathrm{d}\sigma/\mathrm{d}\omega$ for $\varepsilon_1 = 600$ eV electron on Xe [20], and for $\varepsilon_1 = 250$ eV electron on Ba [245, 246] and Eu [244] atoms (as indicated) calculated within the DPWA (*solid curves*) and the BA (*broken curves*). *Curves 1* mark the total BrS, *curves 2*—the OBrS background. *Vertical dashes* mark the 4d ionization thresholds

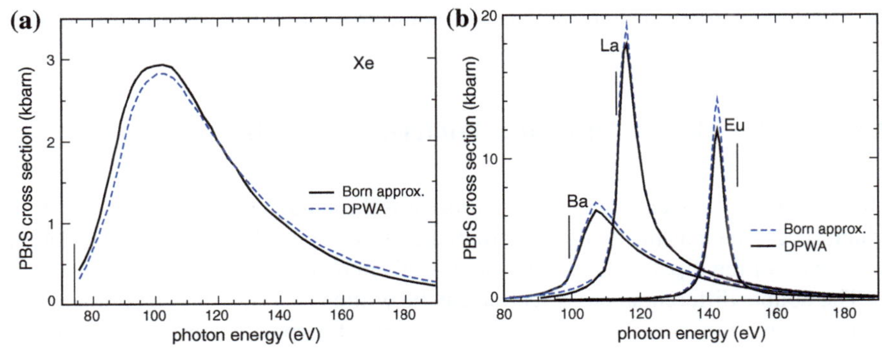

Fig. 4.4 Polarizational BrS cross section $\omega \mathrm{d}\sigma_{\mathrm{pol}}/\mathrm{d}\omega$ calculated within DPWA and BA for 600 eV electrons on Xe [20] and for 250 eV electrons on Ba, La and Eu [244–247] in the vicinity of the 4d ionization thresholds (marked with *vertical dashes*)

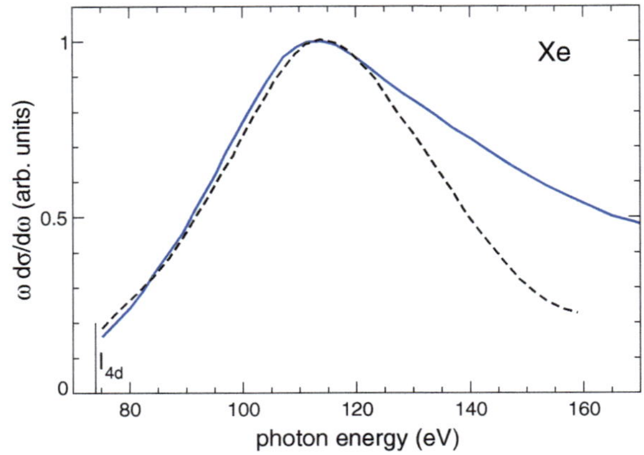

Fig. 4.5 BrS cross section of a 600 eV electron on the Xe atom. *Full curve* stands for the DPWA results [20, 28, 30] *broken curve*—the experimental data [426, 428]. The maximum values of both curves is normalized to one. *Vertical dash* marks the 4d ionization thresholds

Indeed, since the interference of the radiation formed via the two mechanisms can arise only from the region $\rho \lesssim R_{\rm at}$ (i.e. where the OBrS amplitude is noticeable), then to obtain a correct magnitude of $d\,\sigma_{\rm int}$ one must use the distorted wavefunctions rather than plane waves.

To conclude this section we present the comparison of the calculated data on the total BrS spectrum with the experimental data. The full curve in Fig. 4.5 stands for the DPWA cross section $d\sigma - d\sigma_{\rm ord}$ (i.e., the OBrS background subtracted from the total cross section) calculated for a 600 eV collision with a Xe atom [20, 28, 30] The atomic response was treated within the RPAE. The broken curve represents the experimental data [426] (see also [428]). The figure demonstrates quite good agreement between theory and experiment for the position of the maximum. The width of the experimentally measured resonance is somewhat larger than the calculated one.

4.2 Angular Distribution and Polarization of BrS

The important characteristic of the BrS process is the angular distribution of radiation which is described by the double differential cross section $\omega d^2\,\sigma\,/d\omega d\Omega_\gamma$. The latter, in the case of the dipole–photon emission, is given by (2.28) and (2.27).

The double differential cross section can be written as a sum of two components, characterizing the emission of the photons polarized perpendicular ($\perp$) and parallel ($\|$) to the plane $(\mathbf{k}, \mathbf{p}_1)$ [13, 29]:

$$\frac{d^2\sigma}{d\omega d\Omega_{\mathbf{k}}} = \frac{d^2\sigma^{\|}}{d\omega d\Omega_{\mathbf{k}}} + \frac{d^2\sigma^{\perp}}{d\omega d\Omega_{\mathbf{k}}}. \quad (4.4)$$

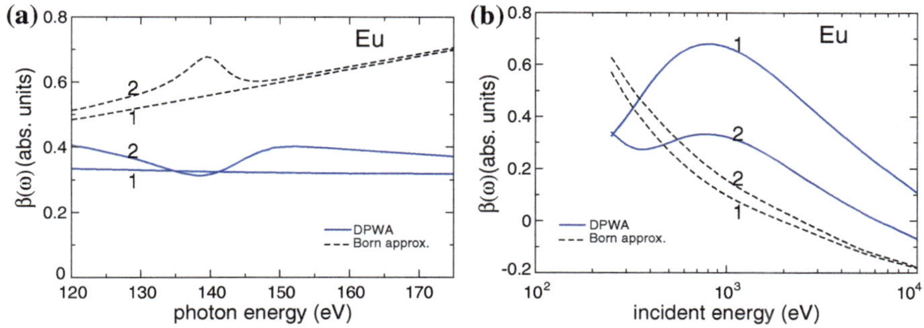

Fig. 4.6 Angular asymmetry parameter versus photon energy for a $\varepsilon_1 = 250\,\mathrm{eV}$ electron on the Eu atom (*left*) and as a function of the incident electron energy ε_1 for the fixed photon energy $\omega = 142.9\,\mathrm{eV}$ (*right*) calculated within the DPWA and BA as indicated. *Curves 1* stand for OBrS, *curves 2*—for the total BrS [244]

Here

$$\frac{d^2\sigma^{\parallel}}{d\omega d\Omega_{\mathbf{k}}} = \frac{1}{8\pi}\frac{d\sigma}{d\omega}\left[1 - \beta(\omega)\Big(2P_2(\cos\theta_{\mathbf{k}}) - 1\Big)\right]$$
$$\frac{d^2\sigma^{\perp}}{d\omega d\Omega_{\mathbf{k}}} = \frac{1}{8\pi}\frac{d\sigma}{d\omega}\Big(1 - \beta(\omega)\Big). \tag{4.5}$$

Results of the DPWA (full curves) and the BA (broken curves) calculations of the angular asymmetry parameters of the total, β_{tot}, and the ordinary, β_{ord}, BrS in electron–Eu collision are presented Fig. 4.6 [244].

Figure 4.6 (*left*) illustrates the differences in the dependencies of the parameters on ω for fixed energy $\varepsilon_1 = 250\,\mathrm{eV}$ of the incident electron. The OBrS anisotropy parameter is a smooth monotonous function, whereas, β_{tot} shows rather rapid variation in the vicinity of the 4d-threshold. Far beyond the resonance in the BrS spectrum β_{tot} tends to β_{ord}. The difference between the DPWA and the Born curves is quite noticeable and it is increasing with the growth of ω.

The right graph in Fig. 4.6 presents the parameters β_{tot} and β_{ord} as functions of the initial electron energy calculated for the photon frequency $\omega = 142.9\,\mathrm{eV}$ which is the position of the giant resonance in the total BrS spectrum (see the right lower graph in Fig. 4.3).

The DPWA curves (full curves) were calculated according to (2.30), the BA dependencies (broken curves) were obtained using the algorithm described in [13].

The discrepancy between the DPWA and the Born results is noticeable over almost the whole region of the projectile's energies. Calculated in BA the angular asymmetry parameters monotonically decrease as ε_1 increases, and both BA curves lie close to each other. More accurate DPWA calculations reveal an oscillation in the ε_1-dependence of both β_{tot} and β_{ord}. An oscillatory, behaviour of β_{ord} as a function of the incident energy was discussed previously in [227], where numerical

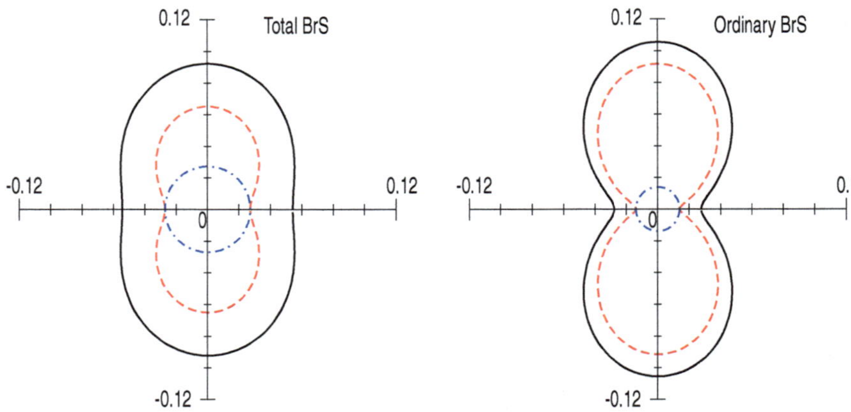

Fig. 4.7 Shape functions of the total (*left*) and ordinary BrS (*right*) for a $\varepsilon_1 = 600\,\text{eV}$ electron on Eu at $\omega = 142.9\,\text{eV}$. *Broken curve* stands for $S^{\parallel}(\omega_{\text{max}}, \varepsilon_1, \theta_{\mathbf{k}})$, *chained curve*—for $S^{\perp}(\omega_{\text{max}}, \varepsilon_1, \theta_{\mathbf{k}})$, *full curve*—for $S(\omega_{\text{max}}, \varepsilon_1, \theta_{\mathbf{k}}) = S^{\parallel}(\omega_{\text{max}}, \varepsilon_1, \theta_{\mathbf{k}}) + S^{\perp}(\omega_{\text{max}}, \varepsilon_1, \theta_{\mathbf{k}})$. The horizontal axis aligned with the initial velocity $\mathbf{v}_1$ of the electron. All shape functions were obtained in DPWA

calculations of the ordinary BrS in a screened Coulomb potential were carried out within the frame of the classical electrodynamics and the classical collision theories. In the cited paper the oscillations of the function $\beta_{\text{ord}}(\varepsilon_1)$ were attributed to a superposition of the electromagnetic waves radiated by a projectile scattered in a screened Coulomb potential with different values of the impact parameter. Full curves in Fig. 4.6 illustrate that a similar effect takes place as well in the processes of both ordinary and total BrS treated in terms of quantum mechanics.

For fixed value of ω the angular distribution of BrS can be described in terms of the shape function $S(\omega, \varepsilon_1, \theta_{\mathbf{k}})$, which allows one to visualize the spatial distribution of the emitted radiation:

$$S(\omega, \varepsilon_1, \theta_{\mathbf{k}}) = \left(\frac{\mathrm{d}^2\sigma}{\mathrm{d}\omega \mathrm{d}\Omega_{\mathbf{k}}}\right) \bigg/ \left(\frac{\mathrm{d}\sigma}{\mathrm{d}\omega}\right). \tag{4.6}$$

Comparing (4.4) and (4.6), one relates the shape function to the angular anisotropy parameter in the case of dipole–photon radiation:

$$S(\omega, \varepsilon_1, \theta_{\mathbf{k}}) = \frac{1 - \beta P_2(\cos\theta_{\mathbf{k}})}{4\pi}. \tag{4.7}$$

Full curves in Figs. 4.7 and 4.8 represent the dependence of $S(\omega, \varepsilon_1, \theta_{\mathbf{k}})$ on the polar emission angle $\theta_{\mathbf{k}}$ (measured with respect to the horizontal axis) for the total and ordinary BrS (as indicated) in collision of $\varepsilon_1 = 600\,\text{eV}$ (Fig. 4.7) and $\varepsilon_1 = 10\,\text{keV}$ (Fig. 4.8) electron on the Eu atom. Both figures refer to $\omega = 142.9\,\text{eV}$ which corresponds to the maximum value of the total spectrum $\omega\mathrm{d}\sigma/\mathrm{d}\omega$. In accordance with (4.4) and (4.5) and the definition (4.6), the shape function $S(\omega, \varepsilon_1, \theta_{\mathbf{k}})$ of unpolarized

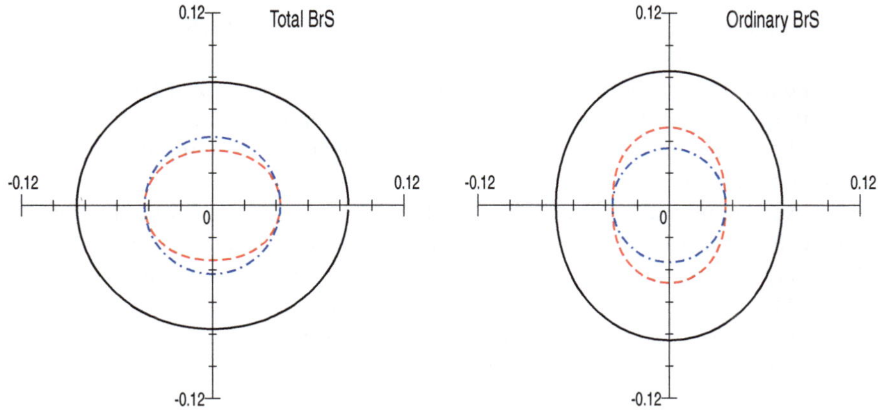

Fig. 4.8 Same as in Fig. 4.7 but for a 10 keV electron

radiation is equal to the sum $S^{\|}(\omega, \varepsilon_1, \theta_{\mathbf{k}}) + S^{\perp}(\omega, \varepsilon_1, \theta_{\mathbf{k}})$. Here the terms stand for the shape functions of the BrS radiation polarized perpendicular, $S^{\perp}$, and parallel, $S^{\|}$ to the plane $(\mathbf{p}_1, \mathbf{k})$.

For the lower energy of the incident electron, Fig. 4.7, the angular distribution of both total and ordinary BrS is noticeably anisotropic. Most of the emitted radiation is polarized parallel to the $(\mathbf{p_1}, \mathbf{k})$ plane. The contribution of the isotropic term, $S^{\perp}(\omega, \varepsilon_1, \theta_{\mathbf{k}})$, to the angular distribution is much smaller (this feature is more pronounced for OBrS).

Figure 4.7 corresponds to the initial energy E_1 for which the angular anisotropy parameters $\beta_{\rm tot}$ and $\beta_{\rm ord}$ reach their maximum values, see right graph in Fig. 4.6. For larger values of ε_1 both anisotropy parameters decrease monotonously resulting on the change of the shape functions. This is illustrated by Fig. 4.8 which corresponds to $\varepsilon_1 = 10\,$keV. In this case, the parameters satisfy the inequalities $|\beta_{\rm tot}|, \beta_{\rm ord} \ll 1$, which result in weak anisotropy of the emitted radiation.

4.3 Approximate Method to Calculate the PBrS Cross Sections with Account for Many-Electron Correlations

It was demonstrated in Sect. 4.1, that for moderate energies of the incoming electron the magnitude of the total BrS cross section is quite sensitive to the distortion of the electron wavefunction by the electrostatic field of a target and to the interference between the two radiative mechanisms. However, for the photon energies in the vicinity of the giant resonance (for example, for ω above the ionization potential of the 4d subshells in Xe, Ba and rare-earth atoms) the magnitude of the cross section is much more affected by the correlation effects which determine, to a great extent, the atomic dynamic response $\alpha(\omega, q)$ to the external electromagnetic and Coulomb fields.

To account for the electron correlations in calculating the PBrS cross sections in the vicinity of the 4d-giant resonances different scheme were used: the Generalized RPAE method [8] for the Ba and La atoms, and the Spin-Polarized RPAE [21] for the Eu atom. It is noted in the cited papers that these approximate schemes provide reasonable agreement between theory and experiment in describing the process of atomic photo effect.

However, numerical computation of of the BrS cross section within the DPWA along with the use of the RPAE (or its above mentioned variants: GRPAE, SP RPAE) is quite complex task. To this end, a simple approximate method, proposed in [244, 245] (see also [242, 243]) for the calculation of $\alpha(\omega, Q)$ and consequently of $d\sigma_{\rm pol}(\omega)$, allows one to avoid rather complicated direct numerical computations of the many-electron correlation effects.

4.3.1 Description of the Method

To construct an approximate expression for generalized polarizability with account for the many-electron correlation effects, let us start with writing $\alpha(\omega, q)$ in the following form:

$$\alpha(\omega, q) = \alpha(\omega) G(\omega, q). \tag{4.8}$$

As written, this equality is a definition of a new function $G(\omega, q)$ equal to the ratio $\alpha(\omega, q)/\alpha(\omega)$ of the *exact* generalized and dipole polarizabilities.

Now let us assume that all the information about the many-electron correlation effects is contained in the dipole polarizability $\alpha(\omega)$, while the factor $G(\omega, q)$ is not that sensitive to them and can be calculated in the simpler approximation, for example within the Hartree–Fock scheme. Hence, instead of (4.8) one writes the following approximate formula:

$$\alpha(\omega, q) \approx \alpha(\omega) \frac{\alpha^{HF}(\omega, q)}{\alpha^{HF}(\omega)} \equiv \alpha(\omega) G^{HF}(\omega, q). \tag{4.9}$$

This relation is the key to the method. The approximate equality in (4.9) reduces a complex problem of the exact computation of $\alpha(\omega, q)$ to a much simpler one: the calculation of the factor $G^{\rm HF}(\omega, q)$ in the Hartree–Fock approximation.

To obtain $\alpha(\omega)$ one can avoid direct calculations by means of the many-body theory. The imaginary part of the polarizability is related to the photoabsorption cross section, $\sigma_\gamma(\omega)$, as

$$\operatorname{Im} \alpha(\omega) = \frac{c}{4\pi\omega} \sigma_\gamma(\omega). \tag{4.10}$$

The real part of $\alpha(\omega)$ can be restored via the dispersion relation:

$$\mathrm{Re}\,\alpha(\omega) = \frac{c}{2\pi^2}\int\limits_0^\infty \frac{\sigma_\gamma(\omega')\,\mathrm{d}\omega'}{\omega'^2 - \omega^2}. \tag{4.11}$$

Relation (4.9) accompanied by (4.10) and (4.11) provide a possibility to get all the necessary information on the atomic dynamic response in the BrS process directly from the experimental data on the photoabsorption. Thus, the general DPWA formalism developed in Sect. 2.2 combined with the proposed semi-empirical method for the calculation of $\alpha(\omega, q)$ could be used, without any alterations, for the calculation of the BrS spectrum formed in collision of an electron with more complex systems, such as molecules, clusters or solids.

In the [242–246] the validity of this method was checked against more rigorous calculations (carried out within various RPA-based schemes) of $\alpha(\omega, q)$ and $\mathrm{d}\sigma_{\mathrm{pol}}(\omega)$ for electron scattering on Ba, La, Eu. The BrS spectrum was calculated in the vicinity of the 4d-subshells ionization potentials where the polarizational mechanism leads to the powerful maximum in the spectrum (see Sect. 4.3.1 for the illustrative numerical results). Recently this approach was used in [365] to calculate, in a broad range of photon energies, the total BrS spectra in e^-–Kr and e^-–La collisions. In the cited paper the function $G(\omega, q)$ (more exactly, its inverse Fourier image, $G(\omega, r)$) was calculated, following [59], within the local spin density approach.

4.3.2 Numerical Results

In this section we present the results of numerical calculations of the BrS cross section formed in collisions of an intermediate energy electron with Ba, La and Eu atoms. The calculations were performed within the frameworks of (a) the HF approximation, (b) the RPAE scheme, and (c) the 'G-function' scheme based on the use of (4.9)–(4.11).

Prior to comparing the cross sections, we illustrate the applicability of the 'G-function' method using the Ba atom as the case study. In Fig. 4.9 (*left*) we compare the dependencies of $|G(\omega, q)|^2$ on the transferred momentum q calculated within the RPAE and the HF frameworks for three photon energies (as indicated) above the 4d threshold ($I_{4d} = 99.3\,\mathrm{eV}$ [276, 364]). Comparing the solid and dashed curves, one can state that in a broad range of q and ω a simple single-electron HF approximation and more sophisticated RPAE approach provide, essentially, the same result for the function $G(\omega, q)$. Thus we conclude that this function is not sensitive to many-electron correlations and can be calculated within the HF scheme.

On the contrast, to obtain the reliable data for the dipole dynamic polarizability $\alpha(\omega)$ one has to go beyond the HF approximation. In Fig. 4.9 (*right*) the curves marked as 'Im_1' and 'Re_1' stand for the, respectively, real and imaginary parts of $\alpha(\omega)$ within the HF approximation. In the vicinity of the 4d threshold, the HF curves by far disagree with the results obtained within the RPAE (the subscript '2' in the Im- and Re-notations) in both the position of the maxima and the height. The curves

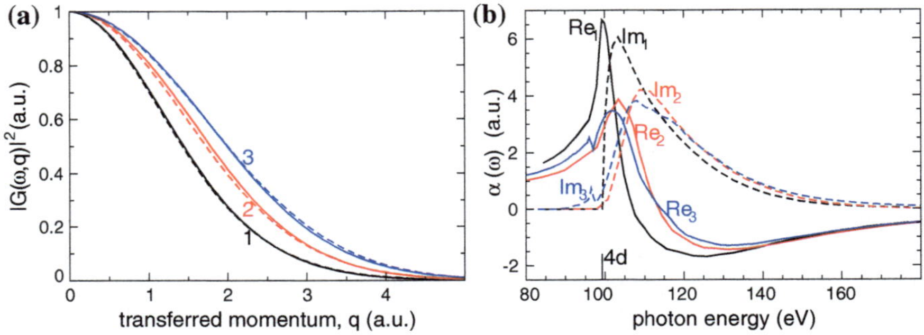

Fig. 4.9 *Left panel* The q dependence of the function $|G(\omega, q)|^2$ calculated for a Ba atom for three different photon energies ω above the 4d threshold $I_{4d} = 99.3\,\text{eV}$: *curves 1, 2, 3* stand for $\omega = 108, 140, 180\,\text{eV}$, respectively. *Dashed curves* stand for the RPAE, the *solid curves*—for HF [246]. *Right panel* Real (*solid curves*) and imaginary (*dashed curves*) parts of $\alpha(\omega)$ for a Ba atom above the 4d threshold (marked with a *vertical line*). Subscripts '*1*' and '*2*' stand for the calculations within the HF and RPAE schemes, respectively, the subscript '*3*' marks the calculations based on (4.10) and (4.11) [246]

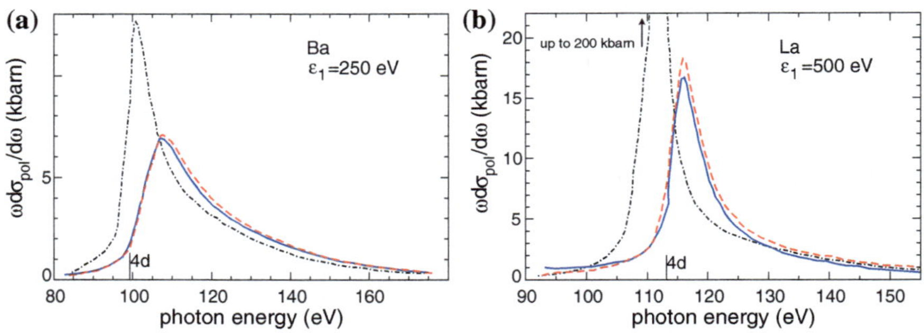

Fig. 4.10 The polarizational part of the cross section, $\omega \mathrm{d}\sigma_{\text{pol}}/\mathrm{d}\omega$, for Ba (*left*) and La (*right*) atoms for the indicated incident electron energies ε_1. *Solid curves* stand for the GRPAE result, *broken curves*—for the 'G-function' approximation (4.9) with $\alpha(\omega)$ calculated within GRPAE, *chained curves* present the HF calculations [245–248]. The *vertical lines* indicate the atomic 4d thresholds

with the subscript '3' stand for the real and imaginary obtained using the relations (4.10) and (4.11). The data on the absorption cross section $\sigma_\gamma(\omega)$ were compiled from [193, 364]. Integration in the dispersion relation (4.11) was carried out over the region of ω from 0.01 to 30 keV, where experimental data are available. It is seen that the RPAE curves '2' and the curves '3' are in much better agreement. Some discrepancy between them can be attributed to the fact that RPAE takes into account only the major part of the many-electron correlations, but not all of them.

Figure 4.10 presents the PBrS cross sections $\omega \mathrm{d}\sigma_{\text{pol}}/\mathrm{d}\omega$ calculated within the DPWA (see Sects. 2.2.1 and 2.2.2) for electron collision with Ba (left graph) and La (right graph) atoms in the vicinity of the 4d ionization potentials: $I_{4d} = 99.3\,\text{eV}$ for Ba and 113.2 eV for La [364]. The incident electron energy is indicated on the graphs. The atomic generalized polarizability $\alpha(\omega, q)$, which enters the PBrS amplitude, (2.16) and (2.21), was calculated within three different frameworks. The chained curves correspond to the HF approximation with no electron correlations taken into account. The solid curves represent the calculations within the GRPAE scheme. Finally, the dashed curves in both graphs stands for the 'G-function' approximation (4.9) with the factor $\alpha(\omega)$ calculated within GRPAE.

The most striking feature in this graphs concerns the inadequacy of the HF approach in describing the atomic dynamic response. Comparison of the chained and solid curves demonstrates, that the the account for the many-electron correlations reduces the maximum value of $\omega \mathrm{d}\sigma_{\mathrm{pol}}/\mathrm{d}\omega$ by a factor of 2 for Ba and by a factor of 20 for lanthanum! Additionally, the maximum is blue-shifted by 8 and 4 eV for Ba and La, respectively.

The discrepancy between the HF and GRPAE results is the largest in the close vicinity of the 4d-thresholds but it diminishes rapidly as ω increase. To a great extend, this is the result of the atomic core relaxation (or, re-arrangement) which is accounted for in the GRPAE scheme [24, 276]. The relaxation of the atomic shells leads to additional screening of the Coulomb field of the hole acting on the excited (ionized), and, therefore, can decrease the photon emission by the electron. The impact of the additional screening is essential only when the characteristic excitation time τ_{e} is larger than the relaxation time τ_{r} [8]. This leads to a decrease in the PBrS cross section for the photon energies close to the ionization threshold, where the motion of the excited electron is slow so that $\tau_{\mathrm{e}} \gg \tau_{\mathrm{r}}$. Far from the threshold the inequality is inverse. As a result, the maximum of $\omega \mathrm{d}\sigma_{\mathrm{pol}}/\mathrm{d}\omega$ is shifted toward larger ω-values.

Another feature to be mentioned in connection with Fig. 4.10 is a remarkable agreement between the GRPAE curves and those based on the approximation (4.9). This agreement demonstrates the 'G-function' scheme works well in the range of photon energies $\omega \sim I_{4d}$ where the electron correlation play essential role in forming a dynamic response of the atoms.

Figures 4.11 and 4.12 present the dependence of the total BrS cross section $\omega \mathrm{d}\sigma_{\mathrm{tot}}/\mathrm{d}\omega$ (2.29) in the vicinity of the 4d-subshell for the $\varepsilon_1 = 250$ eV electron collision with Ba and for the $\varepsilon_1 = 500$ eV electron collision with La atoms, respectively.

In both figures, the solid curves 1 stand for the dependencies obtained within the framework of *GRPAE* (SPRPAE), and the short-dashed curves 2 are the result of the approximation (4.9) with the dipole polarizability $\alpha(\omega)$ calculated within the GRPAE. Comparing the curves 1 and 2 one can state, that the 'G-function' approximation agrees well with the more sophisticated method not only for the PBrS (as in Fig. 4.10) but for the total BrS which includes the OBrS term (the chained curves) and the interference term (not presented in the figures).

The long-dashed curves Figs. 4.11 and 4.12 (in the latter, this curve is marked as "3") were obtained combining (4.9) with relations (4.10) and (4.11) to calculate the dynamic dipole polarizability. The experimental data on the photoabsorption cross sections for Ba and La were compiled from [193, 364, 449]. The discrepancy

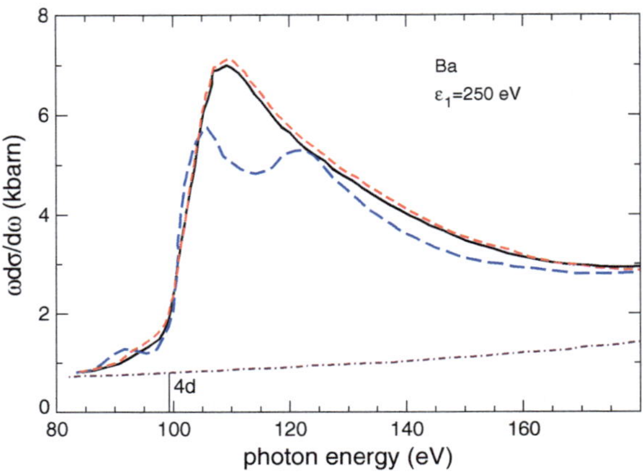

Fig. 4.11 BrS cross section in a $\varepsilon_1 = 250\,\text{eV}$ electron collision with a Ba atom. The total BrS cross section $\omega \mathrm{d}\sigma_{\mathrm{tot}}/\mathrm{d}\omega$ calculated within different frameworks: within GRPAE—the *solid curve*, using the 'G-function' approximation (4.9) with $\alpha(\omega)$ within GRPAE—the *short-dashed* (*red*) *curve*, in the 'G-function' approximation with the use of (4.10) and (4.11) for $\alpha(\omega)$—the *long-dashed* (*blue*) *curve* [245, 246]. The *chained curve* stands for the OBrS background. The *vertical line* indicates the atomic 4d threshold

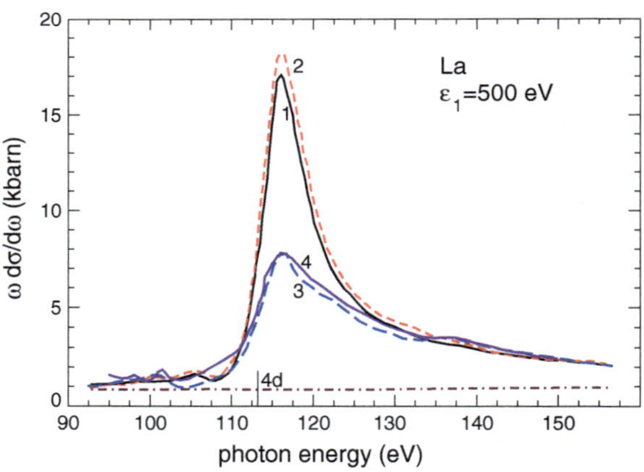

Fig. 4.12 BrS cross section in a $\varepsilon_1 = 500\,\text{eV}$ electron collision with a La atom. The *curves 1*, *2*, *3* denote the total BrS cross section $\omega \mathrm{d}\sigma_{\mathrm{tot}}/\mathrm{d}\omega$ calculated within different frameworks: *curves 1*—within GRPAE, *curves 2*—within the 'G-function' approximation (4.9) with $\alpha(\omega) = \alpha^{GRPAE}(\omega)$, *curves 3*—within the 'G-function' approximation with $\alpha(\omega)$ (4.10) and (4.11) [248]. The *solid curve 4* represents the experimental data for $\omega \mathrm{d}\sigma_{\mathrm{tot}}/\mathrm{d}\omega$ [452]. The *unmarked chained curve* stands for the OBrS background. The *vertical line* indicates the atomic 4d threshold

between curves 1 and 2 (which is more pronounced for La) can be attributed to the difference in the dynamic dipole polarizabilities obtained within GRPAE and from the experimental data on σ_γ. This means that even the GRPAE does not take into account all many-electron correlation effects. To improve the agreement in the vicinity of the maxima one has to perform the calculations of $\alpha(\omega)$ beyond GRPAE.

Curve 4 in Fig. 4.12 represents the experimental BrS spectrum of 500 eV electrons for metallic La [452]. The experimentally measured intensity had no absolute calibration, therefore, in the figure, it is is normalized to the peak value of the curve 3. It is seen that the semi-empirical method (4.9)–(4.11) reproduces quite well the experimental spectrum in the whole ω-region. The influence of the medium on the BrS process in metallic La is discussed in Sect. 4.4.

4.4 Influence of Photon Absorption and Electron Energy Loss on the PBrS Spectrum in a Solid-State Target

To describe the BrS process in a solid we will assume the following. First, the target is considered to be amorphous. The spatial distribution of the atoms is characterized by constant value of volume density $n_{\rm at}$. Second, we assume that the BrS cross section in a solid, $d\sigma_{\rm S}$, represents the incoherent sum of the BrS cross sections $d\sigma_{\rm at}$ of individual atoms. Such an approximation can be justified provided the coherence length $l_{\rm coh} \sim v_1/\omega$ (i.e., the spatial scale within which the radiation of a given frequency ω is formed) is smaller than the (average) interatomic distances l in the solid, and thus the photon emission occurs during independent collisions of a projectile with individual atoms. For the incoming electron and emitted photon energies considered in Sect. 4.4.2 below, the coherence length $l_{\rm coh} \approx 1$ Å whereas the interatomic distances are $l \sim 3 \div 5$ Å. Below we assume that the condition $l_{\rm coh} \ll l$ is met so that so that $d\sigma_{\rm S} \propto n_{\rm at}$.

To account for the photon absorption we assume that the flux of photons of a given energy ω is attenuated by a factor $\propto \exp(-\mu(\omega)L)$ while propagating the distance L in medium. The quantity $\mu(\omega)$ stands for the photon attenuation coefficient.

Within the approximations and assumptions formulated above, the differential cross section of an electron scattered in an infinitely thick target is related to the BrS cross section for an isolated atom as follows:

$$\frac{d^2\sigma_{\rm S}}{d\omega d\Omega_{\mathbf{k}}} = n_{\rm at} \int_0^{\infty}\int_{\omega}^{\varepsilon_1} \frac{d^2\sigma_{\rm at}}{d\omega d\Omega_{\mathbf{k}}} e^{-\mu(\omega)x/\cos\theta_{\mathbf{k}}}\, f(x,\varepsilon)\, dx\, d\varepsilon. \tag{4.12}$$

Here $\theta_{\mathbf{k}}$ is the photon emission angle with respect to the incoming electron beam. The factor $f(x,\varepsilon)$ stands for the distribution function of the electrons scattered in the target. It depends on the electron energy and on the penetration distance x.

4.4.1 Energy Distribution of Scattered Electrons

Determination of the distribution function $f(x,\varepsilon)$ in application in a solid target is a complicated problem.

Within the framework of one of the widely used approach (see, for example, [5, 398]) the energy distribution is ignored. The features of electron deceleration specific for different elements are introduced as follows. It is assumed that the cross section $d\sigma_S/d\omega$ can be obtained by integrating $d\sigma_{at}/d\omega$ over the target thickness:

$$\frac{d^2\sigma_S}{d\omega d\Omega_{\mathbf{k}}} = n_{at} \int\limits_0^{x_{max}} \frac{d^2\sigma_{at}}{d\omega d\Omega_{\mathbf{k}}} e^{-\mu(\omega)x/\cos\theta_{\mathbf{k}}} f(x,\varepsilon)\, dx, \tag{4.13}$$

where x_{max} is the maximum penetration depths for electrons of the initial energy ε_1. The integration over x can be reduced to that over the electron energy: $\int_0^{x_{max}}(\ldots)dx = -\int_{\varepsilon_1}^{\omega}(d\varepsilon/dx)^{-1}\,d\varepsilon$ where $(-d\varepsilon/dx)$ is the mean energy loss rate (or stopping power) of electrons due to inelastic processes in the medium [272, 326, 336, 376].

This approach does not take into account the electrons scattered by large angles, so that their contribution to the spectrum must be accounted for by introducing additional empiric factors. Apart from this, neglecting the energy distribution of electrons imposes a restriction on the upper limit of integration over the penetration depth. All these drawbacks make one to consider formula (4.13) as a quite crude approximation.

In [243], to describe the energy distribution of the electrons scattered in the thick target the model function $f(x,\varepsilon)$, introduced initially in [91], where it was represented as a sum of two terms corresponding to the forward and the backward electron fluxes:

$$f(x,\varepsilon) \to g_f(x,\varepsilon) + g_b(x,\varepsilon). \tag{4.14}$$

where the indices f and b denote the forward and backward fluxes, respectively.

To evaluate the term $g_f(x,\varepsilon)$ in [91] the electron attenuation with respect to the penetration distance was describe by the exponential decay law:

$$N(x) = N_0 \exp\left[-\left(\frac{x}{X}\right)^p\right], \tag{4.15}$$

Here $N(x)$ is the number of electrons at the distance x, N_0 is the number of electrons at the sample entrance, $X = a\varepsilon_1^{\nu}/\rho$, $\nu = 2.44\,(Z/A)^{1/2}$, $p = 1.51\nu/\log_{10}Z$, with ρ, A, Z standing, respectively, for the density, the atomic weight and the atomic number, and $a = 3.33\times10^{-6}$ for ε_1 in keV and x in cm [126].

In terms of the scaled units, $\chi = x/X$ and $\xi = \varepsilon/\varepsilon_1$ the functions $g_f(x,\varepsilon)$ and $g_b(x,\varepsilon)$ can be written as follows [91]. For the forward flux:

$$g_f(\chi,\xi) = pn\chi^p \frac{\xi^{\nu-1}}{(1-\xi^\nu)^{p+1}} \exp\left[-\left(\frac{\chi}{1-\xi^\nu}\right)^p\right]. \tag{4.16}$$

The number of electrons with arbitrary energy (within the interval $[0,\varepsilon_1]$) moving at the scaled distance χ in the forward direction is found as $N(\chi) = N_0\int_0^1 g_\eta(\chi,\xi)d\xi$.

The distribution function $g_b(x,\varepsilon)$ of the backward flux reads:

$$g_{\mathrm{b}}(\chi,\xi)=\frac{2p^2\nu\xi^{\nu-1}}{(1-\xi^\nu)^{p+1}}\int\limits_{\chi}^{\infty}\frac{\mathrm{d}y}{y}\,\frac{y^p(2y-\chi)^p}{\exp\left[(1+(1-\xi^\nu)^{-p})(2y-\chi)^p\right]}. \tag{4.17}$$

The number of electrons with given energy backscattered out from the target, $N_{\mathrm{b}}(\xi)$, and the total number, N_{b}, in the backward electron flux are equal to

$$N_{\mathrm{b}}(\xi)\equiv N_0 g_{\mathrm{b}}(0,\xi)=\frac{p\nu}{2^{p-1}}\frac{\xi^{\nu-1}(1-\xi^\nu)^{p-1}}{[1+(1-\xi^\nu)^p]^2},\qquad N_{\mathrm{b}}=\int\limits_0^1 g_{\mathrm{b}}(0,\xi)d\xi. \tag{4.18}$$

The intensity of characteristic X-rays induced in a target by an electron, calculated in [91] using of the distribution function, defined above by (4.14), (4.16) and (4.17), was in a reasonable agreement with the experimental data. Also, a good agreement with the experimental data [104, 106, 122, 172, 342] was reported for the calculated intensity of the electron flux penetrated through a thin target in the forward direction as well as for the intensity of the backscattered electron flux from a thick target. The comparison was carried out over a wide range of incident electron energies and for various solid targets.

4.4.2 Numerical Results and Experimental Data

The calculation of total BrS cross section $(\mathrm{d}\sigma/\mathrm{d}\omega)_{\mathrm{S}}$ with account for inelastic electron scattering and the attenuation of emitted radiation in metallic La was carried out for the photon energy range ω =100–200 eV [243]. The results of calculations were compared to the BrS cross section on the isolated La atom which was calculated within the "G-function" approximation (see Sect. 4.3). The imaginary part of $\alpha(\omega)$ was obtained from the experimental data on photo-absorption spectra [193, 364, 449]. The real part of $\alpha(\omega)$ was restored using the dispersion relation.

To describe the energy distribution of the electrons scattered in metallic La we used the model (4.14), (4.16) and (4.17). The photon attenuation coefficient $\mu(\omega)$ was calculated as a product of the atomic photoabsorption cross section $\sigma_\gamma(\omega)$ and the volume density n_{at} of the La atoms: $\mu(\omega)=n_{\mathrm{at}}\,\sigma_\gamma(\omega)$.

The atomic cross sections $\mathrm{d}^2\sigma_{\mathrm{at}}/\mathrm{d}\omega\mathrm{d}\Omega_{\mathbf{k}}$ (solid curves) calculated for 0.5 and 4 keV electrons are presented in Fig. 4.13 (left). For the sake of comparison, the ordinary BrS background is also plotted (dashed curves).

In Fig. 4.13 (right) the full curves represent the calculated BrS cross sections for 0.5 (curve 1) and 4 keV (curve 2) electrons bombarding a metallic La target. Comparing the theoretical curves in the left and right panels we note that for the lower incident energy the spectra on the isolated and on the metallic La targets are much alike in their shapes, exhibiting intensive maxima in the vicinity of the 4d threshold ($I_{4\mathrm{d}}$ = 113.2 eV [364]). With the increase of ε_1 up to 4 keV the shape

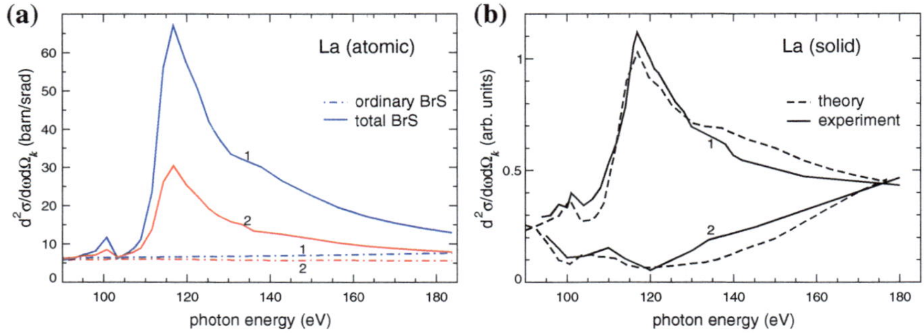

Fig. 4.13 *Left panel* The atomic BrS cross section $d^2\sigma_S/d\omega d\Omega_{\mathbf{k}}$ calculated for $\varepsilon_1 = 0.5\,\text{keV}$ (*full curve 1*) and $\varepsilon_1 = 4\,\text{keV}$ (*full curve 2*) electrons on an isolated La target in the vicinity of the atomic 4d threshold [243]. The angle of the photon emission $\theta_{\mathbf{k}} = 45°$. The *dashed curves 1* and *2* represent the ordinary BrS background radiation. *Right panel* The BrS cross section (4.12) calculated for $\varepsilon_1 = 0.5\,\text{keV}$ (*full curve 1*) and $\varepsilon_1 = 4\,\text{keV}$ (*full curve 2*) electrons on a metallic La target in the vicinity of the atomic 4d threshold [243] versus the experimentally measured spectra (*broken curves*) for the same energies of incident electrons [452, 453]. All curves are normalized to the point $\omega = 176\,\text{eV}$. In *both panels* the angle of the photon emission $\theta_{\mathbf{k}} = 45°$

of the solid La BrS curve changes drastically: the maximum transforms into the minimum in the same ω region. Such an effect is due to the process of radiation absorption in the solid target.

The broken curves in Fig. 4.13 (right) correspond to the experimentally measured [452, 453] emission spectra of metallic La. In the cited papers the experimental data are given in arbitrary units, and the 4 keV spectrum was plotted level with the 0.5 keV curve in some ω-point on the right wing of the spectra. Therefore, to compare the calculated and the experimental cross sections all the the dependencies were matched at the point $\omega = 176\,\text{eV}$.

With an accuracy level of 10 % there is an agreement between the calculated and the measured electron BrS spectra for the metallic La target. Some discrepancy may be attributed to an error of the absolute magnitudes of the photo-absorption cross sections as well as to a mistake introduced in the calculations by the use of the model function (4.14), (4.16) and (4.17).

The calculations performed in [243] have demonstrated that the polarizational mechanism of radiation in collision of an electron with solid La dominates in the region of photon frequencies above the 4d subshell ionization potential for low incident electron energy ($\varepsilon_1 \sim 0.5$ keV). The relative contribution of the polarizational part of the BrS spectrum decreases with the increase of the incident electron energy. It happens due to the increase of the electron penetration distance at which the photons are emitted, and this, in turn, leads to the increase of the photon attenuation effect.

4.5 BrS Spectra in Broad Range of Photon Energies

In Sects. 4.1, 4.3 and 4.4 the main focus was at comparatively narrow regions of the emitted photon energies ω, i.e. in the vicinities of the giant dipole resonances, in elucidating the role of the PBrS process in electron collisions with many-electron atoms. This ω-region is of particular interest because of the collective nature of the giant resonances and, as a result, of the large magnitude of the atomic polarizability which leads to a dominance of the PBrS in the total radiative spectrum.

In this Section we demonstrate that the polarizational mechanism plays an important role in the formation of the total BrS spectrum over the whole region of photon energies rather than only in the vicinity of the giant dipole resonances. The case studies, considered below, include $\varepsilon_1 = 1$ and 25 keV electron scattering from Ne, Ar, Kr and Xe atoms [249, 250, 253, 365]. The spectral distribution of BrS is calculated for the whole range of photon energies: from 10 eV up to $\omega = \varepsilon_1$. Accounting for the PBrS mechanism changes radically the traditional view of the BrS cross section $d\sigma/d\omega$ as a smooth function monotonically decreasing with ω. Instead, the total BrS cross section becomes essentially non-monotonous, bearing in many cases characteristic resonance-like features.

Four graphs in Fig. 4.14 present the total and ordinary BrS spectra, $\omega d\sigma/d\omega$, formed in collisions of $\varepsilon = 1$ keV (the incident velocity $v_1 \approx 8.6$ a.u.) and $\varepsilon = 25$ keV ($v_1 \approx 42.9$ a.u.) with the noble gas atoms, as indicated on the graphs. The vertical dashes denote the HF ionization thresholds of atomic subshells. The solid curves represent the total BrS spectra, the dashed curves—the PBrS ones. The calculations were performed within the framework of DPWA (thick blue line), according to the formulae presented in Sect. 2.2.2, and in the BA (thin black line), (2.1) (the case of Xe is presented only with the DPWA curves). The filled circles in the graphs stand for the data on OBrS spectrum [352].

The graphs presented clearly show that the PBrS mechanism essentially modifies the shape of spectral distribution. Instead of smooth curves, typical for OBrS, the total BrS curves exhibit complicated ω dependencies characterized by wide and powerful maxima and quite narrow cusps in the vicinity of the ionization thresholds. Such a behaviour is totally due to the contributions of the polarizational, $d\sigma_{\rm pol}/d\omega$, as well as the interference, $d\sigma_{\rm int}/d\omega$ terms to the total BrS spectrum (2.29).

First we note that there are maxima in the total spectra in the photon energy range $\omega = 10$–100 eV for Ar, Kr and Xe (up to 900 eV in the case of Ne) which are due to the contribution to the amplitude $f_{\rm pol}$ of virtual dipole excitations of electrons from the outermost s and p shells.

These excitations have a collective nature and are subject to correlation effects [8], the strongest among which are the intershell correlations of the ns- and the np-electrons [25]. In the cited paper the calculations of the photoabsorption cross sections $\sigma_\gamma(\omega)$ of the outer ns-shells of noble gases were carried out within the RPAE framework. It was found that $\sigma_\gamma(\omega)$ is strongly influenced by multielectron correlation effect. For the dipole–photon emission, the total BrS amplitude (2.8) contains only one characteristic, $\alpha(\omega, q)$, dependent on the internal dynamic structure

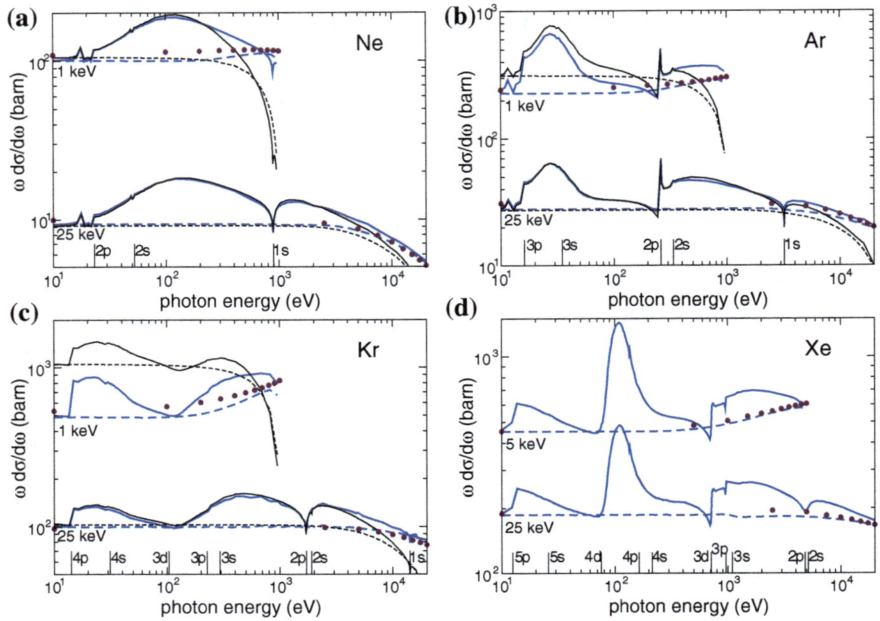

Fig. 4.14 BrS spectra, $\omega d\sigma/d\omega$, formed in collision of electrons with Ne, Ar, Kr and Xe atoms calculated for two incident energies as indicated. *Thick solid* and *dashed curves* describe, respectively, the PBrS and OBrS obtained within the DPWA. *Thin solid* and *dashed curves*—the same but in the BA. In all cases the atomic response was calculated within the RPAE [249, 250, 253]. *Closed circles* present the OBrS data from [352, 376]. *Vertical dashes* mark the HF ionization potentials of the atomic subshells. See also explanations in the text

of a target. The atomic polarizability is closely connected with $\sigma_\gamma(\omega)$ (see (4.10)), so that for its calculation RPAE method accounting for the dipole transitions from all atomic subshells was used in [249, 250, 253].

For the photon energies above the first maximum the total BrS curves for the Ne and Ar atoms exhibit smooth dependencies except for the regions in the vicinity of the ionization thresholds of the intermediate and the inner shells. This behaviour is discussed below in the Section.

In the case of Kr, in the range $\omega = 100$–$1000\,$eV the total spectrum acquires a second maximum, which is related to the dipole excitations into the continuous spectrum of electrons from the intermediate 3s-, 3p-, and 3d-subshells of the atom. Near the ionization potentials of these subshells, the multielectron correlation effects strongly influence the behaviour of $\alpha(\omega, q)$.

Finally, in the case of Xe there a maximum above the 4d-subshell related to the giant dipole resonance in the photoionization cross section. This feature was discussed in detail in Sect. 4.1. The third, less prominent maximum in $\omega d\sigma_{tot}/d\omega$, is due to the correlated excitations of the 3s-, 3p-, and 3d-electrons.

In the graphs for Ne, Ar and Kr the thin solid and dashed curves represent the results obtained within the BA. Comparing these curves with those obtained within the DPWA, one can formulae the general remarks on the applicability of the BA to the total BrS process. For a fixed incident energy the accuracy of the approximation decreases as the atomic number Z increases (compare the $\varepsilon_1 = 1\,\text{keV}$ spectra for the Ne, Ar and Kr targets). Fixing the target atom, one should expect the decrease in the accuracy with lowering the ε_1 values (compare the 25- and 1-keV curves for Ar or/and Kr). Finally, for all targets and any ε_1 value the BA fails in the region of large photon energies $\omega \lesssim \varepsilon_1$.

The difference between the results obtained within the framework of the BA and the DPWA, which is most clearly seen for the $\varepsilon_1 = 1\,\text{keV}$ curves for Ar and Kr, is mainly due to the difference in magnitudes of the OBrS cross sections (dashed curves in the graphs). The BA fails to describe the OBrS since the photon emission occurs mainly inside the atom, where the distorting influence of the atomic potential leads to a significant deviation of a projectiles wavefunction from a plane wave. In contrast, the polarizational BrS, which is formed at distances greater than the atomic radius [45, 454] is much less sensitive to the approximation used for the description of the projectiles movement.

The filled circles in the graphs stand for the data on OBrS spectrum [352] obtained using the relativistic DPWA [410]. The 5–10 % discrepancy between the relativistic and the non-relativistic DPWA (thick dashed curves) results can be attributed to the fact that in the latter scheme only the emission of the dipole photons was accounted for whereas the former approach took into account the contribution of higher multipoles. Another consideration which may explain the discrepancy concerns the accuracy of the data from [352], which is estimated at 10 % by the authors.

To analyze in more detail the general features in the behaviour of the BrS spectrum let us consider the case of a 25 keV electron on krypton atom. Curve 1 in Fig. 4.15 presents the ω-dependence the total BrS cross section $\omega \mathrm{d}\sigma_{\text{tot}}/\mathrm{d}\omega \equiv \mathrm{d}\sigma_{\text{tot}}$, curves 2, 3 and 4 stay for $\mathrm{d}\sigma_{\text{ord}}$, $\mathrm{d}\sigma_{\text{pol}}$ and the interference term $\mathrm{d}\sigma_{\text{int}}$, respectively. The results refer to the DPWA scheme forth projectile electron, and the RPAE approach to calculate the polarizability.

In the photon energy range $\omega = 10\ldots1000\,\text{eV}$ the total BrS cross section has two maxima. Within this spectral range, $\mathrm{d}\sigma_{\text{tot}}$ (see (2.29) is almost equal to the sum of the ordinary and the polarizational terms, whereas the contribution of the interference term is small.

In the vicinity of the 1s, 2s and 2p ionization potentials the total BrS spectral dependence exhibits deep narrow cusps. The value of $\mathrm{d}\sigma_{\text{tot}}$ at the tip of the cusp is smaller than that of $\mathrm{d}\sigma_{\text{ord}}$. As it is seen from Fig. 4.14, similar singularities appear in the spectral dependence for the Ne and Ar in the 1s-thresholds [249].

Such singularities in the total BrS spectrum reflect the behaviour of the real part of the generalized atomic dynamic polarizability $\alpha(\omega, q)$ in the vicinities of the ionization potentials of the inner shells.

The amplitude f_{pol} is expressed in terms of $\alpha(\omega, q)$, (2.16). The behaviour of the latter as a function of ω proves to be the similar for all q values within the range

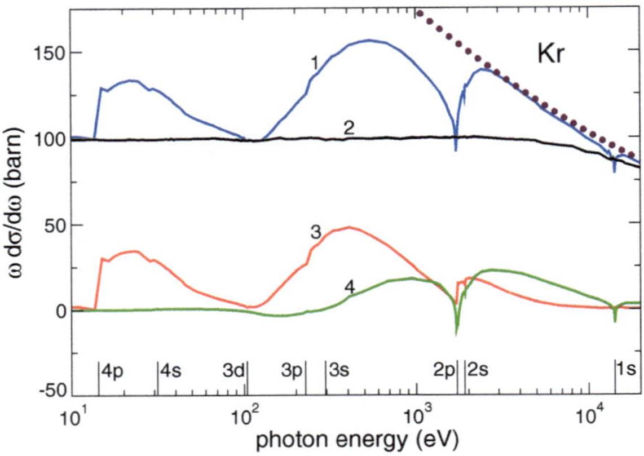

Fig. 4.15 The DPWA cross sections of total (*curve 1*), ordinary (*curve 2*), polarizational (*curve 3*) BrS for 25 keV electrons scattered by the Kr atom [253]. *Curve 4* stands for the interference term in the total cross section, (2.29). The *circles* correspond to the BrS cross section in the point Coulomb field of the charge $Z = 36$. See also explanations in the text

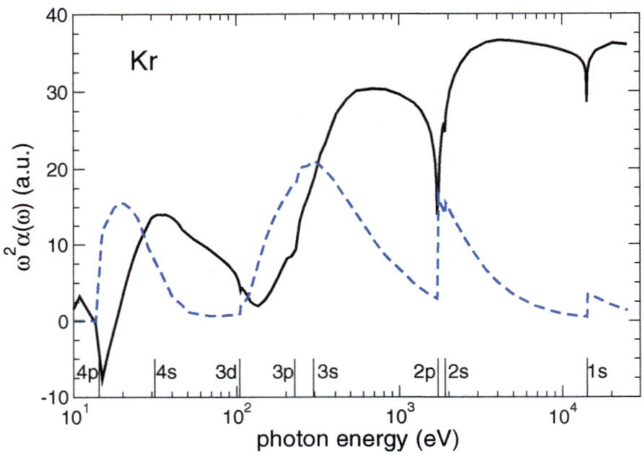

Fig. 4.16 Real (*solid curve*) and imaginary (*dashed curve*) part of dipole dynamic polarizability $\alpha(\omega)$ of the Kr atom multiplied by the factors $(-\omega^2)$ and ω^2, respectively. The results correspond to the calculations within the RPAE [250, 253]. *Vertical dashes* mark the HF ionization potentials of the atomic subshells

$q^2 \lesssim \omega$. Therefore, the ω-dependence of the polarization and interference terms in (2.29) can be estimated as follows:

$$\begin{aligned} \frac{\mathrm{d}\sigma_{\rm pol}}{\mathrm{d}\omega} &\sim \omega^3 \left[\Big(\mathrm{Re}\,\alpha(\omega)\Big)^2 + \Big(\mathrm{Im}\,\alpha(\omega)\Big)^2 \right] g_1(\omega), \\ \frac{\mathrm{d}\sigma_{\rm int}}{\mathrm{d}\omega} &\sim -\omega\,\mathrm{Re}\,\alpha(\omega)\, g_2(\omega), \end{aligned} \tag{4.19}$$

where $g_{1,2}(\omega)$ are smooth, monotonically decreasing functions of ω. This estimate explicitly shows that the interference term can be positive or negative following the sign of $\mathrm{Re}\,\alpha(\omega)$.

Figure 4.16 presents the ω-dependencies of the functions $-\omega^2\mathrm{Re}\,\alpha(\omega)$ (solid curve) and $\omega^2\mathrm{Im}\,\alpha(\omega)$ (broken curve) calculated for a Kr atom within the RPAE scheme accounting for the dipole transitions from all atomic subshells.

The imaginary part of the atomic polarizability of Kr has discontinuities near the ionization potentials of 1s-, 2s-, and 2p-subshells. These discontinuities lead to the cusps in $\mathrm{Re}\,\alpha(\omega)$. In turn, this cusp appears in the shape of the total BrS as a deep narrow minimum, which results from negativeness of the interference term.

The origin of such singularities in the total BrS cross section can be understood from the following considerations. The imaginary part of the dipole polarizability is related to the photoabsorption cross section $\sigma_\gamma(\omega)$, see (4.10). Typically, the behaviour of $\sigma_\gamma(\omega)$ for the inner atomic subshells is hydrogen-like. This means that there are jumps (discontinuities) in the cross section at the points $\omega = I_{\rm in}$ (see, for example, tables presented in [193]). Since the real part and imaginary parts of the polarizability are linked by the dispersion relation (4.11), a discontinuity in $\sigma_\gamma(\omega)$ results in a cusp in $\mathrm{Re}\,\alpha(\omega)$. Both of these features appear in the form discontinuities in the spectral dependence of the total BrS in the regions $\omega \approx I_{\rm in}$. The particular type of discontinuity in $d\sigma_{\rm tot}$ depends on the relative magnitudes of $\mathrm{Re}\,\alpha(\omega)$ and $\mathrm{Im}\,\alpha(\omega)$, as well as on the relative magnitudes of $d\sigma_{\rm pol}$ and $d\sigma_{\rm int}$, and on the sign of the latter.

Finally, let us discuss the ω-region above the 1s ionization potential. For $\omega > I_{1s}$ the total BrS cross section decreases but exceeds the OBrS background a each value of ω. Such a behaviour of $d\sigma_{\rm tot}$ is due to the 'de-screening' effect [12, 236] when the atomic electrons, considered at large photon frequencies as quasi-free ones, do not participate in screening the Coulomb field of the nucleus. As a result, the emission of photons with $\omega \gg I_{1s}$ effectively occurs in the field $-Z/r$ rather than in a neutral atomic field (see Sect. 4.6 for more details). In Fig. 4.15 the filled circles correspond to the BrS spectrum in the point Coulomb potential with charge $Z = 36$. One notices, that for $\omega > I_{1s}$ the point Coulomb results and the results of the exact calculations of the total BrS (curve 1) almost coincide.

4.6 'Stripping' Approximation

4.6.1 De-screening of an Atom in the BrS Process

Below we describe an efficient method for an approximate calculation of the cross section of BrS formed in collisions of electrons with many-electron atomic and ionic targets in the photon energy range exceeding the ionization potentials of the outer subshells. The approach is based on the so-called stripping (or, de-screening) effect which says that with increasing photon energy ω the decrease of OBrS due to screening is partly (or completely in the BA) compensated by the additional polarizational radiation.

The term 'stripping' was introduced in [12] where the BrS process was considered within the framework of the non-relativistic BA and for the dipole–photon range. The shell-by-shell 'stripping' effect was then demonstrated by results of numerical calculations of BrS of fast electrons on many-electron atoms obtained in the BA [13, 69]. It is worth noting that within these approximations stripping had previously been demonstrated for the special case of a hydrogen atom for $\omega \gg I_{1s}$ [100], where the BrS cross section reduces to the spectrum of the bare nucleus. Indeed, as it was mentioned in Chap. 2, in the indicated photon energy range one can use the asymptotic expression (2.4), so that the BrS amplitude of the fast electron on a many-electron atom reduces to that on the bare nucleus. Less formally, this effect is due to the interference between the ordinary and polarizational BrS which, being constructive in the region of high photon frequencies, leads to an increase in the cross section for total BrS.

The physical reason for the 'stripping' effect is that for $\omega \gg I_{1s}$ the electrons of all atomic subshells may be treated as free ones [100]. If the incident electron is also free (the BA) then there is no dipole radiation by a system of free electrons.

These arguments allow one to construct an approximate expression for the total BrS amplitude for photon energies above the first ionization threshold of a many-electron atomic target [12]. For doing this one divides the target electrons into two groups, the 'inner' and the 'outer' electrons, with respect to the photon energy. The inner electrons are those whose binding energies, $I_{\rm in}$, exceeds ω. Correspondingly, the outer electrons have the binding energies $I_{\rm out}$ less than ω:

$$I_{\rm out} < \omega < I_{\rm in}. \tag{4.20}$$

The electromagnetic field of a frequency ω does not distort (polarize) noticeably the cloud of inner electrons and, thus, these electrons do not contribute to the amplitude of the PBrS. On the other hand, the outer electrons behave as free under the action of the field, and their contribution to $f_{\rm pol}$ can be described by (2.4) where one substitutes the quantity $F(q)$ with the form-factor of the outer electrons only, $F_{\rm out}(q)$. As a result the total BrS amplitude acquires the form

$$f_{\rm tot} \approx \frac{4\pi(\mathbf{e}\cdot\mathbf{q})}{q^2}\left(\frac{Z - F(q)}{\omega} + \frac{F_{\rm out}(q)}{\omega}\right) = \frac{4\pi(\mathbf{e}\cdot\mathbf{q})}{q^2}\,\frac{Z - F_{\rm in}(q)}{\omega}, \tag{4.21}$$

where $F_{\rm in}(q)$ is the form-factor of the inner electrons.

The right-hand side of (4.21) demonstrates that the outer electrons do not participate in the screening of the nucleus (or, in other words, the nucleus is 'stripped' by a total number $N_{\rm out}$ of the outer electrons). The physical reason for this partial 'stripping' is as formulated above: for $\omega > I_{\rm out}$ the outer electrons can be considered as free and, therefore, there is no dipole-photon emission by the system 'projectile electron + the outer electrons' [12].

Using these considerations one draws the following qualitative picture of the behaviour of the total BrS cross section as a function of emitted photon energy. Let us label the target subshells with $j = 1, \ldots, n$ so that $j = 1$ corresponds to the outermost subshell and $j = n$ stands for the K-shell. The number of electrons in each subshell is N_j. In the region $\omega < I_1$ all electrons can be considered as the inner ones and, thus, do not contribute to the PBrS amplitude. Therefore, the total BrS cross section σ is equal to σ_0, which is the ordinary BrS cross section for the given atomic/ionic target. With ω increasing more electrons belong to the outer group leading to an effective decrease in the screening of the nuclear field. Within the range $I_j < \omega < I_{j+1}$ the dependence $\sigma(\omega)$ follows the smooth pattern of the ordinary BrS cross section in the Coulomb field of a charge $Z - \sum_{i=j+1}^{n} N_i$. At the threshold of the j-th subshell the cross section experiences a jump, the magnitude of which is proportional (approximately) to N_j^2. For $\omega > I_n \equiv I_{1s}$, where all the electrons are treated as outer, the emission occurs as in the field of the bare nucleus.

This clear physical picture was initially derived [12] and confirmed by numerical calculations [13, 69] within the framework of the non-relativistic BA. Later, the stripping approximation was extended beyond the BA and applied to the case of intermediate energy projectile electrons [236]. Using DPWA the 'stripping' effect was established for the high photon energy region $\omega \gg I_{1s}$. In this limit only the nuclear field contributes effectively to the total BrS.

In the work [251] the approach of [236] was generalized to describe the case of all possible photon energies in the non-relativistic region of the projectile energies. The results of calculations of BrS spectra of 25 keV electrons from Ar and Ne obtained in the framework of this 'stripping' approach were presented, to compare with the results of the partial wave approximation.

To simplify the complex problem of the exact evaluation of the total BrS spectra from atoms and ions at non-Born projectile energies, an alternative approach was also developed in [73], directly applying the result of the 'stripping' approximation to the region of low relativistic energies where the BA was not valid. The advantage of this approach is that it reduces direct calculations of the BrS amplitude to the analysis of the free–free dipole photon matrix elements of ordinary BrS in a point Coulomb field, and in screened Coulomb potentials. For the ordinary BrS spectra from neutral atoms and positive ions, including the point Coulomb case, a simple parametrization was developed [71–73]. Since the accuracy achieved by applying the parametrization is fairly good in a wide range of projectile energies and radiation fraction, these expressions can be successfully used for calculation of the total BrS spectra within the corresponding ranges of photon energies $I_j < \omega < I_{j+1}$, as described above. Note, however, that the parametrization was not applicable either for relativistic incident electron energies or for energies smaller than 10 keV.

In [252] the 'stripping' approaches were compared with each other. Formally, both approaches result in similar-looking formulae for the principal part of the total BrS amplitude, expressing it in terms of a single-electron transition under the action of an effective acceleration operator which depends on the energy of the emitted photon (the details are presented in the subsequent Sects. 4.6.2 and 4.6.3). The major difference between the two approximate formulae is that they utilize different sets of initial/final state wavefunctions of the projectile. In the approach due to [236] the projectile wavefunctions are obtained as solutions of the Schrödinger equation with the potential created by the target nucleus and by all of its electrons. The method of [73] implies that the potential, entering the Schrödinger equation, is dependent on the photon energy.

Both approximate formulae reduce to (4.21) in the Born limit. To analyze to what extent they coincide/deviate if one considers different targets, incident electron energies and ranges of ω the BrS spectra were computed within the frameworks of both approaches over a wide range of photon energies and for various projectile energies and atomic targets [252].

To conclude the introductory part we mention that both approaches, as well as the 'stripping' approximation itself, become inadequate for relativistic projectiles, where the effects of retardation (see [35, 57]) and the radiation of higher multipoles strongly influence the BrS spectrum. However, in this case it is possible to represent the BrS total amplitude in terms of the free–free matrix element of some effective operator [258].

4.6.2 The 'Stripping' Effect Beyond the Born Approximation

In what follows we assume the initial and final energies of the projectile electron to be high enough compared to the energies of the target electrons. In this case the exchange effects are negligent, and the motion of the projectile can be considered in the atomic (ionic) frozen-core potential

$$U_{\rm at}(\mathbf{r}) = -\frac{Z}{r} + U_0(\mathbf{r})\,, \tag{4.22}$$

where in addition to the Coulomb potential of the nucleus the potential due to the electron cloud is accounted for:

$$U_0(\mathbf{r}) = \int d\mathbf{r}_1 \dots d\mathbf{r}_N \sum_{a=1}^{N} \frac{|\Psi_0(\{\mathbf{r}_a\})|^2}{|\mathbf{r}-\mathbf{r}_a|}\,. \tag{4.23}$$

Here, the sum is carried out over atomic electrons, $\mathbf{r}_a$ stands for the position vector of the ath electron, $\Psi_0(\{\mathbf{r}_a\})$ is the ground state atomic wavefunction dependent on the positions of all electrons, $\{\mathbf{r}_a\} \equiv \mathbf{r}_1 \dots \mathbf{r}_N$.

The projectile wavefunctions of the initial, $\psi^{(+)}_{\mathbf{p}_1}(\mathbf{r})$, and the final, $\psi^{(-)}_{\mathbf{p}_2}(\mathbf{r})$, states are found from the Schrödinger equation

$$\left(\frac{\hat{p}^2}{2} + U_{\rm at}(\mathbf{r})\right) \psi^{(\pm)}_{\mathbf{p}}(\mathbf{r}) = \varepsilon\, \psi^{(\pm)}_{\mathbf{p}}(\mathbf{r})\,. \tag{4.24}$$

The superscripts correspond to the outgoing ('+') and the incoming ('$-$') boundary conditions imposed on the asymptotic behaviour of the wavefunctions.

In the lowest order of perturbation theory in the interaction between the projectile and the atomic electrons which leads to virtual atomic excitations,

$$\hat{V} = \sum_{a=1}^{N} \hat{v}^{(a)}, \qquad \hat{v}^{(a)} = \frac{1}{|\mathbf{r}-\mathbf{r}_a|}\,, \tag{4.25}$$

the OBrS and PBrS amplitudes are written as follows:

$$f_{\text{ord}} = \langle \mathbf{p}_2^{(-)} \, | \mathbf{e} \cdot \hat{\mathbf{p}} | \, \mathbf{p}_1^{(+)} \rangle \,, \tag{4.26}$$

$$f_{\text{pol}} = f_{\text{pol}}^{(\text{r})} + \text{i} \, f_{\text{pol}}^{(\text{i})} \,, \tag{4.27}$$

where

$$f_{\text{pol}}^{(\text{r})} = - \sum_{a=1}^{N} \text{v.p.} \sum_n \left[\frac{\langle 0 | \mathbf{e} \cdot \hat{\mathbf{p}}_a | n \rangle \langle n | \hat{v}_{\mathbf{p}_2 \mathbf{p}_1}^{(a)} | 0 \rangle}{\omega_{n0} - \omega} + \frac{\langle 0 | \hat{v}_{\mathbf{p}_2 \mathbf{p}_1}^{(a)} | n \rangle \langle n | \mathbf{e} \cdot \hat{\mathbf{p}}_a | 0 \rangle}{\omega_{n0} + \omega} \right], \tag{4.28}$$

$$f_{\text{pol}}^{(\text{i})} = - \pi \sum_n \langle 0 | \mathbf{e} \cdot \hat{\mathbf{p}}_a | n \rangle \langle n | \sum_{a=1}^{N} \hat{v}_{\mathbf{p}_2 \mathbf{p}_1}^{(a)} | 0 \rangle \delta \left(\omega_{n0} - \omega \right) \,. \tag{4.29}$$

Here '0' and 'n' denote the ground and excited states of the target, ω_{n0} is the transition energy, $\hat{\mathbf{p}}_a$ and $\hat{\mathbf{p}}$ are the momentum operators of the a-th atomic electron and of the projectile electron, respectively, and $\hat{v}_{\mathbf{p}_2 \mathbf{p}_1}^{(a)} = \langle \mathbf{p}_2^{(-)} \, | \hat{v}^{(a)} | \, \mathbf{p}_1^{(+)} \rangle$. The sum $\sum_n$ is carried out over all excited states including the excitations into the continuum.

The real part of the polarizational amplitude, $f_{\text{pol}}^{(\text{r})}$, accounts for the virtual excitations $0 \longleftrightarrow n$.

The imaginary part of the polarizational amplitude, $f_{\text{pol}}^{(\text{i})}$, describes a two-step process: the excitation $0 \longrightarrow n$ under the action of the Coulomb interaction $\sum_{a=1}^{N} \hat{v}^{(a)}$ and the subsequent de-excitation $n \longrightarrow n$ accompanied by the photon emission (the energy conservation, $\varepsilon_n = \omega + \varepsilon_0$, is implied). The factor $\langle 0 | \mathbf{e} \cdot \hat{\mathbf{p}}_a | n \rangle$ is proportional to the photoabsorption amplitude, while the factor $\langle n | \sum_{a=1}^{N} \hat{v}_{\mathbf{p}_2 \mathbf{p}_1}^{(a)} | 0 \rangle$ is related to the amplitude of atomic excitation (ionization) via electron impact.

To construct f_{pol} within the 'stripping' approximation one considers the following transformations of $f_{\text{pol}}^{(\text{r})}$. First, for given ω, in accordance with (4.20), the atomic electrons are split into two groups, of inner and outer electrons. Assuming that the strong inequality $\omega \ll I_{\text{in}}$ is fulfilled, one neglects the contribution of the virtual excitations of the inner electrons to the sum in (4.28). Then, the amplitude $f_{\text{pol}}^{(\text{r})}$ is defined by the contribution of the virtual excitations of the outer electrons only:

$$f_{\text{pol}}^{(\text{r})} \approx - \sum_{a_{\text{out}}} \text{v.p.} \sum_n \left[\frac{\langle 0 | \mathbf{e} \cdot \hat{\mathbf{p}}_{a_{\text{out}}} | n \rangle \langle n | \hat{v}_{\mathbf{p}_2 \mathbf{p}_1}^{(a_{\text{out}})} | 0 \rangle}{\omega_{n0} - \omega} + \frac{\langle 0 | \hat{v}_{\mathbf{p}_2 \mathbf{p}_1}^{(a_{\text{out}})} | n \rangle \langle n | \mathbf{e} \cdot \hat{\mathbf{p}}_{a_{\text{out}}} | 0 \rangle}{\omega_{n0} + \omega} \right]. \tag{4.30}$$

Using the inequality $\omega > I_{\text{out}}$ one expands the denominators in powers of the small parameter ω_{n0}/ω: $(\omega_{n0} \pm \omega)^{-1} = \pm \omega^{-1} - \omega_{n0} \, \omega^{-2} + O(I_{\text{out}} \omega^{-3})$. It can be demonstrated that the terms containing $\omega_{n0} \, \omega^{-2}$ do not contribute to the sum on the right-hand side of (4.30). The leading term, proportional to ω^{-1}, is evaluated with the help of the completeness relation $\sum_n \, |n\rangle \, \langle n| = 1$. The result is

$$f_{\mathrm{pol}}^{(\mathrm{r})} = \frac{\mathbf{ie}}{\omega} \langle \mathbf{p}_2^{(-)} \left| \nabla_{\mathbf{r}} U_0^{(\mathrm{out})}(\mathbf{r}) \right| \mathbf{p}_1^{(+)} \rangle + O\left(\frac{I_{\mathrm{out}}^2}{\omega^3} \right), \tag{4.31}$$

where

$$U_0^{(\mathrm{out})}(\mathbf{r}) = \left\langle 0 \left| \sum_{a_{\mathrm{out}}} \frac{1}{\left| \mathbf{r} - \mathbf{r}_{a_{\mathrm{out}}} \right|} \right| 0 \right\rangle \tag{4.32}$$

is the part of the potential (4.23) due to the outer electrons.

Therefore, the real part of the total BrS amplitude can be represented as follows:

$$\mathrm{Re}\, f_{\mathrm{tot}} \approx \langle \mathbf{p}_2^{(-)} \left| \mathbf{e} \cdot \hat{\mathbf{p}} \right| \mathbf{p}_1^{(+)} \rangle + \frac{\mathbf{ie}}{\omega} \langle \mathbf{p}_2^{(-)} \left| \nabla_{\mathbf{r}} U_0^{(\mathrm{out})}(\mathbf{r}) \right| \mathbf{p}_1^{(+)} \rangle . \tag{4.33}$$

To derive the final expression for f_{tot} let us introduce the operator of the total acceleration $\hat{\mathbf{a}}_{\mathrm{at}}$ of the projectile in the field $U_{\mathrm{at}}(\mathbf{r})$ (see (4.22)):

$$\hat{\mathbf{a}}_{\mathrm{at}} = -\nabla_{\mathbf{r}} U_{\mathrm{at}}(\mathbf{r}) = -Z \frac{\mathbf{r}}{r^3} + \hat{\mathbf{a}}^{(\mathrm{in})} + \hat{\mathbf{a}}^{(\mathrm{out})}, \tag{4.34}$$

where $\hat{\mathbf{a}}^{(\mathrm{out})}$ and $\hat{\mathbf{a}}^{(\mathrm{in})}$ are the operators of the accelerations due to the potentials of the outer and the inner electrons.

Expressing the OBrS amplitude (the first term in (4.33)) in terms of the matrix element of $\hat{\mathbf{a}}_{\mathrm{at}}$ one obtains the following expression for the total BrS amplitude within the 'stripping' approximation [252]:

$$f_{\mathrm{tot}} \approx \frac{\mathrm{i}}{\omega} \langle \mathbf{p}_2^{(-)} \left| \mathbf{e} \cdot \hat{\mathbf{A}} \right| \mathbf{p}_1^{(+)} \rangle - \mathrm{i}\pi \sum_{a=1}^{N_{\mathrm{out}}} \sum_{n} \langle 0 \left| \mathbf{e} \cdot \hat{\mathbf{p}}_a \right| n \rangle \langle n \left| \hat{v}_{\mathbf{p}_2\mathbf{p}_1}^{(a)} \right| 0 \rangle \, \delta\left(\omega_{n0} - \omega \right) \tag{4.35}$$

where the sum is carried out over the outer electrons, and the operator $\hat{\mathbf{A}}$ is given by

$$\hat{\mathbf{A}} = -Z \frac{\mathbf{r}}{r^3} + \hat{\mathbf{a}}^{(\mathrm{in})} . \tag{4.36}$$

Equation (4.35) generalizes the result obtained in [236], where the focus was on the high-frequency limit of the BrS spectrum, $\omega \gg I_{1s}$, so that all target electrons can be considered as 'inner' ones. The scheme outlined above allows one to introduce the stripping effect for each atomic subshell by comparing its ionization potential with the photon energy.

Note that expression (4.35) is exact in the sense that in its derivation no approximate schemes have been used to describe the many-electron target. The result is applicable for any type of target (atomic, ionic, molecular).

Two assumptions have been made to evaluate (4.35):

(1) the dynamics of the projectile is subject to the action of the frozen-core potential (see (4.22)) created by the target nucleus (or nuclei, if a many-atom molecule is considered) and electrons;
(2) the process of photon emission can be described within the frame of the dipole approximation.

The formula (4.35) indicates that for photon energies ω within the range defined by (4.20) the total BrS of an electron effectively consists of two terms.

The first term is a free-free transition matrix element of the operator $\hat{\mathbf{A}}$ which is the acceleration of the projectile due to the action of the Coulomb field of the nucleus partly screened by the inner-shell electrons. This matrix element has to be evaluated between the exact initial, $\psi^{(+)}_{\mathbf{p}_1}(\mathbf{r})$, and the final, $\psi^{(-)}_{\mathbf{p}_2}(\mathbf{r})$, projectile wavefunctions satisfying the Schrödinger equation (4.24) with the frozen-core potential (4.22). This term involves a joint action of the two mechanisms of BrS formation: the ordinary BrS and the polarizational BrS, where only virtual atomic excitations are taken into account. The sum of these terms results, effectively, in the renormalization of the vertex describing the electron–photon interaction, but it leaves unchanged the projectile's wavefunctions.

The second term, $f^{(i)}_{\rm pol}$, is due to the real excitations of the atom during the collision. This describes a two-step process which corresponds to the excitation of an outer-shell atomic electron under the action of the Coulomb interaction with the projectile and the subsequent radiative de-excitation. This term is of most importance for photon energies in vicinities of the outer-shell ionization thresholds.

To conclude this section let us mention that the Born limit of (4.35) is rather straightforward. Indeed, substituting in the first term for the wavefunctions $\psi^{(+)}_{\mathbf{p}_1}(\mathbf{r})$ and $\psi^{(-)}_{\mathbf{p}_2}(\mathbf{r})$ the corresponding plane waves, one obtains the formula (4.21). The contribution of the real excitations to the total BrS amplitude (the second term in (4.24)) was not accounted for when evaluating (4.21).

4.6.3 Alternative Approach to the 'Stripping' Effect

Different physical arguments were used in [73] to formulate the 'stripping' approximation beyond the framework of the BA for a projectile electron. This alternative approach can be summarized as follows.

Suppose that the photon energy lies between the ionization potential of two neighbouring atomic subshells,

$$I_{j-1} < \omega < I_j \,, \tag{4.37}$$

where $j = 1, \dots, n$ and n stands for the $1s$ subshell. In this case the Born formula (4.21) suggests that the total BrS process can be interpreted in terms of the radiative transition of an electron scattered by the field of a partly screened nucleus, $U_{\rm ion}(r) = -Z/r + U_{\rm in}(r)$, with $U_{\rm in}(r)$ being the potential due to the inner electrons only.

As $r \longrightarrow \infty$ the potential $U_{\rm ion}(r)$ has the form of an ionic field with net charge $Z_{\rm ion} = Z - N_{\rm out}$, where $N_{\rm out} = \sum_{i=j+1}^{n} N_i$.

Let this clear physical picture (which is justified within the BA) be extrapolated to the range of projectile energies where the BA is not applicable to describe the scattering process. Thus, one assumes that within the 'stripping' approximation the real part of the total BrS amplitude can be expressed in terms of the matrix element

$$\mathrm{Re}\, \tilde{f}_{\rm tot} = \left\langle \tilde{\mathbf{p}}_2^{(-)} \,|\, \mathbf{e}\cdot\hat{\mathbf{p}} \,|\, \tilde{\mathbf{p}}_1^{(+)} \right\rangle , \tag{4.38}$$

where the tilde signs are written to distinguish between the quantities used here and the quantities introduce in the preceding section.

The wavefunctions $\tilde{\psi}_{\mathbf{p}}^{(\pm)}(\mathbf{r})$, which correspond to the scattering states $|\tilde{\mathbf{p}}^{(\pm)}\rangle$, satisfy the Schrödinger equation in the ionic field $U_{\rm ion}(r)$:

$$\left(\frac{\hat{p}^2}{2} + U_{\rm ion}(\mathbf{r})\right) \tilde{\psi}_{\mathbf{p}}^{(\pm)}(\mathbf{r}) = \varepsilon\, \tilde{\psi}_{\mathbf{p}}^{(\pm)}(\mathbf{r}) \,. \tag{4.39}$$

Thus, the calculation of the total BrS spectrum requires calculation of the spectrum of the neutral atom (for $\omega < I_1$) and for a number of ions (of the ionicities $Z - \sum_{i=j}^{n} N_i$, $j = 1, \dots, n$), as well as for the pure Coulomb spectrum (when $\omega > I_n \equiv I_{1s}$). To obtain accurate data on BrS from neutral atoms for low relativistic energies of the projectile $1\,\mathrm{keV} < \omega < 2000\,\mathrm{keV}$ one can use tabulated results obtained in the numerical relativistic DPWA [352]. Results for a few atomic ions, obtained in a similar way for a number of projectile energies and fraction of energy radiated are also available [282]. Simple analytic expressions for BrS spectra of atoms and their ions for the energy range $10\,\mathrm{keV} < \omega < 2\,\mathrm{MeV}$ have been developed in [71–73]. The accuracy of these expressions is usually better than 5 %, becoming worse with decreasing electron energy.

Despite the relative simplicity of its practical implementation, this approach is less justified from the formal viewpoint than the formulation of the 'stripping' approximation outlined in Sect. 4.6.2. To indicate explicitly the difference between the matrix element (4.38) and its counterpart (4.35) let us reduce the matrix element of the momentum operator on the right-hand side of (4.38) to that of the acceleration operator. Then, taking into account equation (4.39) and recalling the definition (4.36), one notices that the operator $\hat{\mathbf{A}}$ is the total acceleration of the projectile in the field $U_{\rm ion}(\mathbf{r})$. Therefore

$$\mathrm{Re}\, \tilde{f}_{\rm tot} = \frac{\mathrm{i}}{\omega} \langle \tilde{\mathbf{p}}_2^{(-)} \left| \mathbf{e}\cdot\hat{\mathbf{A}} \right| \tilde{\mathbf{p}}_1^{(+)} \rangle \,. \tag{4.40}$$

Comparison of this result and the first term on the right-hand side of (4.35) illuminates the difference between the two approaches: both amplitudes are the matrix elements of the same operator $\hat{\mathbf{A}}$ but calculated using different initial/final state wavefunctions. In (4.35) the wavefunctions $\psi_{\mathbf{p}}^{(\pm)}(\mathbf{r})$ are the solutions of the Schrödinger equation

(4.24) with the static potential (4.22) due to the target's nucleus and all electrons, whereas the wavefunctions $\tilde{\psi}_{\mathbf{p}}^{(\pm)}(\mathbf{r})$ are obtained from (4.39), where the potential $U_{\text{ion}}(\mathbf{r})$ depends (implicitly) on the photon energy. Indeed, for $\omega > I_n = I_{1s}$ this potential reduces to the point Coulomb field $-Z/r$, for $\omega < I_1$ (no outer electrons) $U_{\text{ion}}(\mathbf{r})$ coincides with (4.22), and for $I_1 < \omega < I_n$ the potential is given by $U_{\text{ion}}(r) = -Z/r + U_{\text{in}}(r) = U_{\text{at}}(\mathbf{r}) - U_{\text{out}}(r)$.

Let us point out the cases in which both approaches to the 'stripping' approximation produce identical results. First, it is the region of photon energies below the ionization threshold of the target. In this case (4.40) and the first term on the right-hand side of (4.35) coincide, since $U_{\text{ion}}(\mathbf{r}) = U_{\text{at}}(\mathbf{r})$. The second case refers to the region of high photon energies, $\omega \gg I_{1s}$, where the amplitude f_{tot} reduces to the BrS amplitude of the fast electron on a bare nucleus. Finally, both approaches give identical results in the Born limit. Both formulae reduce to (4.21) if the scattering states are described by plane waves.

4.6.4 Numerical Results

Aiming to carry out a quantitative comparison of the two approaches the 'stripping' approximation, in this section we present the results of numerical calculations of the BrS spectra formed in the collision of an electron with $\varepsilon = 5$–50 keV electron with neutral Ar, Kr, Xe, Au atoms ($Z = 18, 36, 54, 79$, respectively) [252]. The data shown in Figs. 4.17 and 4.18 refer to the differential cross section $\omega\, \mathrm{d}\sigma/\mathrm{d}\omega$ integrated over the angles of the emitted photon and the scattered electron:

$$\mathrm{d}\sigma \equiv \omega \frac{\mathrm{d}\sigma}{\mathrm{d}\omega} = \frac{\omega^2}{c^3}\frac{p_2}{p_1} \int\limits_{4\pi}\int\limits_{4\pi} \frac{\mathrm{d}\Omega_{\mathbf{p}_2}\,\mathrm{d}\Omega_{\mathbf{k}}}{(2\pi)^4} \sum_{\lambda} |f|^2 \,, \tag{4.41}$$

where the sum is carried out over the photon polarizations.

To calculate the radiative amplitudes the non-relativistic DPWA was used for the wavefunctions of a projectile electron. The radial parts of the wavefunctions were obtained by solving the Schrödinger equation with a frozen-core potential. The atomic wavefunctions were considered within the HF approximation. More details on the DPWA formalism and the corresponding algorithms can be found in Sect. 2.2 and in [29, 244, 249].

Three panels in Fig. 4.17 present the BrS data calculated for 5 and 25 keV electrons in collisions with Ar, Kr and Xe atoms as indicated. The differential cross sections of the following four types are compared:

(1) The OBrS cross section $\mathrm{d}\sigma_{\text{ord}}$ (dashed blue curves), which was calculated using the OBrS amplitude (4.26) in (4.41).

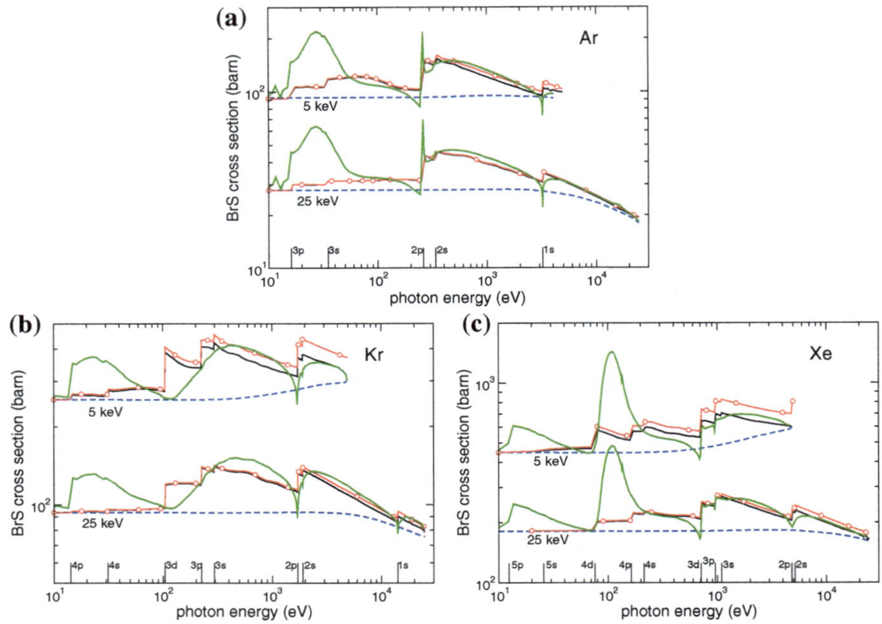

Fig. 4.17 BrS spectra, $\omega d\sigma/d\omega$, formed in collision of $\varepsilon = 5$ and 25 keV electrons with Ar, Kr and Xe atoms [252]. The *dashed curves* describe the OBrS, the (*black*) *thin solid lines* correspond to the 'stripping' approximation as outlined in Sect. 4.6.2. The (*red*) *thin solid curves* with *open circles* represent the cross sections obtained within the scheme described in Sect. 4.6.3. The (*green*) *thick solid curves* represent the total BrS spectra calculated within the distorted partial waves approximation without the use of the 'stripping' approximation. *Vertical lines* mark the HF ionization potentials of the atomic subshells. See also explanations in the text

(2) The total BrS cross section, $d\sigma_{\rm tot}$, (thick solid green curves) was computed using the exact expressions (4.27)–(4.29) for the amplitude of PBrS. The interaction of atomic electrons with the dipole photon was treated within the frame of RPAE.
(3) The total BrS cross section in the 'stripping' approximation, $d\sigma_{\rm s}$, (thin solid black curves) calculated using the first term on the right-hand side of (4.35) to construct the amplitude.
(4) The total BrS cross section $d\tilde{\sigma}_{\rm s}$ (thin solid red curves with open circles) calculated within the framework of the alternative 'stripping' approximation (4.40).

In Fig. 4.18 the cross sections $d\,\sigma_{\rm ord}$, $d\,\sigma_{\rm s}$ and $d\tilde{\sigma}_{\rm s}$ are presented for an 5–50 keV electron–Au collision.

Figure 4.17 demonstrates that, apart from the region of comparatively low ω (which are $\omega < I_{2s} \approx 335$ eV for Ar, $\omega < I_{3s} \approx 295$ eV for Kr and $\omega < I_{4s} \approx 214$ eV for Xe), the exact cross sections are reproduced quite well if one accounts only for the real part of the total BrS amplitude within the 'stripping' approximations. The noticeable deviations for low photon energies are not surprising. Indeed, the first maxima in the total BrS spectrum for Ar and Kr, and the first two maxima for Xe are

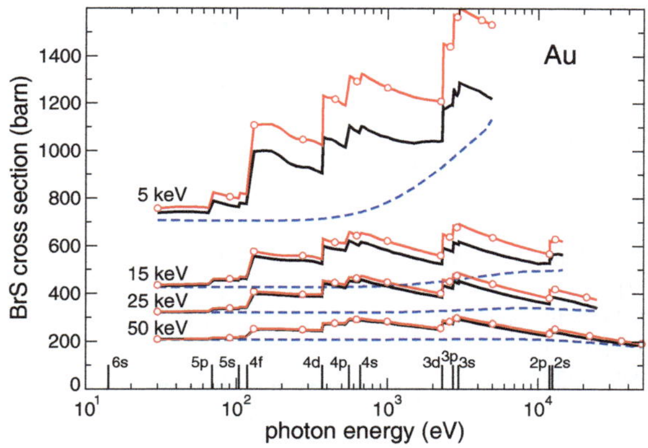

Fig. 4.18 BrS spectra, $\omega d\sigma/d\omega$, formed in collision of electrons of various initial energies (as indicated) with an Au atom [252]. The *dashed curves* describe the OBrS, the *solid lines* correspond to the 'stripping' approximation as outlined in Sect. 4.6.2. The *curves with open circles* represent the cross sections obtained within the scheme described in Sect. 4.6.4. *Vertical lines* mark the HF ionization potentials of the atomic subshells

due to the contribution of the real excitations of atomic electrons (see (4.29) and the second term in (4.35)). The use of the amplitude $f^{(i)}_{pol}$ in (4.41) results in an additive term in the BrS cross section proportional to the photoabsorption cross section [12, 249]. In Fig. 4.17 the quantity $f^{(i)}_{pol}$ was not taken into account when calculating the BrS spectra in the 'stripping' approximation(s).

Comparing the data obtained by means of the two different approaches to the 'stripping' approximation as discussed in Sects. 4.6.2 and 4.6.3 one concludes that both methods agree quite well with each other. In the worst case, for the hard photon emission by low energy electron scattered from Au, the difference between data obtained in the two approximations does not exceed 13 %. In other cases the difference is less than 10 %.

The agreement between the approaches is very good if the parameter $\nu = Z/p_1$ is less than one. It becomes worse for $\nu \gg 1$. For a fixed target and for a given photon energy the ratio of the two cross sections, $\delta(Z, \varepsilon, \omega) = d\tilde{\sigma}_s/d\sigma_s$, increases with decrease of the incident energy ε. For a fixed incident energy $\delta(Z, \varepsilon, \omega)$ is an increasing function of Z.

As a function of ω the ratio goes to one in the low-frequency part of the spectrum. Thus, in the limit of small ω both approaches to the 'stripping' approximation produce the same result. This is clear since both matrix elements characterizing the 'stripping' amplitudes, the first term in (4.35) and the one given by (4.40), have the same limit as $\omega \longrightarrow 0$. The agreement becomes poorer as the fraction of radiated energy ω/ε increases.

We would like to stress that the 'stripping' approximation represents by itself a very efficient method for calculating the characteristics of the BrS process (the spectral and spectral-angular distributions) in the whole range of photon energies.

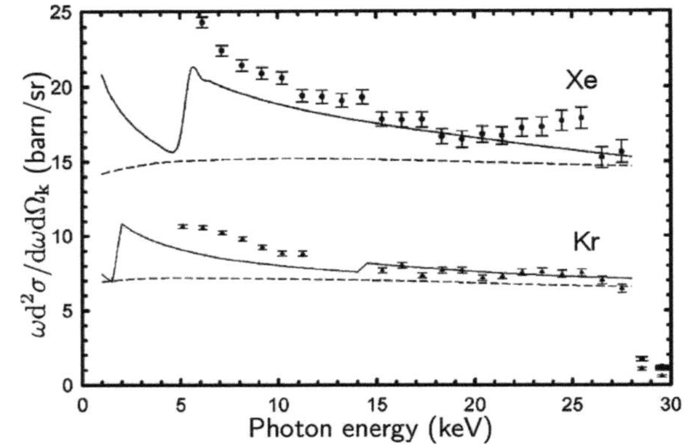

Fig. 4.19 Double differential BrS cross section, $\omega d^2\sigma/d\omega d\Omega_{\mathbf{k}}$, at $\theta_{\mathbf{k}} = 90°$ formed in collision of $\varepsilon_1 = 28\,\text{keV}$ electron with Xe and Kr atoms. *Filled circles*—experimentally measured data, *broken curve*—OBrS, *full curves*—estimation based on the "stripping" approximation. See explanations in the text. The graph represents Fig. 1 from [346] with minor editorial changes

Indeed, this approach allows one to treat the total BrS amplitude as a single-electron free–free transition under the action of a local operator of the effective acceleration. This results in a considerable reduction of the computational efforts in comparison with those needed for the exact calculations see Sects. 4.1–4.3.

4.6.5 Experimental Results

Absolute doubly differential BrS cross sections $d^2\sigma/d\omega d\Omega_{\mathbf{k}}$ for $\varepsilon_1 = 28$ and $50\,\text{keV}$ electrons from Xe, Kr, Ar and Ne atoms were measured recently in a broad range of the emitted photon energies by Portillo and Quarles [345, 346] (see also [356, 357]). The emitted photon was detected at $\theta_{\mathbf{k}} = 90°$ with respect to the incident electron beam.

In the cited papers the experimental data were compared to the prediction of the OBrS theory as well as to the estimates based on the model account for the PBrS channel. The results presented in Figs. 1–3 from [346] are reproduced in Figs. 4.19 and 4.20 with minor editorial changes. The filled circles with error bars are the experimental data. The broken curves stand for the OBrS cross sections and the full curves represent the total BrS cross sections calculated in the cited paper as described below.

In [230] the tabulation of the shape function

$$S_{\text{ord}}(\theta_{\mathbf{k}}) = (d^2\sigma_{\text{ord}}/d\omega d\Omega_{\mathbf{k}}) \Big/ (d\sigma_{\text{ord}}/d\omega) \tag{4.42}$$

and of the spectrum $d\sigma_{\text{ord}}/d\omega$ is presented for neutral atomic targets with the charge number $1 \le Z \le 92$ and for the incident electron energies $1 \le \varepsilon_1 \le 500\,\text{keV}$. The calculations were performed within the relativistic DPWA formalism of OBrS and with account for the multipole expansion of the emitted radiation [281, 409, 410].

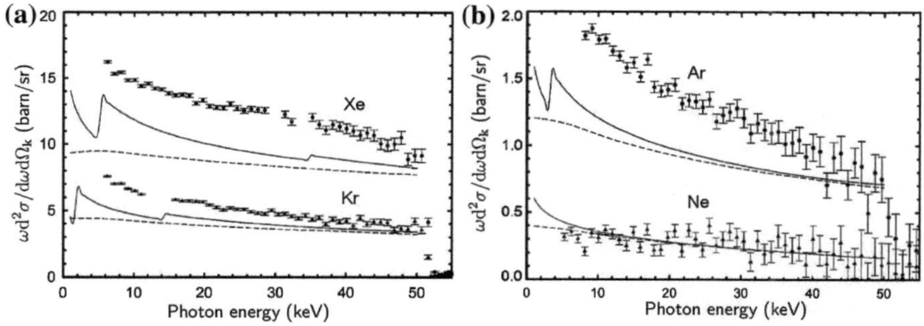

Fig. 4.20 Same as in Fig. 4.19 but for $\varepsilon_1 = 50$ keV electron incident on Xe and Kr atoms (*left graph*) and Ar and Ne atoms (*right graph*). The graphs represent Figs. 2 and 3 from [346] with minor editorial changes

It is clearly seen that apart from the case of Ne atom and for the photon energies well below the kinematic end point $\omega = \varepsilon_1$ the experimental data noticeably exceed the OBrS background $d^2\sigma_{ord}/d\omega d\Omega_{\mathbf{k}}$. In [346], this excess was attributed to the the PBrS mechanism and it was stated, that their measurements are "...the first direct evidence of the contribution of PBrS in electron bremsstrahlung is important over the whole range of radiated photon energy".[3]

To estimate the spectral dependence of the double differential cross section of the total BrS, $d^2\sigma_{tot}/d\omega d\Omega_{\mathbf{k}} = S_{tot}(\theta_{\mathbf{k}})d\sigma_{tot}/d\omega$ with the PBrS channel accounted for, the authors of [346] calculated the total cross section $d\sigma_{tot}/d\omega$ as the sum of two terms, $d\sigma_{ord}/d\omega + d\sigma_{pol}/d\omega$, neglecting thus the interference part of the cross section. The PBrS single differential cross section was estimated within the framework of the "stripping" approximation due to the scheme proposed in [73] (see also Sect. 4.6.3). The shape function was taken from fully relativistic OBrS calculations [230]. Thus, it was assumed that $S_{pol}(\theta_{\mathbf{k}})$ equals to $S_{ord}(\theta_{\mathbf{k}})$ from (4.42). The result of these calculations are presented in Figs. 4.19 and 4.20 with full curves.

In the subsequent publications of Quarles's group [190, 356, 357] the quantitative analysis was carried out of the contributions of various background processes which might have resulted in the excess of the measured photon yield over the OBrS spectrum. In particular, one if the processes which was estimated is due to the Compton scattering from the target gas of the photons produced by electron beam interaction with beam-line collimators. Another processes is also related to the photons produced by the beam which then interact with the gas in the target cell via the photoeffect. The resulting secondary electrons can scatter into the detector window or target cell walls and produce the additional photon spectrum. The estimates carried out in the cited papers indicated that the overall contribution of the additional photon background is small enough to be responsible for the observed discrepancies between experimental data and results of the OBrS theory.

[3] As have been already mentioned, the experimental evidence of the PBrS contribution in the photon energy range of a giant photoabsorption resonance was provided earlier for La and rare earth atoms [451, 453] and Xe atom [160–162, 425–428].

It was noted in [346] that while "stripping" approximation agrees well with the trend of the experimental data at both incident energies, the absolute discrepancies are noticeably higher for $\varepsilon_1 = 50\,\text{keV}$. Later, the validity of several assumptions made in [346] was examined in [335]. It was noted that although the stripping approximation produces reasonable predictions for the total spectrum in the non-relativistic dipole–photon domain there is, however, no argument for its validity for the angular distributions, in which higher multipoles are important. Thus, the assumption $S_{\text{pol}}(\theta_{\mathbf{k}}) = S_{\text{ord}}(\theta_{\mathbf{k}})$ for relativistic projectiles was questioned. In [335] a more reasonable estimate of the PBrS term $\mathrm{d}\sigma_{\text{pol}}/\mathrm{d}\omega$ was made based on the used of the non-relativistic dipole-photon shape function along with the stripping approximation for 50 keV electrons on Ar. The resulting values of $\mathrm{d}\sigma_{\text{pol}}/\mathrm{d}\omega$ turned out to be somewhat higher than obtained in [346] although not large enough to resolve the problem. The contribution of the inelastic BrS channels, estimated in [335] within the framework of the logarithmic approximation [37], turned out to be small enough to disregard them completely in the given experimental circumstances.

In all the cited papers it was stated that a more accurate calculations are needed to produce theoretical data to be compared with the measured spectra.

4.7 Bethe Ridge in the PBrS Process

In this Section we discuss a peculiar feature in the PBrS emitted in the collision of a fast heavy charged particle with a many-electron atom. This feature originates from the kinematics of the atomic electrons in the process and manifests itself as an additional maximum in the velocity dependence of the PBrS cross section.

A similar peculiarity, known as the Bethe ridge, appears in the differential cross section of the impact ionization of an atom by a charged particle [279], where there is a ridge in the dependence of the cross section on the transferred momentum q and the momentum of the outgoing electron p_e at $q \sim p_e$. This ridge in the differential cross section is a result of the momentum transfer to one of the target electrons in the collision process.

The peculiarity in the polarizational BrS process arises from the similar dynamics of atomic electrons [241]. However, in this process after virtual excitation the atomic electron returns to its initial state radiating a photon. The Bethe-type virtual excitations of electrons in the PBrS process give rise to the additional maximum in the velocity dependence of the cross section.

4.7.1 Contributions of Various Radiative Processes to the Total Spectrum

As demonstrated below, the Bethe peculiarity arises in the velocity dependence of PBrS cross section in collisions of fast heavy charged projectiles with many-electron atoms at the photon energies close to the ionization potentials of inner atomic shells.

In this ω-range PBrS dominates over other radiative mechanisms such as OBrS, secondary electron BrS (SEB) [386], Radiative Ionization (RI) or 'inelastic' BrS [36, 37, 339] (see also Sects. 5.3 and 6.4) and molecular orbital X-rays [173].

For a heavy projectile, $m \gg 1$, the photon yield via the OBrS mechanism is suppressed by the factor m^2 as compared with the intensity of PBrS. Hence, the OBrS channel can be disregarded in the wide range of the photon energies except for very small ones and very large ones. Indeed, matching the polarizational and ordinary BrS contributions to the total amplitude (2.1) for a heavy projectile, one derives that for $\omega \lesssim (Z/m\alpha(0))^{1/2}$ (where Z is the nucleus charge and $\alpha(0)$ is the static polarizability of the atom) the OBrS amplitude, being divergent (the infrared divergence), dominates. As ω becomes much larger than the K-shell ionization threshold so that the minimum momentum transfer $q_{\min} \approx \omega/v_1$ exceeds the momenta $p_{\rm at}$ of all atomic electrons, then $\alpha(\omega, q) \approx -F(q)/\omega^2 \to 0$, leading again to the dominance of OBrS.

SEB is produced by secondary electrons in the target media. The intensity of SEB is proportional to the density of atoms in the target. For gaseous targets SEB is usually negligible due to their low density.

One can also neglect the contribution of the molecular orbital radiation to the total photon emission spectrum when considering collisions of fast particles (see [388] and Sect. 3.3).

In contrast to the SEB and molecular orbital radiation, the process of RI ('inelastic' BrS) is of importance in collisions of charged particles with atoms. As is it demonstrated in Sects. 5.3 and 6.4, the coherence effect of radiation in 'elastic' PBrS (i.e., the process in which the colliding particles remain in their initial internal states after the collision) leads to the dominance over the RI channels in the wide range of photon energies: $I < \omega < v_1 R_{\rm at}^{-1}$, which which we call the region of coherence. Here I and $R_{\rm at}$ are the mean ionization potential and atomic radius, v_1 is the collision velocity, which is assumed to be large, $v \gg 1$. Beyond the region of coherence RI dominates over PBrS one in the total photon emission spectrum. An exception to this rule occurs in collisions of fast heavy charged particles with atoms or ions or in atomatom (ionatom or ionion) collisions in the region of high photon energies. In this region the inelastic process has a threshold, which is equal to $v_1^2/2$. However, PBrS takes place at higher frequencies, up to $\omega \approx 2v_1^2$, dominating in this region in the total photon emission spectrum. Therefore, the photon frequency range $v_1^2/2 \le \omega \lesssim 2v_1^2$ is convenient for the observation of PBrS and thus for studying the Bethe peculiarity.

Fast collisions involving a heavy projectile imply the applicability of the BA. In this approximation the PBrS amplitude follows from (2.1) where one omits the term on the right-hand side inversely proportional to m. The PBrS cross section reads:

$$\omega \frac{\mathrm{d}\sigma_{\rm pol}}{\mathrm{d}\omega} = \frac{16 Z_0^2 \omega^4}{3c^3 v_1^2} \int\limits_{q_{\min}}^{q_{\max}} \frac{\mathrm{d}q}{q} \left|\alpha(\omega, q)\right|^2 . \tag{4.43}$$

The transferred momentum q varies from $q_{\min} = p_1 - p_2$ to $q_{\max} = p_1 + p_2$ which can be expressed via v_1, ω and the reduced mass μ of the projectile and the target atom:

$$q_{\max\,/\,\max} = \mu v_1 \left(1 \pm \sqrt{1 - \frac{2\omega}{\mu v_1^2}}\right). \tag{4.44}$$

Here, the upper/lower sign corresponds to $q_{\max}/\, q_{\min}$.

In the region of photon energies comparable and higher than the ionization potentials of inner atomic shell, which is of interest in this section, the main contribution to the generalized polarizability $\alpha(\omega, q)$, and, consequently, to the cross section (4.43), comes from the inner-shell electrons.

In the same approximation the RI cross section reads as [37]

$$\omega \frac{\mathrm{d}\sigma_{\mathrm{RI}}}{\mathrm{d}\omega} = \frac{16 Z_0^2}{3 v_1^2 c^3} \sum_{\alpha} \int_{q_{\min}^{\alpha}}^{q_{\max}^{\alpha}} \frac{\mathrm{d}q}{q} \left(N_\alpha - \sum_{i}^{N_\alpha} \sum_{j}^{N} \left| W_{ij}(\mathbf{q}) \right|^2 \right), \tag{4.45}$$

where

$$W_{ij}(\mathbf{q}) = \left\langle i \left| \mathrm{e}^{\mathrm{i}\mathbf{q}\cdot\mathbf{r}} \right| j \right\rangle. \tag{4.46}$$

Indices i and j denote the occupied one-electron states, and index α marks the quantities associated with the atomic subshell α; N is the total number of atomic electrons and N_α is the number of electrons in the subshell α. The summation in is performed over all atomic subshells α.

The minimum and the maximum transferred momenta in the RI process are equal to

$$q_{\min}^{\alpha} = v_1 \left(1 - \sqrt{1 - \frac{2(\omega + I_\alpha)}{v_1^2}}\right), \quad q_{\max}^{\alpha} = v_1 \left(1 + \sqrt{1 - \frac{2(\omega + I_\alpha)}{v_1^2}}\right), \tag{4.47}$$

where I_α is the ionization potential of the the atomic subshell α from which the outgoing electron is ejected.

In Fig. 4.18 we compare contributions to the total radiation spectrum originating from different radiative mechanisms for the collision of a 1 MeV proton with the Al atom [241]. The figure demonstrates that PBrS dominates in the total radiation in a wide frequency range. The RI process becomes visible in the total photon emission spectrum only at low frequencies. This occurs since the RI cross section contains the infrared divergence, $\mathrm{d}\sigma_{\mathrm{RI}}/\mathrm{d}\omega \propto \omega^{-1}$ as $\omega \to 0$, while the PBrS one tends to zero. The OBrS channel defines the spectrum at $\omega > 10$ keV, where the PBrS contribution practically vanishes (Fig. 4.21).

The PBrS cross section $\mathrm{d}\sigma_{\mathrm{pol}}/\mathrm{d}\omega$ was calculated according to (4.43) and (4.44). The generalized polarizabilities of atomic subshells were calculated using the HF wave functions. Figure 4.22 illustrates good agreement of the calculated double

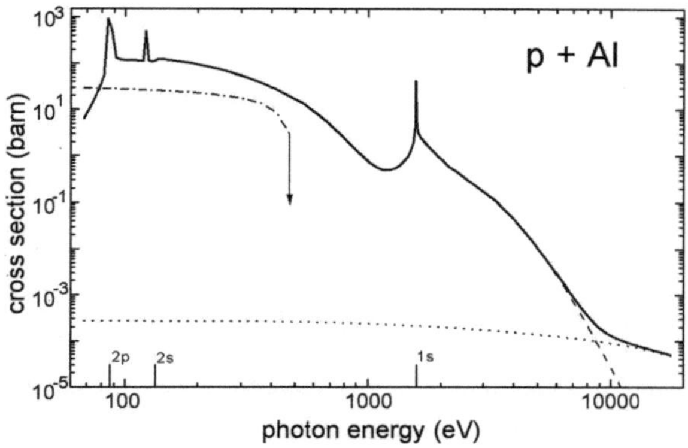

Fig. 4.21 Contributions of various radiative mechanisms to the total cross section (the *solid curve*) in the collision of a 1 MeV proton (the incident velocity $v_1 = 6.3$ a.u.) with the Al atom. The *dashed curve* stands for PBrS (clearly visible only for large ω), the *dotted one*—for OBrS. RI is represented by the *dashed-dotted curve*. The *vertical dashes* mark the ionization potentials of atomic subshells. The RI threshold, $\omega = v_1^2/2$, is marked by the *vertical arrow*. The PBrS and RI cross sections have been calculated within the HF approximation

Fig. 4.22 The PBrS cross section versus photon energy for the collision of a 1 MeV proton with the Al atom. The photon emission angle is fixed at $\theta_{\mathbf{k}} = 90°$. The *solid curve* stands for the cross section calculated within the HF approximation [241]. *Filled circles* represent the experimental data [204, 208]

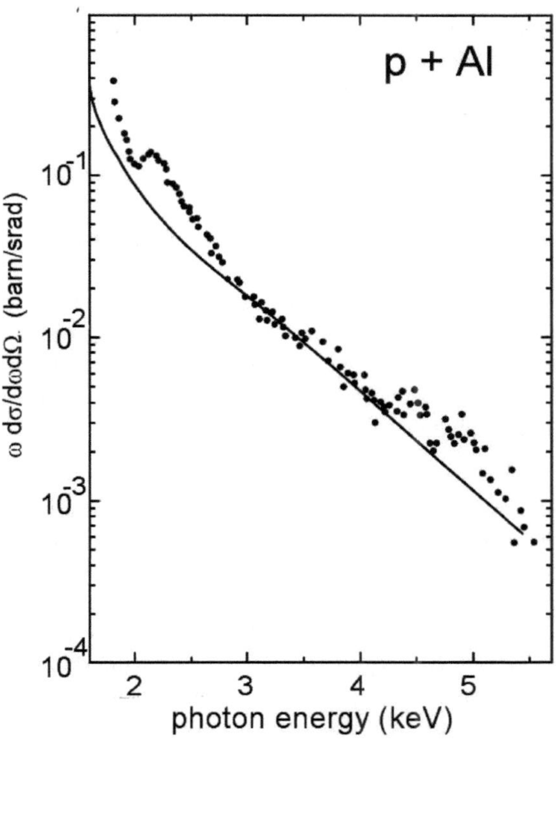

differential PBrS cross section, $\omega d^2\sigma/d\omega d\Omega_{\mathbf{k}}$ with the experimental results from [204, 208].

It is worth noting, that for the photon energies comparable or higher than the K- and L-shell ionization potentials, the cross section obtained within the HF framework is very close to that calculated using the hydrogen-like approximation (see Sects. 3.1 and 5.4).

4.7.2 Analysis of the Bethe Peculiarity in PBrS

Let us now discuss the peculiarity in the velocity dependence of the PBrS cross section, which arises due to the specific behaviour of the generalized polarizability as a function of transferred momentum q. The matrix element of the type $\langle 0|\exp(i\mathbf{q}\cdot\mathbf{r})|n\rangle$ in (2.14) as a function of the energy ε_n of the intermediate state $|n\rangle$ is maximal at $\varepsilon_n = q^2/2$, when the oscillatory character of the wave function $\psi_n(\mathbf{r})$ is partly compensated by the exponential factor. This means that the momentum is mainly transferred to the electron in the intermediate state. Similar behaviour of matrix elements is known from electron atom impact ionization, where it is called 'the Bethe ridge'. The differential inelastic scattering cross section has a sharp maximum corresponding to the momentum transfer from the projectile particle to the ejected electron [279]. In the PBrS process the Bethe peculiarity results in the following relationship between the photon energy and the transferred momentum:

$$\omega \approx \frac{q^2}{2} + I_\alpha, \tag{4.48}$$

where I_α is the ionization potential of the subshell from which the electron is excited (ionized).

The relation (4.48) defines the curve in the plane (ω, q) in the vicinity of which the generalized polarizability $\alpha(\omega, q)$ as a function has resonance character. The increase becomes more prominent with the increase of the principal quantum number of the ionized subshell [257]. Figure 4.23 illustrates the resonance behaviour of the real and imaginary parts of $\alpha(\omega, q)$ as well as that of the integrand, $|\alpha(\omega, q)|^2/q$ in (4.43).

The resonant-like behavior of $\left|\alpha(\omega, q)\right|^2$ results in a peculiarity in the PBrS cross section $d\sigma_{pol}$ as a function of v_1. Indeed, when calculating the integral in (4.43), the main contribution comes from the domain $q \gtrsim q_{min}$. The value q_{min} varies as a function of v_1. Therefore, the integral and thus, the PBrS cross section, varies rapidly when q_{min} as a function of v_1 crosses the region of the Bethe peculiarity (4.48).

For a fast heavy projectile the following relations are valid: $q_{max} \gg q_{min} \approx \omega/v_1$ and $\alpha(\omega, q_{max}) \ll \alpha(\omega, q_{min})$. Using this value of q_{min} in (4.48) results in a simple relationship, which defines the position of the Bethe peculiarity in the velocity dependence of the PBrS cross section at a certain value of ω:

$$v_1 \approx \sqrt{(\omega + I)/2}. \tag{4.49}$$

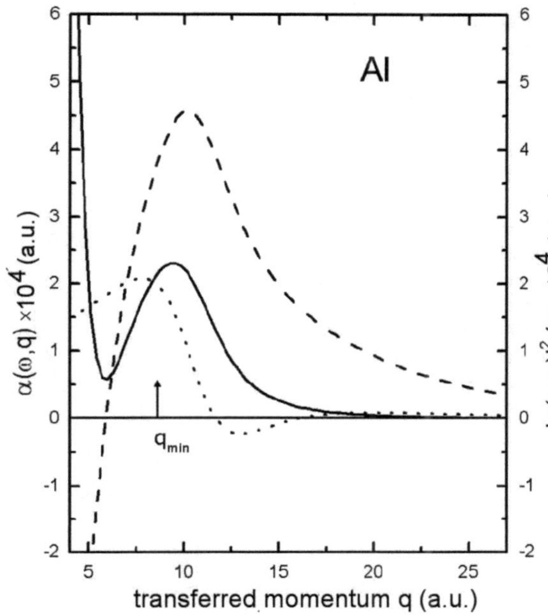

Fig. 4.23 Generalized polarizability of the Al atom versus the transferred momentum q at fixed photon energy $\omega = 1483$ keV [241]. *Dashed* and *dotted curves* represent $\mathrm{Re}[\alpha(\omega, q)]$ and $\mathrm{Im}[\alpha(\omega, q)]$ respectively. The *solid curve* shows the behaviour of the integrand in (4.43)

Differentiating (4.43) with respect to v_1 one derives the exact equation determining the position of the extrema in the velocity dependence of the cross section:

$$2 \int_{q_{\min}}^{q_{\max}} |\alpha(\omega, q)|^2 \frac{dq}{q} = \left| \alpha \left(\omega, \frac{\omega}{v_1} \right) \right|^2 . \qquad (4.50)$$

Equation (4.50) can easily be solved numerically. Such a calculation shows that (4.49) turns out to be a reliable estimate.

4.7.3 Numerical Results

Figure 4.24 presents the PBrS cross section $\omega d\sigma_{pol}/d\omega \equiv d\sigma_{pol}$ in the collision of a proton on atomic aluminum as a function of projectile velocity v_1 calculated at various photon energies. Normally, there is only one maximum in the velocity dependence of $d\sigma_{pol}$. Indeed, at low values of v_1 the polarization effect of the target becomes small, because the collision takes place at small impact parameters. At high collision velocities, the cross section decrease is a result of the decrease of the collision time. Therefore, there is a certain optimal value $v_1 \sim \omega R_{at}$ which ensures the largest yield of radiation. However, the dependencies shown in the figure by solid curves exhibit a more complicated structure.

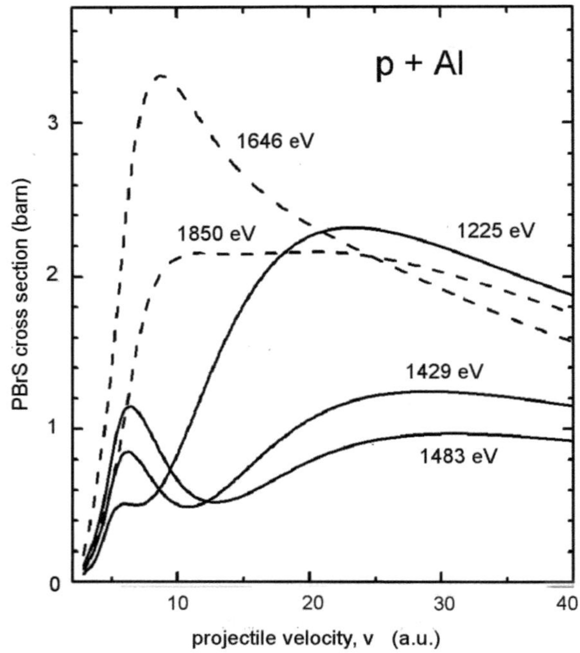

Fig. 4.24 The velocity dependence of the PBrS cross section $d\sigma_{pol}$ at various photon energies, as indicated, in the collision of a proton with the Al atom [241]. The HF ionization threshold of the atomic K-shell is equal to $I_{1s} = 1591$ eV. The Bethe peculiarity, i.e. the additional maximum, arises in the velocity dependencies at $\omega < I_{1s}$ shown by *solid curves*. *Dashed curves* correspond to the energies above I_{1s}

It is seen from Fig. 4.24, that the dependence of $d\,\sigma_{pol}$ on v_1 is modifies essentially with the increase of the photon energy. At $\omega = 1225$ eV there appears the additional maximum which manifests the Bethe peculiarity. Its location is in good agreement with the estimate (4.49). The magnitude of $d\sigma_{pol}$ at the maximum gradually increases with ω up to the K-shell ionization threshold, $I_{1s} = 1591$ eV. Beyond the threshold (the dashed curves) the additional maximum disappears.

The formation of the additional maximum in the velocity dependence and its evolution with ω are determined mainly by the properties of the generalized polarizability of a target atom. Figure 4.24 demonstrates that the Bethe peculiarity appears in the form of a resonance in the q-dependence of $\mathrm{Re}\,\alpha(\omega, q)$. However, such a behaviour is typical only for photon energies slightly lower than the ionization threshold I of an inner shell. In this region the imaginary part of the polarizability, formed mainly by the electrons of the nearest outer shell, is greatly reduced. At photon energies slightly above I, the magnitude of $\mathrm{Im}\,\alpha(\omega, q)$ increases greatly due to the electron transitions, not linked to the kinematic condition (4.48), and thus it masks the Bethe peculiarity. The latter, therefore, appears as an additional maximum in the velocity dependence of $d\sigma_{pol}$ for ω below the ionization threshold, and becomes more pronounces as ω goes closer to I. These features of the polarizability qualitatively explain the dependencies presented in Fig. 4.24.

In Fig. 4.25 the velocity dependencies of the PBrS (dashed curve), RI (dotted curve) and the total (solid curve) cross sections are presented. The RI process has

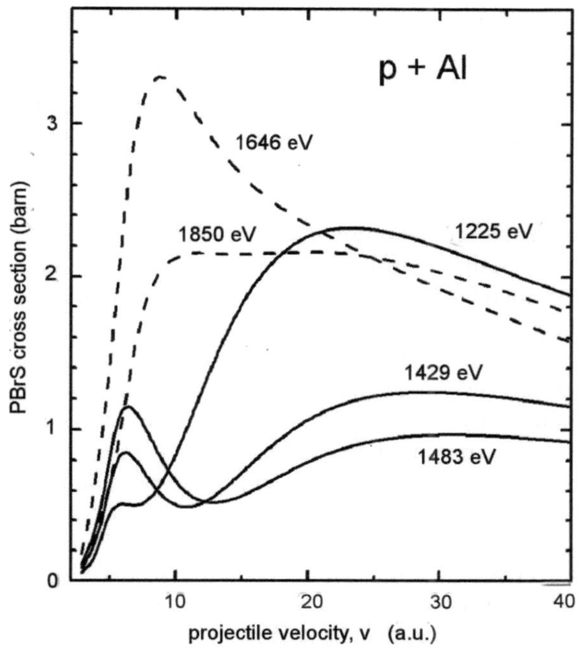

Fig. 4.25 The velocity dependencies of the polarizational BrS (*dashed curve*) and RI (*dotted curve*) cross sections in the proton–Al collision at $\omega = 1483\,\text{eV}$. The *solid curve* is the total cross section, PBrS + RI [241]

the threshold at $\omega \sim v_1^2/2 - I$, where I is the atomic ionization potential. Therefore, at the fixed photon energy ω, the RI process takes place only if $v_1 > \sqrt{2(\omega + I)}$. However, the PBrS of the same energy can be generated at lower collision velocity. In accordance with the estimate (4.49), the Bethe peculiarity in the velocity dependence of the polarizational BrS cross section arises at $v_1 \sim \sqrt{\omega/2} < \sqrt{2\omega}$. This means that the feature can be observed in the total photon emission spectrum, because in this region the RI does not mask the manifestation of the Bethe ridge.

The presented results show, that the Bethe peculiarity appears in the velocity dependence of the PBrS cross section. The peculiarity can be observed in the total photon emission spectrum at frequencies lying below the ionization thresholds of the inner (K- or L-) atomic shells. The relationship (4.49) elucidates the connection between the photon frequency and the location of the peculiarity.

4.8 Polarizational Mechanism in Electron–Ion Radiative Capture

In this Section we investigate the role of dynamic polarization of the target electrons in the process of radiative capture of an electron by a multicharged ion ('polarizational recombination', PR).

The radiative capture and the BrS processes are quite similar, differing only in the type of radiative transition involved: the former corresponds to a free-free transition whereas the latter requires a free-bound one.

Initially, the PR mechanism was studied in [115] in application to electron–metallic-cluster collision. It was demonstrated that the PR mechanism gives rise to a giant resonance in the cross section of radiative capture. In the vicinity of the resonance, which is related to the giant resonance in the photoionization of the cluster, the cross section is dominated by PR (see Sect. 7.4 for more details).

In connection with electron–ion collisions the PR process was considered in [102, 103] and further discussed in [62, 64, 65] Using the arguments of classical mechanics they derived a simple approximate formula for the contribution of PR, and claimed that the contribution of PR to the total recombination rates may be comparable or exceed (by more that an order of magnitude) the standard ('ordinary') contribution of radiative recombination. More accurate analysis [238, 239] showed, that, intrinsically, the proposed approximate formula grossly overestimates the PR contribution. The results of numerical calculations carried out in the cited papers indicated that the effect of PR can enhance the recombination cross sections for photon energies outside the regions dominated by dielectronic recombination. The magnitude of the effect is on the level of 10–30 % for Ne-like and Ni-like ions [238]. However, for multi-electron ionic ions with lower ionicities, the non-resonant photorecombination background can be enhanced by up to the order of magnitude [239]. This happens, as we show below in the section, due to the positive interference between the 'ordinary' and the 'polarization' parts of the non-resonant amplitude. This interference is additionally enhanced due to the coherence effect in the polarization of the multi-electron shell. For the ionic targets with small number of electrons the effect is much less pronounced, as it was demonstrated by recent calculations of the electron capture rates to heavy hydrogen-like ions [442].

Recently, it was noted [331] that the PR channel can be responsible for the discrepancy between the measured cross sections [97] and the available theoretical data. In the cited paper, the electron-impact excitation cross sections $\sigma_{\rm exc}$ for dominant X-ray lines from Fe XVII were measured together with those, $\sigma_{\rm RR}$, for radiative recombination of the beam electrons into the M-shell ionic levels. The measured $\sigma_{\rm exc}$ cross sections were determined by normalizing to the $\sigma_{\rm RR}$ values. The latter, in turn, were independently normalized to the theoretical RR values. It was claimed in [97] that the available theoretical σ cross sections were higher by $\gtrsim$25 % than the experimentally determined ones. The estimate made in [331] of the enhancement factor derived analytically in [238, 239] showed that the RR background is affected by approximately 25 % if one accounts for the PR mechanism. The tabulation of the smooth background with account for the PR channel, based on the calculations performed within the relativistic Dirac-Fock approximation [402], is now available [403].

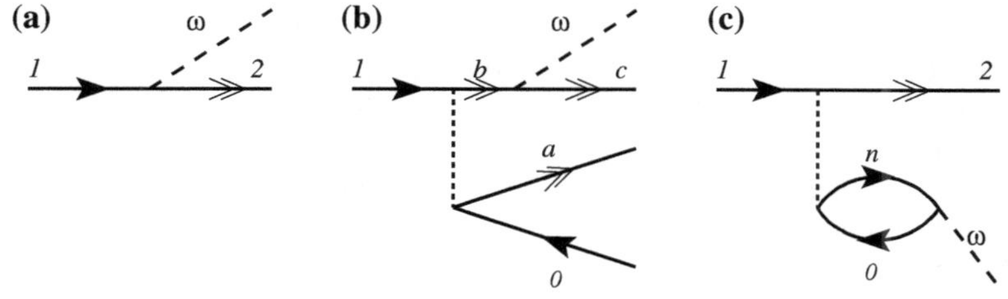

Fig. 4.26 Diagrammatic representation of RR (**a**), DR (**b**), and PR (**c**). PR includes the part of DR which interferes with RR, while (**b**) stands for the non-interfering part of DR. *Lines with double arrows* are discrete electron states in the ionic field, *lines with the arrows* to the *left* are holes in the target ground state, *dashed line* is the emitted photon, and *vertical short-dashed line* is the Coulomb interaction [239]

4.8.1 Qualitative Description of the Radiative Capture Channels for Multi-Electron Ionic Targets

Theoretically, the process of electron–ion photorecombination is usually analyzed in terms of two mechanisms: radiative recombination (RR) and dielectronic recombination (DR) [184,324].

RR is a direct process in the course of which an unbound electron with energy ε_1 undergoes a free-bound radiative transition in the static field of the target ion A^{k+} of the ionicity k: $e^- + A^{k+} \to A^{(k-1)+} + \omega$. In the final state $|2\rangle$ the electron is bound to the target. The excess of energy is taken away by the emitted photon: $\omega = \varepsilon_1 - \varepsilon_2$. The diagrammatic representation of the RR process, see Fig. 4.26a, is identical to that of the OBrS process (the first diagram in Fig. 2.2).

DR stands for a two-step process, $e^- + A^{k+} \to \left(A^{(k-1)+}\right)^* \to A^{(k-1)+} + \omega$, which is illustrated by the diagram Fig. 4.26b [239]. In the first stage, the incoming electron is captured into a doubly excited state $\left(A^{(k-1)+}\right)^*$ with the energy $\varepsilon \approx \varepsilon_a + \varepsilon_b - \varepsilon_0$. The second stage of the process is the radiative decay of the excited state. Of course, the photon can also be emitted by the other electron, b, and exchange diagrams should be added. Also, to obtain correct DR resonance energies and widths, configuration mixing between doubly excited states must be taken into account.

The main difference between RR and DR is that the latter is a resonant process. Its contribution to the emission spectra appears as a set of peaks placed atop the smooth RR background. The positions and the widths of the peaks are related to the energies and widths of the intermediate excited states.

Another difference is that the final state of RR, $|2\rangle$, is assumed to be that of an electron moving in the field of the (ground-state) target ion. In contrast, the final states populated in DR include doubly-excited states $|\bar{0}ac\rangle$ of the compound ion. Of course, the final states formed in RR can also result from the DR process. However, their contribution to the DR cross section is by no means dominant, since most DR resonances have many radiative decay channels. As a result, one can often neglect the interference between RR and DR and obtain the total recombination cross section as a sum of the smooth RR background and DR peaks [184, 324].

Quantitatively, the recombination process can be characterized by the cross section of selective recombination, $\sigma(\nu_2)$, when the electron is captured into the final state, $|\nu_2\rangle = |n_2 l_2\rangle$ (here n_2, l_2 are the principal and the orbital quantum numbers, respectively, if one adopts a non-relativistic description), and the total cross section, $\sigma = \sum_{\nu_2} \sigma(\nu_2)$, when the final state is not specified.

As a function of ε_1, the cross section varies rapidly in those regions where DR resonances occur, and where DR, as a rule, dominates over RR, leading to $\sigma(\nu_2) \approx \sigma_{\rm DR}(\nu_2)$. For electron energies outside such regions $\sigma(\nu_2) \approx \sigma_{\rm RR}(\nu_2)$, and the cross section exhibits smooth behaviour. For electron capture accompanied by the target excitation from the jth subshell, the range of incident electron energies ε_1 where DR resonances occur, is given by

$$I_j - 2I_e < \varepsilon_1 < I_j, \tag{4.51}$$

where I_j is the ionization potential of subshell j, and I_e is the ionization energy of the lowest unoccupied (excited state) orbital.

In multicharged many-electron ions the ionization potentials of different subshells are well separated. Therefore, there exist wide ranges of incident electron energies where the cross section is free from the DR resonances. In [238, 239] it was demonstrated that for electron energies outside the DR resonance regions, the cross section $\sigma(\nu_2)$ is enhanced compared to $\sigma_{\rm RR}(\nu_2)$. This increase is due to the contribution of another distinct mechanism of recombination, the polarizational recombination (PR).[4] The presented PR diagram Fig. 4.26c describes a process in which the photon is emitted by the target ion polarized by the incident electron. Its amplitude $f_{\rm PR}$ contains the sum over all allowed virtual transitions $0 \to n$. Figure 4.26c shows only part of the diagrammatic representation of $f_{\rm PR}$: the time-reversed diagram as well as those including the exchange between the electrons are not drawn. Comparing Fig. 4.26c with the second diagram in Fig. 2.2), one concludes that the PR and PBrS processes are similar, differing only in the type of radiative transition involved: the latter corresponds to a free-free transition $1 \to 2$ whereas the former to a free-bound one.

The PR mechanism is somewhat similar to "radiative dielectronic recombination" [184], characterized by off-shell excitations of the target electrons. It was estimated in the cited paper but the general comment was that 'the computation of this amplitude is difficult...'.

At first glance it may appear that Fig. 4.26c is only a small correction to Fig. 4.26a, since the acceleration of the target electrons due to the incident electron is much smaller than that of the incident electron in the field of the ion. However, in $f_{\rm PR}$ the contributions of all electrons in the valence or inner shell are added coherently, leading thus to $f_{\rm PR} \propto N_\nu$ where N_ν is the number of electrons in a particular target orbital ν.

[4] We use this term following [102, 115].

Also, important is that both RR and PR processes result in the same final state. Therefore, their sum $f_{\rm RR}+f_{\rm PR}$ may be enhanced due to the constructive interference. It was demonstrated in [238, 239] that PR leads to an increase of the cross section, mostly due to the interference between the RR and PR amplitudes. This enhancement is much smaller than predicted in [62, 64, 65, 102, 103] but more significant than estimated in [184].

4.8.2 Formalism

To evaluate the RR and PR amplitudes, as well as the resulting formulae for the cross section, we follow the paper [238].

Considering the incoming electron moving in the static spherically symmetric potential of an ion in the ground state, one represents its non-relativistic wavefunction $\Psi^{(+)}_{\mathbf{p}_1}(\mathbf{r})$ in the form of the DPWA series (2.11).

The total amplitude corresponding to the electron transition from the initial scattering state $|1\rangle \equiv |\mathbf{p}_1^{(+)}\rangle$ to the final bound orbital $|2\rangle \equiv |n_2 l_2 m_2\rangle$, accompanied by the photon emission $\omega = \varepsilon_1 - \varepsilon_2$, consists of two terms:

$$f_{\rm tot} = f_{\rm RR} + f_{\rm PR}\,. \tag{4.52}$$

Taking into account close resemblance of RR and PR with OBrS and PBrS, respectively, one writes analytical expressions for $f_{\rm RR}$ and $f_{\rm PR}$ by introducing the substitution $\rangle\mathbf{p}_2^{(-)}| \rightarrow \rangle 2|$ on the right-hand sides of (2.9) and (2.10):

$$f_{\rm RR} = \langle 2\,|\mathbf{e}\cdot\mathbf{r}|\,1\rangle\,, \tag{4.53}$$

$$f_{\rm PR} = -\sum_n\left[\frac{\langle 0\,|\mathbf{e}\cdot\mathbf{D}|\,n\rangle\left\langle 2;n\left|\hat{V}\right|1;0\right\rangle}{\omega_{n0}-\omega-{\rm i}0} + \frac{\left\langle 2;0\left|\hat{V}\right|1;n\right\rangle\langle n\,|\mathbf{e}\cdot\mathbf{D}|\,0\rangle}{\omega_{n0}+\omega}\right]. \tag{4.54}$$

Here $\mathbf{D} = \sum_a \mathbf{r}_a$ is the operator of the dipole moment of the target ion, and $\hat{V} = \sum_a 1/|\mathbf{r}-\mathbf{r}_a|$ is the Coulomb interaction between the incident and the ionic electrons. In (4.54), the sum $\sum_n$ is carried out over the whole spectrum of the (dipole) excitations $|0\rangle \rightarrow |n\rangle$ of the target, including the continuous spectrum of excitations.

When the photon energy ω is close to the excitation energy ω_{n0} of a discrete spectrum state n, the PR amplitude is resonantly enhanced. The enhancement corresponds to a dielectronic resonance at $\varepsilon_1 \approx \varepsilon_2 + \varepsilon_n - \varepsilon_0$, where $f_{\rm PR}$ amplitude is dominated by the single term

$$f_{\rm PR} \approx -\left.\frac{\langle 0|\mathbf{e}\cdot\mathbf{D}|n\rangle\langle n;2|V|1;0\rangle}{\varepsilon_n-\varepsilon_0-\omega-{\rm i}\Gamma_n/2}\right|_{\omega\approx\varepsilon_n-\varepsilon_0}\,. \tag{4.55}$$

Here the total width Γ_n of the dielectronic state is introduced.

The structure of the right-hand side of (4.55) indicates that the resonant part of the PR amplitude accounts for the DR processes leading to the final state with one excited electron. Therefore, in the region of dielectronic resonances the PR mechanism is indistinguishable from the DR one.

The cross section is proportional to the modulus squared of the amplitude f_{tot}. Hence, it can be written as a sum of three terms: (i) $\sigma_{\mathrm{RR}} \propto |f_{\mathrm{RR}}|^2$ and (ii) $\sigma_{\mathrm{PR}} \propto |f_{\mathrm{PR}}|^2$ standing for the cross sections of the RR and PR processes, and (iii) $\sigma_{\mathrm{int}} \propto \mathrm{Re}(f^*_{\mathrm{RR}}\, f_{\mathrm{PR}})$ being the interference term which can be of either sign.

Introducing the DPWA series (2.11) into (4.53) and (4.54), one derives the following expression for the cross section of selective radiative capture in the $\nu_2 \equiv (n_2 l_2)$ state [238]:

$$\begin{aligned}\sigma_{\mathrm{tot}}(\nu_2) &= \sigma_{\mathrm{RR}}(\nu_2) + \sigma_{\mathrm{int}}(\nu_2) + \sigma_{\mathrm{PR}}(\nu_2)\\ &= \frac{16\pi\omega^3}{3p_1^3c^3} \sum_{l_1=l_2\pm 1} \max\{l_1, l_2\}\, \left(\mathscr{R}_{\nu_2\nu_1} + \mathscr{P}_{\nu_2\nu_1}\right)^2 .\end{aligned} \tag{4.56}$$

Here $\nu_1 = (\varepsilon_1, l_1)$. The quantities $\mathscr{R}_{\nu_2\nu_1}$ and $\mathscr{P}_{\nu_2\nu_1}$ are the partial RR and PR amplitudes:

$$\begin{aligned}\mathscr{R}_{\nu_2\nu_1} &\equiv r_{\nu_2\nu_1} = \int_0^\infty \mathrm{d}r\, P_{\nu_2}(r)\, r\, P_{\nu_1}(r)\,,\\ \mathscr{P}_{\nu_2\nu_1} &= -\sum_{\nu_0} \frac{N_{\nu_0}}{2l_0+1} \sum_{\nu_n} \max\{l_0, l_n\}\, \frac{2\omega_{n0} r_{\nu_0\nu_n}\, V^{(1)}_{\nu_n\nu_2,\nu_1\nu_0}}{\omega_{n0}^2 - \omega^2}\end{aligned}, \tag{4.57}$$

where $V_{\nu_n\nu_2,\nu_1\nu_0}$ denotes the radial matrix element of the dipole part of the Coulomb interaction (see, e.g., [287, 385]). The sum $\sum_{\nu_0}$, is carried out over the subshells $\nu_0 = (n_0 l_0)$ occupied in the ionic ground state, N_{ν_0} stands for the number of electrons in subshell. The inner sum, $\sum_{\nu_n} \equiv \sum_{\varepsilon_n l_n}$, is over the excited state orbitals (summation over ε_n also implies integration over the continuous spectrum of excitations).

4.8.2.1 Crude Estimate of the PR Cross Section

In [102,103] a simple approximate formula for the ratio of the PR and RR cross sections was derived by means of the classical mechanics:

$$\frac{\sigma_{\mathrm{PR}}(\nu_2)}{\sigma_{\mathrm{RR}}(\nu_2)} \approx \left|\frac{\omega^2\alpha(\omega)}{Z_{\mathrm{eff}}}\right|^2 . \tag{4.58}$$

Here $\alpha(\omega)$ is the dynamic dipole polarizability of the ionic target, and Z_{eff} is the charge seen by the projectile at some effective distance, r_0, from the nucleus (see [102, 234]). The radius r_0 depends on ε_1 and ω. It is found as the root of the [102]:

$$\varepsilon_1 = -\frac{Z}{r_0} + U_{\rm el}(r_0) + \frac{\omega^2 r_0^2}{2}\,, \tag{4.59}$$

where Z is the nuclear charge, and $U_{\rm el}(r)$ is the electrostatic potential of the target electrons. By definition, $Z_{\rm eff} = Z - r_0\, U_{\rm el}(r_0)$. For low electron energies $Z_{\rm eff}$ reduces to the net ionic charge, $Z_{\rm ion}$, while for large ε_1 and ω it becomes equal to Z. Calculations, based on the estimate (4.59), indicated strong enhancement effect (an order of magnitude and larger) due to the PR mechanism [65, 102, 103].

Let us demonstrate, following [238], that in application to the electron capture by multi-charged ion the estimate (4.59) should be used with caution.

To do this, we first derive (4.59) from the general quantum-mechanical formalism outline above. In the course of derivation the limits of validity of this estimate are also established.

Equation (4.59) contains the dynamic dipole polarizability. This implies that effective distances r between the projectile and the target, which are important for the PR mechanism, are greater than the average radius R of the ion. As has been already mentioned, this condition is well-known in the theory of dipole PBrS and has clear physical justification [454]. The polarization of the electron cloud is most pronounced if the Coulomb field of the projectile is uniform on the scale of R. Thus, distant collisions when the projectile is outside the target, $r \gg R$, are of importance. For close collisions, $r \ll R$, the field, being almost spherically symmetric, induces a small dipole moment on the target.

An extension of the 'large-distance approximation' to the PR process should be done with care [238]. In particular, its direct application prescribes the wavelength of the projectile to be much greater than R (i.e., a low-energy incoming electron) and the final state ν_2 must be a Rydberg state, whose effective radius also exceeds R.

Under these assumptions, the operator $\hat{V}$ on the right-hand side of (4.53) can be approximated with its dipole part $\hat{V} \approx \sum_a \mathbf{r}_a \cdot \mathbf{r}/r^3 = \mathbf{r}\cdot\mathbf{D}/r^3$. This allows one to express the PR amplitude in terms of the dynamic dipole polarizability $\alpha(\omega)$ of the target ion:

$$f_{\rm PR} \approx -\left\langle 2\left|\frac{\mathbf{e}\cdot\mathbf{r}}{r^3}\right|1\right\rangle \alpha(\omega)\,. \tag{4.60}$$

The operator in (4.60) is proportional to $-Z_{\rm ion}\mathbf{r}/r^3$, which is the acceleration operator in the field $-Z_{\rm ion}/r$, i.e. in the electrostatic field of the target in the region $r \gg R$ ($Z_{\rm ion}$ stands for the net ionic charge). If one assumes that the initial- and final-state wavefunctions of the electron can be treated as solutions of the Schrödinger equation in the field of a point charge $Z_{\rm ion}$, then the matrix element of the acceleration can be converted into the matrix element of the operator $\mathbf{r}$. This allows one to relate $f_{\rm PR}$ to the RR amplitude

$$f_{\rm PR} \approx -\left\langle 2\left|\mathbf{e}\cdot\mathbf{r}\right|1\right\rangle \frac{\omega^2\alpha(\omega)}{Z_{\rm ion}} = -f_{\rm RR}\,\frac{\omega^2\alpha(\omega)}{Z_{\rm ion}}\,. \tag{4.61}$$

As a result, the ratio of the PR to RR cross sections is given by

$$\frac{\sigma_{PR}(\nu_2)}{\sigma_{RR}(\nu_2)} \approx \left|\frac{\omega^2\alpha(\omega)}{Z_{ion}}\right|^2 . \tag{4.62}$$

This formula coincides with (4.58) if one substitutes Z_{eff} with Z_{ion} in the latter.

The approximations tacitly made in [102] in deriving the estimate (4.58) (see Sect. 2 of the cited paper) are equivalent to those made above. Hence, the distance r_0 which follow from (4.59) must satisfy the condition $r_0 \gg R$. This means that in the range of validity of (4.58) one can put $Z_{eff} = Z_{ion}$, while for $r_0 < R$ (where $Z_{eff} > Z_{ion}$) the estimate is not valid.

The condition $r_0 > R$ imposes restrictions on the ε_1 and ω values for which one can rely on both (4.58) and (4.62). when estimating the relative contribution of the PR channel. To establish these restrictions, one considers (4.59) in the region $r_0 > R$ replacing $Z/r_0 - U_{el}(r_0)$ with Z_{ion}/r:

$$\varepsilon_1 + \frac{Z_{ion}}{r_0} = \frac{\omega^2 r_0^2}{2} . \tag{4.63}$$

It is clear that r_0 is a decreasing function of ω. Therefore, for a fixed ε_1 the relation $r_0 = R$ defines the maximum photon energy consistent with the large-distance approximation:

$$\omega_{max} = \frac{I}{n_*}\left(1 + \frac{\varepsilon_1}{I}\right)^{1/2} , \tag{4.64}$$

where I is the ionization potential of the target and n_* is the effective principal quantum number of the outermost occupied ionic orbital. In deriving (4.64) the estimate $I = Z_{ion}^2/2n_*^2$ was used, and R was treated as the classical turning point of the outer-shell electron in the ionic field, $R \approx Z_{ion}/I$.

The energy ω_{max} must satisfy the conditions

$$\varepsilon_1 \le \omega_{max} < \varepsilon_1 + I . \tag{4.65}$$

In the extreme case $\omega_{max} = \varepsilon_1$ corresponds to capture into the state with $\varepsilon_2 = 0$, whereas $\omega_{max} \sim \varepsilon_1 + I$ results from capture into the lowest vacant orbital of the ion. For many-electron ions ($n_* \gg 1$) relations (4.64) and (4.65) lead to

$$\varepsilon_1 \ll I , \qquad \omega \ll I . \tag{4.66}$$

Hence, estimates (4.58) and (4.62) are only applicable to the capture of low-energy electrons into Rydberg states with principal quantum numbers $n_2 \gg n_*$.

However, for such ε_1 and n_2 the contribution of PR is small. Indeed, for $\omega \ll I$ the dynamic polarizability can be replaced by its static value $\alpha(0) \sim R^3/Z_{\rm eff}$. Then, using $R \approx Z_0/I$, one obtains: $\omega^2\alpha(\omega)/Z_{\rm eff} \sim (\omega/I)^2(R/a_0) \ll 1$ (a_0 is the Bohr radius).

4.8.2.2 "Stripping" Approximation

To estimate quantitatively the influence of the PR mechanism on the total cross section of $\sigma_{\rm tot}$ in non-resonant energy regions (4.51), one can use the approach developed in the PBrS theory [12, 100, 236, 252] (see also Sect. 4.6) which is known as the "stripping" phenomenon [12].

For the selective capture to the state ν_2 with the energy ε_2, the energy of the emitted photon is fixed at $\omega = \varepsilon_1 - \varepsilon_2$. In a many-electron target, the ionization potentials of different subshells are well separated, so that the electrons can be divided into two groups, the 'inner' and the 'outer' electrons, with respect to ω. This subdivision is carried out in accordance with (4.20).

The cloud of strongly-bound inner electrons is weakly polarized by an external electromagnetic field of the frequency ω. Therefore, assuming the strong inequality $\omega \ll I_{\rm in}$, one neglects the contribution of the inner electrons to the sum in (4.54) and approximates the amplitude $f_{\rm PR}$ by the contribution of the outer-shell electrons (cf. (4.30) [238]:

$$f_{\rm PR} \approx -\sum_n \left[\frac{\langle 0|\mathbf{e}\cdot\mathbf{D}_{\rm out}|n\rangle\langle n;2|\hat{V}_{\rm out}|1;0\rangle}{\omega_{n0}-\omega} + \frac{\langle 0;2|\hat{V}_{\rm out}|1;n\rangle\langle n|\mathbf{e}\cdot\mathbf{D}_{\rm out}|0\rangle}{\omega_{n0}+\omega} \right]. \quad (4.67)$$

Here $\mathbf{D}_{\rm out} = \sum_{a=1}^{N_{\rm out}} \mathbf{r}_a$ is the dipole momentum of the the outer electrons, and $\hat{V}_{\rm out} = \sum_{a=1}^{N_{\rm out}} 1/|\mathbf{r}-\mathbf{r}_a|$ is the Coulomb interaction between the incident and the outer electrons; $N_{\rm out}$ is the number of such electrons.

Evaluation of the right-hand side in (4.67) is similar to how it was done in derivation of (4.31). The result reads

$$f_{\rm PR} \approx \frac{1}{\omega^2}\langle 2|\mathbf{e}\cdot\widehat{\mathbf{a}}_{\rm out}|1\rangle\,, \quad (4.68)$$

where $\widehat{\mathbf{a}}_{\rm out}$ is the acceleration due to the static field of the outer electrons.

To obtain the final expression for $f_{\rm tot}$ one uses (4.31) to introduce the operator of the total acceleration $\widehat{\mathbf{a}}$ the electron in the ionic field. Then, relating the dipole matrix element (4.53) to that of the operator $\mathbf{e}\cdot\widehat{\mathbf{a}}$ (see, e.g., [385]), one writes the RR amplitude in the form

$$f_{\rm RR} = -\frac{1}{\omega^2}\,\langle 2\,|\mathbf{e}\cdot\widehat{\mathbf{a}}|\,1\rangle\,. \quad (4.69)$$

Using (4.68) and (4.69), one constructs the total amplitude of the selective radiative capture as follows [238]:

$$f_{\text{tot}} \approx -\frac{1}{\omega^2} \left\langle 2 \left| \mathbf{e} \cdot \widehat{\mathbf{A}} \right| 1 \right\rangle , \tag{4.70}$$

where the operator $\widehat{\mathbf{A}}$ (the effective acceleration) is defined in (4.36). The operator $\widehat{\mathbf{A}}$ does not contain the term due to the outer electrons. Hence, in F_{tot} the outer electrons do not participate in the screening of the nucleus (or, in other words, the ion is 'stripped' of its N_{out} outer electrons). As a result $|f_{\text{tot}}| > |f_{\text{RR}}|$, so that the total cross section is enhanced compared with σ_{RR}, since the effective acceleration of the projectile becomes larger. A physical origin of the stripping effect 'stripping' is explained in Sect. 4.6.

It was demonstrated in [238] (see also Sect. 4.8.3) below), that the 'stripping' approximation provides a correct estimate of the PR effect away from the thresholds. This feature is similar to what is seen from Fig. 4.17 where the BrS spectra obtained within the 'stripping' approximation are compared with the results of more accurate calculation.

Thus, the role of the target polarization can be described in terms of 'stripping'. However, there exists an even simpler way of estimating the ratio $\sigma_{\text{tot}} / \sigma_{\text{RR}}$ [238], based on the combination of the 'stripping' approximation and of the arguments of the so-called Kramers electrodynamics [234].

Equation (4.68) can be re-written as follows:

$$f_{\text{PR}} \approx \frac{1}{\omega^2} \left\langle 2 \left| \mathbf{e} \cdot \mathbf{r}\, \frac{N_{\text{out}}(r)}{r^3} \right| 1 \right\rangle, \tag{4.71}$$

where $N_{\text{out}}(r) = 4\pi \int_0^r \rho_{\text{out}}(r') r'^2 \mathrm{d}r'$ is the number of outer-shell electrons within the sphere of radius r, and ρ_{out} is their density.

To compare f_{PR} with F_{RR} from (4.53), one extracts the factor $N_{\text{out}}(r)/r^3$, taken at some point $r = r_0$, from the matrix element (4.71). The choice of r_0 is determined by the condition that the distances $r \sim r_0$ contribute most effectively to the matrix element. The quasi-classical theory of radiative transitions [234] states, that the radiation of frequency ω is emitted most effectively in the vicinity of the turning points of classical trajectories for which the projectile's angular velocity is close to ω. Therefore, the distance r_0 can be estimated as a root of the equation (4.59). Then, the PR amplitude f_{PR} is expressed in terms of f_{RR}, and the total amplitude reads

$$f_{\text{tot}} \approx f_{\text{RR}} \left(1 + \frac{N_{\text{out}}(r_0)}{\omega^2 r_0^3} \right). \tag{4.72}$$

This gives the following estimate for the ratio of the cross sections:

$$\frac{\sigma_{\rm tot}}{\sigma_{\rm RR}} \approx 1 + \frac{2N_{\rm out}(r_0)}{\omega^2 r_0^3}, \tag{4.73}$$

where we have assumed that the 2nd term in brackets in equation (4.72) is a small correction.

4.8.3 Numerical Results

In [238] the numerical calculations of $\sigma_{\rm RR}$, $\sigma_{\rm PR}$ and the interference term $\sigma_{\rm int}$ were performed for $\varepsilon_1 = 0.01$–5 keV electrons incident on a number of Ni-like and Ne-like ions. The calculations were aimed at the elucidation of the role of the PR mechanism in modifying the smooth RR background in the photon energy regions away from the DR peaks.

The two main features of the PR contribution were noted. First, the relative magnitude of $\sigma_{\rm PR}$ is very small, varying between 10^{-3} and 10^{-1} of $\sigma_{\rm RR}$. Accordingly, the ratio $\sigma_{\rm PR}\,/\,\sigma_{\rm RR}$ in non-resonant regions is much smaller than the values which can be inferred from (4.58).

The second feature is that the total cross section systematically exceeds the RR cross section in the non-resonant regions, although this effect is not large. Taking into account the relation $\sigma_{\rm PR}\,/\,\sigma_{\rm RR} \ll 1$ and recalling equation (4.56), one realities that it is the interference term, $\sigma_{\rm int}$, which leads to this enhancement of the cross section. Hence, one can write

$$\frac{\sigma_{\rm tot}}{\sigma_{\rm RR}} \approx 1 + \frac{\sigma_{\rm int}}{\sigma_{\rm RR}} \tag{4.74}$$

For photon energies well above the ionization thresholds of many-electron subshells, the interference between the radiation emitted via RR and PR mechanisms is constructive and results in $\sigma_{\rm int} > 0$ and, consequently, in $\sigma_{\rm tot} > \sigma_{\rm RR}$. A similar effect is also known in BrS (see the data and discussion presented in Sect. 4.6.4).

These features are illustrated by Figs. 4.27 and 4.28.

In Fig. 4.27 the ratios $\sigma_{\rm tot}(4p)/\,\sigma_{\rm RR}(4p)$ are presented for the selective recombination of electrons to the 4p-subshells of various Ni-like ions, as indicated. Full curves stand for the ratios value obtained by calculating the exact amplitudes $f_{\rm RR}$ and $f_{\rm PR}$, (4.53), (4.54) using the HF radial wavefunctions to calculate the partial amplitudes (4.57). Broken curves correspond to the 'stripping' approximation, (4.70). In implementing this equation, the contributions of the outer subshells were successively dropped, as the photon energy crosses their ionization thresholds. The latter are shown in the figure with vertical (red) bars.

As mentioned, the 'stripping' approximation provides a correct estimate of the PR effect away from the thresholds, in the energy intervals $I_{3d,3p,3s} < \omega < I_{2p,2s}$ and $\omega > I_{2p,2s}$. Close to the thresholds the polarizational amplitude has a strong energy dependence, because of the resonant denominators, see (4.55). This feature is obviously not accounted for by the 'stripping' approximation.

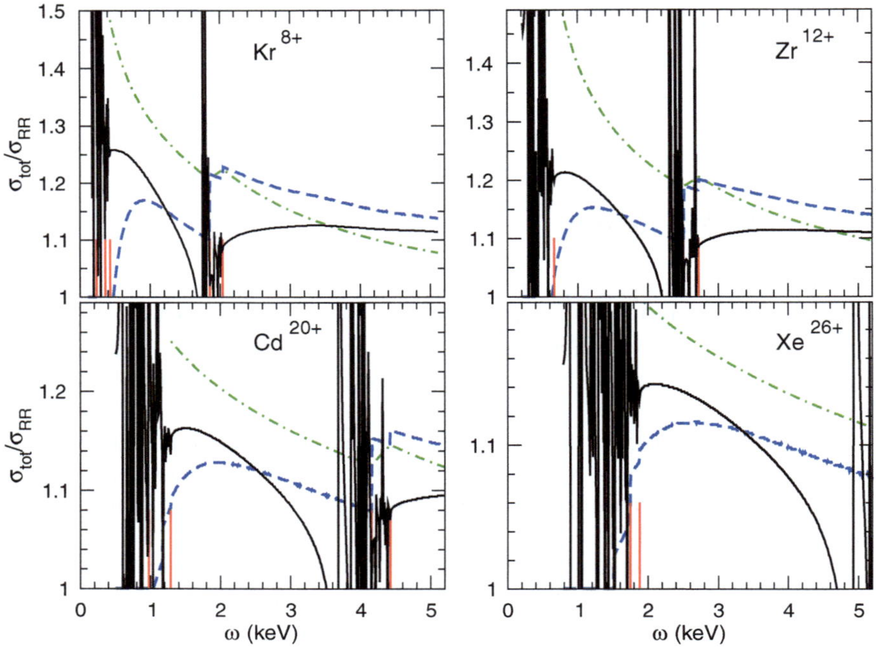

Fig. 4.27 Ratio $\sigma_{\rm tot}\,/\,\sigma_{\rm RR}$ for the selective capture into the $4p$ states of Ni-like ions as a function of the photon energy ω. The ratios shown by *full* (*black*) *curves* were obtained using the exact formulae, (4.53) and (4.54). *Broken* (*blue*) *curves* correspond to the stripping approximation (4.70), *while chain (green) curves* show the results obtained from (4.73). *Vertical bars* (in some cases obscured by the resonances) indicate the HF ionization potentials [238]

The results obtained from a simple estimate (4.73) are shown by chained (green) curves. Comparing these to the accurate numerical results one concludes, that in spite of its simplicity, the estimate yields a reasonable estimate for the ratio $\sigma_{\rm tot}\,/\,\sigma_{\rm RR}$ in the non-resonant regions.

The actual value of the enhancement due to PR is about 10 to 30%. It depends on the energy but changes little along the isoelectronic sequence. This ratio characterizes both the selective recombination cross sections $\sigma(nl)$ (as illustrated by Fig. 4.27 for the 4p states) and the total recombination cross sections $\sigma = \sum_{nl} \sigma(nl)$ [238].

In the cited paper, similar results were obtained in for a number of Ne-like ions. As discussed above, the increase is due to constructive interference between the RR and PR amplitudes. However, for a Ne-like target with a given Z, the ratio $\sigma_{\rm tot}\,/\,\sigma_{\rm RR}$ is smaller than that for the corresponding Ni-like ion. The decrease of the relative role of polarization and the interference term $\sigma_{\rm int}$ with the ionicity $Z_{\rm ion}$ could be expected. One can deduce an explicit dependence of the ratio $\sigma_{\rm int}\,/\,\sigma_{\rm RR}$ on $Z_{\rm ion}$ from (4.73). Approximating the factor $N_{\rm out}(r_0)/r_0^3$ by $N_{\rm out}/R_{\rm out}^3$, and estimating the (average) radius and ionization potential of the outer electrons as $R_{\rm out} \sim Z_{\rm ion}/I_{\rm out}$ and $I_{\rm out} \sim Z_{\rm ion}^2$, one derives $\sigma_{\rm int}\,/\,\sigma_{\rm RR} \sim (N_{\rm out}/Z_{\rm ion})(I_{\rm out}^2/\omega^2)$. Hence, for a fixed

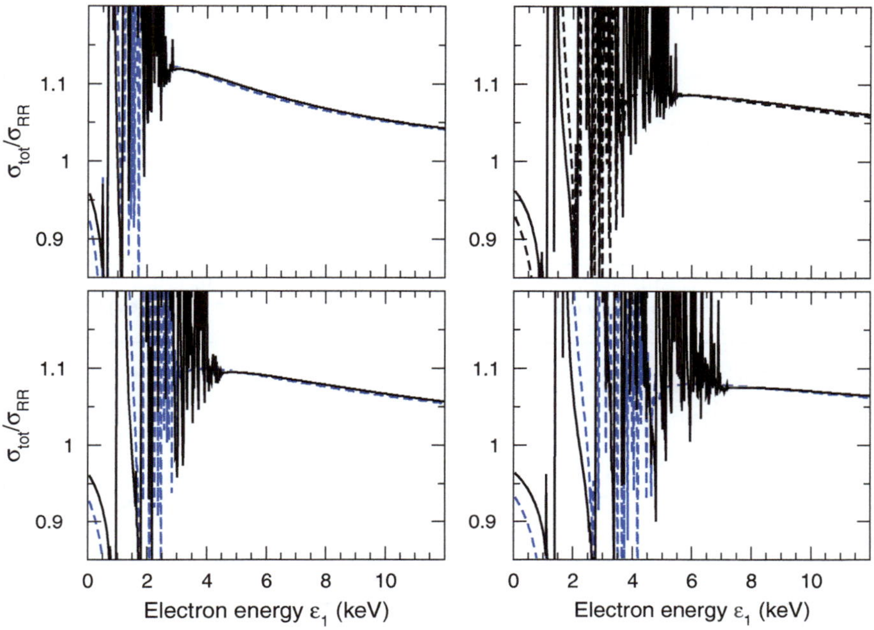

Fig. 4.28 Dependence of the ratios $\sigma_{\rm tot}\,/\,\sigma_{\rm RR}$ on the electron energy for Ne-like ions. *Broken (blue) curves* depict $\sigma_{\rm tot}(3)/\,\sigma_{\rm RR}(3)$, while *full (black) curves* correspond to the ratios of cross sections summed over the final states, $\sigma_{\rm tot}(3\text{–}10)/\,\sigma_{\rm RR}(3\text{–}10)$ [238]

number of outer electrons and for a given $I_{\rm out}/\omega$ ratio, the relative magnitude of $\sigma_{\rm int}$ decreases as $Z_{\rm int}^{-1}$.

This conclusion is illustrated by the results presented in Fig. 4.28 where the ratios $\sigma_{\rm tot}(3)/\,\sigma_{\rm RR}(3)$ and $\sigma_{\rm tot}(3\ldots 10)/\,\sigma_{\rm RR}(3\ldots 10)$ versus the electron energy ε_1 are plotted. Here the notation $\sigma(n)$ indicate the sum of the selective recombination cross sections $\sigma(nl)$ over allowed values of orbital momentum, $l = 0, 1, \ldots, n-1$. The notation $\sigma(n_1 \ldots n_2)$ stand for $\sum_{n=n_1}^{n_2} \sigma(n)$. The actual size of the enhancement varies between 5 and 15 %. Similar to the Ni-like case, the ratio $\sigma_{\rm tot}\,/\,\sigma_{\rm RR}$ depends weakly on the principal quantum number of the final electron state. It is also worth noting that the polarization contribution results in a prominent dip in $\sigma_{\rm tot}$ at the energy just below the dielectronic resonance region.

The manifestation of PR can be much stronger for the multi-electron ionic targets with lower ionicities and in the energy range of the (possible) giant resonance in the photoionization cross section. This is illustrated by Fig. 4.29 the four graphs in which show the photorecombination cross section versus the photon energy calculated for different ionic targets as indicated [239].

In each graph, the broken line stands for $\sigma_{\rm RR}$ calculated as a sum of *all* selective cross sections $\sigma_{\rm RR}(nl)$. The terms with $n \le 10$ were obtained via numerical evaluation of the partial RR amplitudes defined in (4.57). Contributions of higher

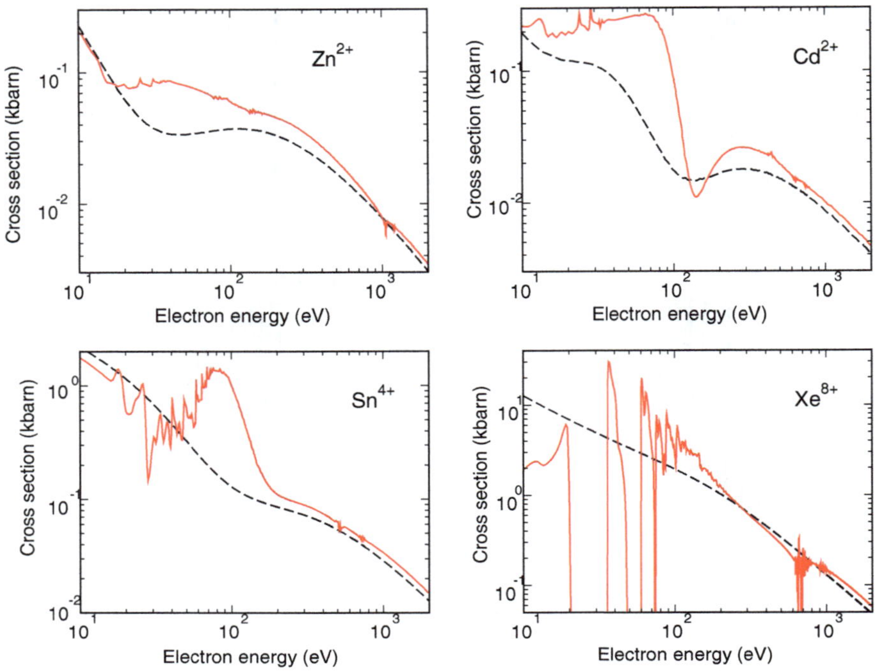

Fig. 4.29 Photorecombination cross sections for Zn^{2+}, Cd^{2+}, Sn^{4+}, and $Xe2^{8+}$: RR, *broken curves*; RR+PR without resonant contributions, (4.55), *full curves* [239]

Rydberg states, which are important at low projectile energies, were added to the recombination rates by using the Kramers formula (see. e.g., [385]).

The full curves stand for the total cross section σ_{tot} which is the sum of all selective cross sections $\sigma_{tot}(nl)$, (4.56). The PR and interference terms were computed numerically with the help of (4.57) for the principal quantum numbers up to $n = 10$, and were nullified for the higher n-values. Important is, that these dependencies stay for the *non-resonant* part of σ_{tot}. This was obtained by omitting the squared moduli of all discrete excitation terms, (4.55), from the cross section. Such calculation preserves all interference and continuum contributions. Thus, the full curves in Fig. 4.29 represent the RR background modified by the non-resonant part of PR. Note, that the inclusion of PR does not always increase the cross sections: the interference term σ_{int} can be positive (as a rule, above thresholds) and negative (below the thresholds).

Three ions, Cd^{2+}, Sn^{4+}, and Xe^{8+} belong to the Pd-like sequence. Here the target polarization at low energies is dominated by the $4d \to nf, \varepsilon f$ transitions from the $4d^{10}$ subshell. The increase in the cross section due to PR in Cd^{2+} and Sn^{4+} is even greater than in Zn^{2+} [239]. It can be related to the giant resonance in the $4d$ photoabsorption found in atoms and low-charged ions from Pd to Ba (see, e.g., [304] and references therein).

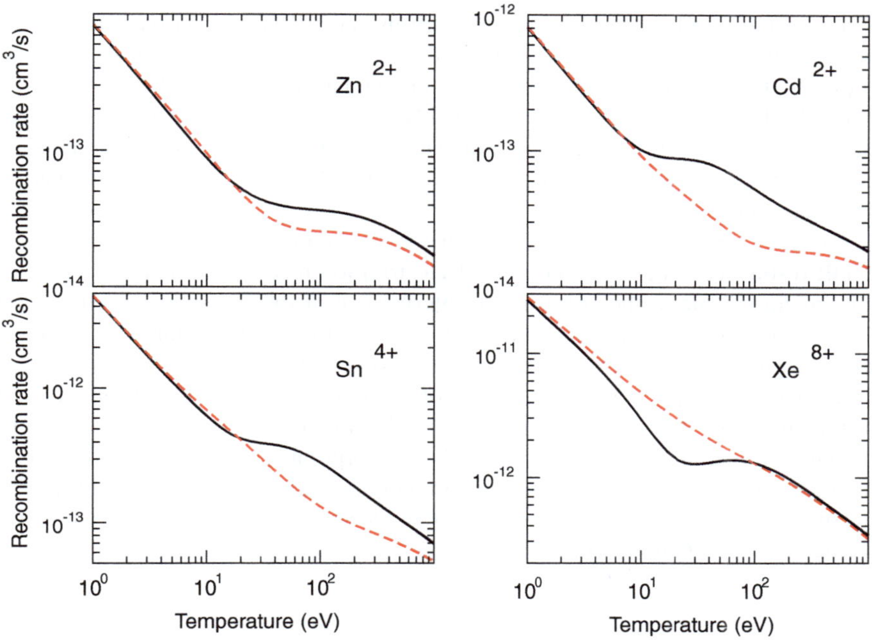

Fig. 4.30 Maxwellian recombination rates for Zn^{2+}, Cd^{2+}, Sn^{4+}, and $Xe2^{8+}$ obtained from the RR cross sections (*dashed curves*), and from the RR+PR cross sections (*solid curves*), omitting any DR resonance contribution [239]

As the nuclear charge increases, the $4f$ orbital "collapses", and the bulk of the dipole strength of the $4d$ subshell is shifted from the giant resonance in the continuum to the $4d-nf$ discrete spectrum (see, e.g., [86] and the discussion in Sect. 4.1). This transition is clear in Fig. 4.29, when comparing Cd^{2+}, Sn^{4+} and Xe^{8+}. Simultaneously, the interference between the discrete excitations and the background becomes very large [239].

The photorecombination cross sections, $\sigma(\varepsilon)$, are used to calculate so-called recombination rates, which are further used for ionization-balance calculations and the correct estimations of elemental abundance in plasmas, e.g., [102, 312, 324, 337]. At a given velocity of the collision, v, the recombination rate $r(v)$ is related to the cross section as $v\,\sigma(\varepsilon)$. However, for the plasmas applications a more instructive quantity is obtained by averaging $r(v)$ over v using the Maxwellian distribution, $w_T(v)$. Thus, the recombination rate corresponding to a given temperature T is defined as follows

$$\langle r_T \rangle = \int_0^\infty v\,\sigma(\varepsilon)\,w_T(v)\,dv \qquad (4.75)$$

The averaged recombination rates $\langle r_T \rangle$ as a function of temperature calculated for the ions Zn^{2+}, Cd^{2+}, Sn^{4+}, and $Xe2^{8+}$ are presented in Fig. 4.30 [239]. The broken curve stand for the rates obtained from the RR cross sections, full curves were calculated by inserting 'σ_{tot} minus the resonant part of PR' into the integral in (4.75). Thermal averaging washes out all resonant interference features, but, as shown by Fig. 4.30, the overall effect of PR remains important. Also to be noted, that inclusion of the PR mechanism can lead to both the increase (see the graphs for Zn^{2+}, Cd^{2+}, Sn^{4+}) and the decrease (the $Xe2^{8+}$ graph) in the smooth recombination rate background.

To conclude this Section we state that for electron energies outside the regions of DR resonances, the cross sections of electron-ion recombination can be noticeably enhanced due to the constructive interference of the photons emitted via the radiative recombination and the polarizational recombination mechanisms. For multiply charged targets this enhancement, which is characterized by the ratio $\sigma_{int} / \sigma_{RR}$, varies from a few per cent to a few tens of percent, depending on the ionicity and on the energy of the incident electron. The effect can become much larger (up to the order of magnitude) for the ions the photoabsorption spectrum of which exhibits the giant resonance lying in the continuum. For such targets the inclusion of the PR mechanisms leads to the detectable change in the smooth background for thermally averaged recombination rates.

Chapter 5
PBrS in Non-Relativistic Collisions of Structural Particles with Atoms and Ions

5.1 Introductory Notes

The qualitative picture of the PBrS process in a collision of two structural atomic particles is as follows. The colliding particles become mutually polarized and, thus, acquire induced dipole moments. The induced dipole moments of the projectile and the target form the total dipole moment of the system which alters during the collision due to the motion of the colliders. The alteration of the induced dipole moment is the source of the PBrS radiation. PBrS process in collisions of fast (but non-relativistic) ions and/or atoms was considered for the first time in [33, 34] where the main features of the process were described and analyzed.

Experimental observations of PBrS in collisions of fast protons with various atomic targets have demonstrated good agreement between theoretical predictions and the measured data [204, 205]. The spectral and spectral-angular distribution of PBrS calculated independently [167] also agree fairly well with the experimental results.

In slow collisions of atomic particles neither ordinary nor polarizational mechanisms lead to an adequate description of the radiation process. In this limit, another mechanism, known as molecular-orbital X-ray emission (see, for example, [173]) plays a dominant role. This radiation originates from electronic transitions between different levels of a quasi-molecule (QM) created during the collision. The first observation of the molecular-orbital X-ray emission was reported in [369]. Subsequent theoretical [95, 284, 305] and experimental [139, 316, 317] work showed that this kind of continuous spectrum radiation can be understood as a radiative decay in QM.

Despite the considerable intensity of the molecular-orbital emission in slow collisions of atomic particles, the polarizational mechanism of radiation is important in asymmetric collisions at large impact parameter [388]. The intensity of OBrS is negligibly small due to the large masses of the colliding particles.

A slow atomic/ionic collision proceeds adiabatically and, as a result, a photon can be emitted by QM only at the instants when the splitting between the QM levels

A. V. Korol and A. V. Solov'yov, *Polarization Bremsstrahlung*,
Springer Series on Atomic, Optical, and Plasma Physics 80,
DOI: 10.1007/978-3-642-45224-6_5, © Springer-Verlag Berlin Heidelberg 2014

coincides with the photon frequency ω, i.e. $\omega = \omega_{fi}(t)$, where $\omega_{fi}(t)$ is the transition frequency between the initial and the final states at the instant t. This condition can be met several times during the collision. For instance, a photon of the same frequency can be emitted on the incoming and outgoing parts of the ion trajectory. These events interfere, giving rise to oscillations of the cross section as a function of ω and the impact parameter. This effect was predicted theoretically [96, 284, 305], and later measured experimentally [315, 377, 413].

The target electron can be captured by the field of the projectile during the collision. The electron, moving from one ion to another, radiates rather intensively. In [388] it was shown that the radiation intensity in such process is the product of the PBrS intensity and the probability of the electron capture. This result is valid for large impact parameters.

5.2 Collisions of Fast Atomic Particles

In this section we present the formalism and the results of numerical calculations of the BrS spectra formed in collisions of two fast non-relativistic heavy particles.

5.2.1 BrS Amplitude

Let us consider a collision of two fast but non-relativistic atoms[1] (the masses and nuclear charges of which are denoted as $M_{1,2}$ and $Z_{1,2}$, respectively), resulting in the emission of a photon. The translational momenta of the atoms before and after the collision are notated as $\mathbf{p}_{1,2}$ and $\mathbf{p}'_{1,2}$, respectively. Apart from the translation momentum the state of each particle is characterized by the set of quantum numbers which refer to its internal (electronic) state. We assume that each atom before and after the collision is in its grounds state, '0'.

The interaction between the atoms is described by the potential $U(\{\mathbf{r}\}_1, \{\mathbf{r}\}_2) \equiv \hat{U}$ (here $\{\mathbf{r}\}_j, j = 1, 2$, denote the coordinates of all the particles in atom '1' and atom '2'), which is the sum of the pair interactions between the constituents of the two atoms. For each one of the colliders, the interaction with the field of radiation is described by the operator $\hat{V}$ which has the general form $\hat{V} = \sum_{i=1}^{N}(e_i/m_i)\exp(-\mathrm{i}\mathbf{k}\cdot\mathbf{r}_i)\,(\mathbf{e}\cdot\hat{\mathbf{p}}_i)$. The sum is carried out over the atom's constituents of the total number N, the quantities e_i and m_i stand for the charges and the masses of the constituents, and $\hat{\mathbf{p}}_i$ is the momentum operator.

In the lowest orders of the perturbation theory in $\hat{U}$ and $\hat{V}$ the process of photon emission is represented by four diagrams drawn in Fig. 5.1. The solid lines denote the atoms, the dashed lines stand for the emitted photon, whose momentum is $\mathbf{k}$, energy

[1] Unless specified further the term 'atom' is used for any heavy projectile (neutral atom, ion, bare nucleus, etc).

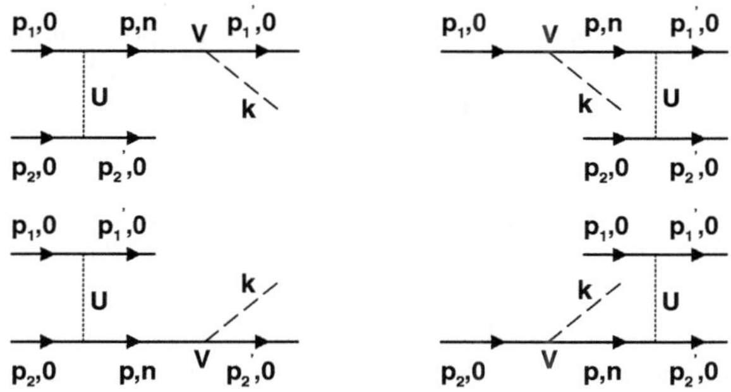

Fig. 5.1 Diagrammatic representation of the BrS amplitude in atom–atom collisions

ω and polarization **e**, the vertical dotted lines represent the interaction $\hat{U}$. The two upper diagrams correspond to the photon emission by a projectile which is virtually polarized by the target. The amplitude of this process is written as follows:

$$f_1 = \int \frac{d\mathbf{p}}{(2\pi)^3} \sum_n \left[\frac{\langle \mathbf{p}_1', 0 \left| \hat{V} \right| \mathbf{p}; n \rangle \langle \mathbf{p}, n;\ \mathbf{p}_2', 0 \left| \hat{U} \right| \mathbf{p}_1, 0;\ \mathbf{p}_2, 0 \rangle}{\varepsilon_n - \varepsilon_0 + E_{\mathbf{p}} - E_{\mathbf{p}_1} + E_{\mathbf{p}_2'} - E_{\mathbf{p}_2}} \right.$$
$$\left. + \frac{\langle \mathbf{p}_1', 0;\ \mathbf{p}_2', 0 \left| \hat{U} \right| \mathbf{p}, n;\ \mathbf{p}_2, 0 \rangle \langle \mathbf{p}, n \left| \hat{V} \right| \mathbf{p}_1; 0 \rangle}{\varepsilon_n - \varepsilon_0 + \omega + E_{\mathbf{p}} - E_{\mathbf{p}_1}} \right]. \tag{5.1}$$

Here $E_{\mathbf{p}} = p^2/2M$ stands for the kinetic energy of the translation motion, ε denotes the energy of the electronic state of the projectile, the integration is carried out over the translation momentum of the projectile in the intermediate (virtual) state, and the sum $\sum_n$ is evaluated over the whole spectrum (including the excitations into the continuum) of the internal degrees of freedom of the atom 1.

The process of the photon emission by the target, is described by the lower pair of diagrams, and defines the amplitude f_2. The analytic expression for f_2 one obtains from (5.1) by exchanging the atomic subscripts '1' and '2'.

The total BrS amplitude is given by

$$f = f_1 + f_2. \tag{5.2}$$

To evaluate the terms $f_{1,2}$ one starts with the factorization of the center-of-mass motion in the definition of the wavefunction of a complex system:

$$\Psi_{\mathbf{p}n}(\{\mathbf{r}_i\} = e^{i\mathbf{p}\cdot\mathbf{R}} \psi_n (\{\mathbf{r}_i - \mathbf{R}\}), \tag{5.3}$$

where $\mathbf{p}$ is the momentum of the center-of-mass, $\mathbf{R}$ is its position radius, ψ_m $(\{\mathbf{r}_i - \mathbf{R}\})$ is the wavefunction describing the internal degrees of freedom of the system's constituents with the subscript 'n' staying for all the necessary quantum numbers.

Let us carry out the integration over the position vectors of the atomic center-of-masses in the matrix elements of the operators $\hat{V}$ and $\hat{U}$. To start with we consider the matrix element describing the photon emission by a moving atom

$$\left\langle \mathbf{p}'; m \left| \hat{V} \right| \mathbf{p}; n \right\rangle = - \left\langle \mathbf{p}'; m \left| \sum_{i=1}^{N+1} \frac{e_i}{m_i c} \mathrm{e}^{\mathrm{i}\mathbf{k}\cdot\mathbf{r}_i} \left(\mathbf{e} \cdot \hat{\mathbf{p}}_i \right) \right| \mathbf{p}; n \right\rangle . \tag{5.4}$$

The sum is carried out over all atomic constituents including the nucleus; $\hat{\mathbf{p}}_i$, e_i, m_i are the momentum operator, the charge and the mass of a ith atomic constituent.

Using the wavefunctions (5.3), one arrives at

$$\langle \mathbf{p}'; m \,|\, \hat{V} \,|\, \mathbf{p}; n \rangle = -(2\pi)^3 \delta(\mathbf{p} - \mathbf{p}' - \mathbf{k}) M_{mn}(\mathbf{k}, \mathbf{e}, \mathbf{p}), \tag{5.5}$$

where the factor $\delta(\mathbf{p} - \mathbf{p}' - \mathbf{k})$ ensures the momentum conservation. The matrix element $M_{mn}(\mathbf{k}, \mathbf{e}, \mathbf{p})$ is defined as follows

$$M_{mn}(\mathbf{k}, \mathbf{e}, \mathbf{p}) = \left\langle m \left| \sum_{i=1}^{N+1} e_i \mathrm{e}^{\mathrm{i}\mathbf{k}\cdot\mathbf{r}_i} \right| n \right\rangle \frac{\mathbf{e}\cdot\mathbf{p}}{Mc} + \left\langle m \left| \sum_{i=1}^{N+1} \frac{e_i}{m_i c} \mathrm{e}^{\mathrm{i}\mathbf{k}\cdot\mathbf{r}_i} (\mathbf{e}\cdot\hat{\mathbf{p}}_i) \right| n \right\rangle .$$

Here M is the total mass of the atom. For $\mathbf{p} = 0$ the quantity $M_{mn}(\mathbf{k}, \mathbf{e}, 0)$ stands for the matrix element of the photon emission calculated in the center-of-mass of the atom.

Within the framework of non-relativistic dipole approximation, $\omega R_{\mathrm{at}}/c \ll 1$, $v/c \ll 1$, $(v = p/M)$, which is applicable even for the photon energies comparable with the ionization potential of the inner shells, the expression for $M_{mn}(\mathbf{k}, \mathbf{e}, \mathbf{p})$ is simplified:

$$M_{mn}(\mathbf{k}, \mathbf{e}, \mathbf{p}) = \delta_{mn} \frac{\mathbf{e}\cdot\mathbf{p}}{Mc} \sum_{i=1}^{N+1} e_i + \mathrm{i} \frac{\omega_{mn}}{c} (\mathbf{e}\cdot\mathbf{d}_{mn}) . \tag{5.6}$$

Here $\mathbf{d}_{mn}$ is the dipole matrix element between the electron states m and n, $\omega_{mn} = \varepsilon_m - \varepsilon_n$ is the transition energy.

Separating the center-of-mass coordinates in the matrix element of the scattering process (i.e.,of the operator $\hat{U}$) and integrating over them, one obtains:

$$\langle \mathbf{p}'_1, m'_1; \mathbf{p}'_2, m'_2 | \hat{U} | \mathbf{p}_1, m_1; \mathbf{p}_2, m_2 \rangle = (2\pi)^3 \delta(\mathbf{p}_1 + \mathbf{p}_2 - \mathbf{p}'_1 - \mathbf{p}'_2) \frac{4\pi}{Q_2^2} \times S^{(1)}_{m'_1 m_1}(\mathbf{Q}_1) S^{(2)}_{m'_2 m_2}(\mathbf{Q}_2) . \tag{5.7}$$

Here the matrix elements $S_{mn}(\mathbf{Q})$ are defined as follows

$$S_{m'm}(\mathbf{Q}) = \left\langle m' \left| \sum_{i=1}^{N+1} e_i e^{-i\mathbf{Q}\cdot\mathbf{r}_i} \right| m \right\rangle, \tag{5.8}$$

where $\mathbf{Q} = \mathbf{p} - \mathbf{p}'$ stands for the momentum transferred to the atom.

Let us note that the matrix element (5.7) is similar to the scattering amplitude of two charged particles calculated within the Born approximation. The functions (5.8) play the role of the charges of the scattering particles.

As it will be demonstrated further in this section, to describe the BrS of structural particles it is sufficient to consider the function $S_{m'm}(\mathbf{Q})$ in the region of not too large transferred momenta: $QR_{\text{at}} \ll M/Zm_e$ (here Z is the nucleus charge and m_e is the electron mass). In this limit the expression $S_{m'm}(\mathbf{Q})$ can be transformed to

$$S_{m'm}(\mathbf{Q}) = Z\delta_{m'm} - \left\langle m' \left| \sum_{i=1}^{N} e_i e^{-i\mathbf{Q}\cdot\mathbf{r}_i} \right| m \right\rangle + i\,\frac{Zm_e}{M}\,\mathbf{Q}\cdot\mathbf{d}_{m'm}\,. \tag{5.9}$$

To derive this formula one makes use of the following expansion for the exponent from (5.8):

$$Ze^{-i\mathbf{Q}\cdot\mathbf{r}_{N+1}} \approx Z - i\,\mathbf{Q}\cdot\mathbf{r}_{N+1}, \qquad \mathbf{r}_{N+1} = -\sum_{i=1}^{N} \frac{m_i\mathbf{r}_i}{M}, \tag{5.10}$$

where $\mathbf{r}_{N+1}$ denotes the position vector of the nucleus in the atomic center-of-mass frame expressed via the coordinates of N atomic electrons.

Introducing (5.5) and (5.7) into (5.1), one integrates over the translation momentum of the first atom in the intermediate state and derives the following expression for the amplitude f_1:

$$f_1 = -\frac{4\pi}{Q_2^2}\,S_2(Q_2)\sum_n \left(\frac{M_{0n}^{(1)}(\mathbf{k},\mathbf{e},\mathbf{p}_1')\,S_{n0}^{(1)}(\mathbf{q}_1)}{\omega - \omega_{n0} + p_1'2/2M_1 - (\mathbf{p}_1' + \mathbf{k})^2/2M_1} \right.$$
$$\left. - \frac{S_{0n}^{(1)}(\mathbf{q}_1)M_{n0}^{(1)}(\mathbf{k},\mathbf{e},\mathbf{p}_1)}{\omega + \omega_{n0} - p_1^2/2M_1 + (\mathbf{p}_1' - \mathbf{k})^2/2M_2} \right). \tag{5.11}$$

Here $\mathbf{Q}_2 = \mathbf{p}_2 - \mathbf{p}_2'$, $\mathbf{q}_1 = \mathbf{p}_1 - \mathbf{p}_1' - \mathbf{k}$. In (5.10), as well as in the subsequent formulae, we omitted the delta-function $\delta(\mathbf{Q}_1 + \mathbf{Q}_2 - \mathbf{k})$ which expresses the momentum conservation law. The factor $4\pi S_2(Q_2)/Q_2^2$ (for brevity, the notation $S_2(Q_2) \equiv S_{00}^{(2)}(\mathbf{Q}_2)$ has been introduced) on the right-hand side of (5.11) is the Fourier image of the static potential of the second atom. The factor $\sum_n(\ldots)$ describes dynamic response of the first atom to the external field of the second atom and to that of the emitted photon.

Polarization of the atoms is most pronounced in the photon energy region $\omega \geq I$, where the BrS process can be described within the framework of the dipole approximation. Introducing (5.6) into (5.11), and omitting the terms containing small parameters $v_1/c \ll 1$, $Q/M_1v_1 \ll 1$ and $\omega/M_1v_1^2 \ll 1$ in the denominators, one obtains the following expression for the dipole amplitude f_1:

$$f_1 = \frac{4\pi}{Q_2^2}\frac{\omega}{c}\,(\mathbf{e}\cdot\mathbf{Q}_2)\,G_1(\omega,\mathbf{Q}_2)\,S_2(Q_2). \tag{5.12}$$

Here the following notation was introduced:

$$G_1(\omega,\mathbf{Q}_2) = -\frac{S_1(Q_2)}{M_1\omega^2}\sum_{i=1}^{N+1} e_i + \alpha_1(\omega,\mathbf{Q}_2). \tag{5.13}$$

The generalized polarizability of the first atom, $\alpha_1(\omega,\mathbf{Q}_2)$, is defined in accordance with (2.14). In the region of small transferred momenta ($Q_2R_{\text{at}} \ll 1$) $\alpha_1(\omega,\mathbf{Q}_2)$ reduces to the dipole dynamic polarizability $\alpha_1(\omega)$ (see (2.15). Generalized polarizability of a spherically symmetric atom depends only on the absolute value of the transferred momentum. We will use this feature below when calculating the BrS cross section. For a neutral atom, $\sum_{i=1}^{N+1} e_i = 0$, and, thus, the first term on the right-hand side of (5.13) disappears. For ions $G_1(\omega,\mathbf{Q}) \approx \alpha_1(\omega,\mathbf{Q})$ due to $M_1 \gg m_e$. Within the same approximation $S_2(Q_2) = Z_2 - F_2(Q_2)$ where $F_2(Q_2)$ is the form-factor of the second atom.

The formula for the amplitude f_2 in (5.2) one obtains interchanging the atomic indices 1 and 2. Hence, for the total BrS amplitude one obtains:

$$f = \frac{4\pi\omega}{c}\frac{\mathbf{e}\cdot\mathbf{Q}}{Q^2}\Big(G_1(\omega,\mathbf{Q})\,S_2(Q) - G_2(\omega,\mathbf{Q})\,S_1(Q)\Big), \tag{5.14}$$

where $\mathbf{Q} \equiv \mathbf{Q}_1$. To derive (5.14) we used the condition $\mathbf{Q}_2 = -\mathbf{Q}_1 + \mathbf{k} \approx -\mathbf{Q}_1$, which follows from the relation $Q_2 \geq \omega/v_1 \gg \omega/c = k$ where v_1 is the velocity of the first atom in the reference frame of the second atom.

In the region $\omega \gg I$ one can neglect the energies $\omega_{n0} \sim I$ in comparison with ω in the denominators in (5.11). Additionally, taking into account $M \gg m_e$, the BrS amplitude can be written in the following simple form:

$$f = \frac{4\pi\omega}{c}\frac{\mathbf{e}\cdot\mathbf{Q}}{Q^2}\Big(Z_2F_1(Q) - Z_1F_2(Q)\Big). \tag{5.15}$$

It follows from (5.14) and (5.15) that there is no dipole BrS in the collision of two identical atoms: $f = 0$ [33, 34]. In this case, the radiation which appear is due to the relativistic corrections (see [38, 39, 43] and Sect. 6.2).

5.2.2 BrS Cross Section

To calculate the cross section which corresponds to the amplitude (5.14) let us choose the following independent unit vectors of the photon polarization:

$$\mathbf{e}_1 = \mathbf{p}_1 \times \mathbf{k} / |\mathbf{p}_1 \times \mathbf{k}|, \qquad \mathbf{e}_2 = \mathbf{k} \times \mathbf{e}_1 / k,$$

The spectral-angular distribution of the radiation is described by the formula:

$$\frac{\mathrm{d}^2\sigma(\xi_i)}{\mathrm{d}\omega \mathrm{d}\Omega_{\mathbf{k}}} = \frac{\omega^3}{2\pi c^3 v_1^2} \int\limits_{\omega/v_1}^{2\mu v_1} \frac{\mathrm{d}Q}{Q} \left| G_1(\omega, Q) S_2(Q) - G_2(\omega, Q) S_1(Q) \right|^2$$
$$\times \left[1 + \cos^2\theta_{\mathbf{k}} + \xi_3 \sin^2\theta_{\mathbf{k}} + \left(\frac{\omega}{v_1 Q}\right)^2 (1 - 3\cos^2\theta_{\mathbf{k}} - 3\xi_3 \sin^2\theta_{\mathbf{k}}) \right]. \tag{5.16}$$

Here $\theta_{\mathbf{k}}$ is the angle between $\mathbf{k}$ and $\mathbf{p}_1$, $\mu = M_1 M_2/(M_1 + M_2)$ is the reduced mass of the colliding atoms. The lower and upper limits of integration over the transferred momentum Q are found from the energy and momentum conservation laws.

The Stokes parameter ξ_3 characterizes the degree of the linear polarization of the emitted radiation along the vectors $\mathbf{e}_1$ and $\mathbf{e}_2$. Namely, the probability to be polarized along $\mathbf{e}_1$ is equal to $1/2(1 + \xi_3)$, and along $\mathbf{e}_2$ it is $1/2(1 - \xi_3)$. Hence, the emitted radiation is linearly polarized. Setting $\xi_3 = 0$ in (5.16) and multiplying the resulting expression by 2, one derives the cross section $\mathrm{d}^2\sigma/\mathrm{d}\omega\mathrm{d}\Omega_{\mathbf{k}}$ averaged over the photon polarizations.

Equation (5.16) indicates that to calculate the BrS cross section in collision of two neutral atoms, it is necessary to determine the amplitudes in the transferred momentum domain $QR_{\mathrm{at}} \sim 1$ since outside this region the amplitudes fall off rapidly. On the contrary, in ion–atom (or ion–ion) collisions the integral increases as $1/Q$ in the region $QR_{\mathrm{at}} \ll 1$, provided these transferred momenta are kinematically allowed. The latter is fulfilled if $Q_{\min} R_{\mathrm{at}} = \omega R_{\mathrm{at}}/v_1 \ll 1$. It is easy to check this using the relations:

$$S(Q) = Z - N + O(QR_{\mathrm{at}}), \qquad G(\omega, Q) = \alpha(\omega)\,[1 + O(QR_{\mathrm{at}})].$$

The difference in the cross sections behaviour for small Q is due to the long-range Coulomb interaction which attends ion–atom and ion–ion collisions but is absent in collisions of two neutral atoms. The long-range character of the interaction is important in the region $Q \sim Q_{\min} \ll R_{\mathrm{at}}^{-1}$, since in this case the dominant contribution to the BrS spectrum arises from the distant collisions. Provided the condition $\omega R_{\mathrm{at}}/v_1 \geq 1$ is fulfilled, the radiation is formed at the distances comparable with the atomic radii. As a result, the BrS cross sections for neutral atoms and for charged ions are of the same order of magnitude.

In the region $\omega R_{\rm at}/v_1 \ll 1$ the formula (5.16) is simplified by omitting the terms containing the parameter $(\omega/v_1 Q)^2$. This is justified since the most important region for the BrS formation in a collision of neutral atoms is $QR_{\rm at} \sim 1$ where the parameter is small. For neutral colliders, setting the lower limit of integration in (5.16) to zero (this can be done since the contribution of the region $Q \sim Q_{\rm min}$ is negligibly small), one derives:

$$\frac{d^2\sigma(\xi_i)}{d\omega d\Omega_{\mathbf{k}}} = \frac{\omega^3}{2\pi c^3 v_1^2} J(\omega) \left(1 + \cos^2\theta_{\mathbf{k}} + \xi_3 \sin^2\theta_{\mathbf{k}}\right),$$

where $J(\omega)$ stands for the integral over Q carried out from zero to infinity. This formula shows, that the angular distribution of non-polarized photons is proportional to $(1 + \cos^2\theta_{\mathbf{k}})$, and thus coincides with that of the rotating dipole [278].

The angular distribution of radiation, proportional to $(1 + \cos^2\theta_{\mathbf{k}})$, formed in ion–atom (ion–ion) collisions in the photon frequency domain $\omega R_{\rm at}/v_1 \ll 1$ can be obtained from (5.16) in the framework of so-called logarithmic approximation. Namely, expanding the integrand in powers of Q and integrating in the limits from $Q_{\rm min}$ up to $Q \sim 1/R_{\rm at}$, one retains only the main term, containing the logarithm of $(R_{\rm at}Q_{\rm min})^{-1}$, which is considered as a large quantity.

Averaging (5.16) over the photon polarizations and integrating over $d\Omega_{\mathbf{k}}$, one finds the BrS spectrum:

$$\frac{d\sigma}{d\omega} = \frac{16\omega^3}{3c^3 v_1^2} \int_{\omega/v_1}^{2\mu v_1} \frac{dQ}{Q} \left| G_1(\omega, Q) S_2(Q) - G_2(\omega, Q) S_1(Q) \right|^2. \tag{5.17}$$

The comments made above concerning the peculiarities in evaluation and possibilities to simplify the spectra-angular distribution $d^2\sigma/d\omega d\Omega_{\mathbf{k}}$ in the domain $\omega R_{\rm at}/v_1 \ll 1$, are applicable to (5.17) as well.

Most simple expression for the spectrum can be obtained in the photon energy range $\omega \gg I$:

$$\frac{d\sigma}{d\omega} = \frac{16\omega^3}{3c^3 v_1^2} \int_{\omega/v_1}^{\infty} \frac{dQ}{Q} \left| Z_1 F_2(Q) - Z_2 F_1(Q) \right|^2. \tag{5.18}$$

The upper limit of integration is made infinite since the domain $Q \geq Q_{\rm max}$ does not contribute noticeably due to a rapid decrease of the integrand.

The structure of the right-hand side of (5.18) is quite clear. In the region of large photon frequencies, $\omega \gg I$, the BrS spectrum is formed due to the deceleration of electrons of each collider in the field of the bare nucleus of the partner. The BrS amplitude of the quasi-free electrons of one of the atoms (for example, the first one) in the nucleus field of the second atom is proportional to the factor $Z_2 F_1(Q)$, which is analogous to the product of two charges in the OBrS amplitude of two structureless charges. Therefore, the total BrS amplitude for the two atoms is proportional to $Z_1 F_2(Q) - Z_2 F_1(Q)$. The minus sign shows that the induced dipole moments of the colliders are oriented in the opposite directions.

The functions $S^{(1)}_{n0}(Q)$ and $S^{(2)}_{n0}(Q)$ in the cross sections (5.16) and (5.17) take into account the effect of nucleus recoil (see the third term on the right-hand side of (5.9)). The recoil manifests itself in the region of large transferred momentum Q where $F(Q) \leq QR_{\rm at}(Zm_e/M)$. Fixing the value of Q one can derive the following expression for the double differential cross section $\mathrm{d}^2\sigma/\mathrm{d}\omega\mathrm{d}Q$ in the region where the recoil terms from (5.9) are dominant:

$$\frac{\mathrm{d}\sigma^2}{\mathrm{d}\omega\mathrm{d}Q} = \frac{16\omega^3}{3c^3v_1^2}\frac{Z_1^2Z_2^2}{Q}\left|\frac{\alpha^{(1)}(\omega)}{M_1} - \frac{\alpha^{(2)}(\omega)}{M_2}\right|^2, \tag{5.19}$$

valid for

$$\frac{M}{Zm_e}F(Q) \ll QR_{\rm at} \ll \frac{M}{Zm_e} \tag{5.20}$$

Integrating (5.19) over the transferred momentum Q within the limits followed from (5.20), one notices that the expression obtained contains the small parameter $(Z_1Z_2/M_{(1,2)}) \ll 1$. Therefore, the contribution of the recoil terms to the cross sections (5.16) and (5.17) is negligibly small.

For a structureless projectile (5.17), reduces to the cross section obtained in [45, 454] for the BrS on many-electron atom with account for its dynamic polarizability. Indeed, setting $G_2(\omega, Q) = \alpha(\omega, Q)$ and $S_2(Q) = Z - F(Q)$ for the target, and $G_1(\omega, Q) = -1/M\omega^2$ and $S_1(Q) = -1$ for the projectile particle (with M being its mass), one derives from (5.17):

$$\frac{\mathrm{d}\sigma}{\mathrm{d}\omega} = \frac{16\omega^3}{3c^3p^2}\int_{\omega/v_1}^{2p_1}\frac{dq}{q}\left|\frac{Z - F(Q)}{\omega^2} - M\alpha(\omega, q)\right|^2. \tag{5.21}$$

Here the first term in the modulus describes the radiation by a projectile in the atomic field, whereas the second represents the radiation due to the atom polarization (see also Sect. 2.1).

5.2.3 Numerical Example: BrS in He + Xe Collision

It was mentioned in Chap. 2 that for the same initial velocity of a projectile the intensity of BrS formed in the collision of a heavy particle with a many-electron target can be comparable or even higher than that in the electron-atom collision. To demonstrate that the BrS intensity in collision of two neutral atoms can be of the same order as in collision involving a charged projectile. we present the results of numerical calculation [33] of the BrS spectra formed in $e^- + \mathrm{Xe}$, $\alpha + \mathrm{Xe}$ and $\mathrm{He} + \mathrm{Xe}$ collisions for the range of photon energies beyond the ionization potential of the Xe 4d subshell, $I_{4d} = 73.4$ eV. The calculated differential cross sections $\mathrm{d}\sigma/\mathrm{d}\omega$ are presented in Fig. 5.2. All curves correspond to the initial velocity $v_1 = 5$ a.u.

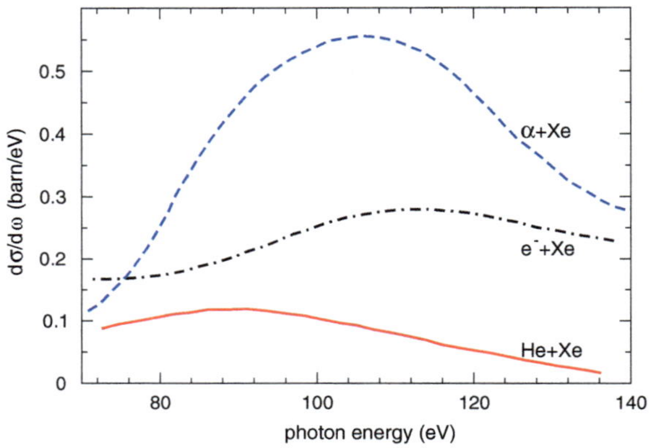

Fig. 5.2 Differential cross section $d\sigma/d\omega$ of BrS in collisions of an electron, an α-particle and a He atom (as indicated) with a Xe atom [33]. In all cases the initial velocity of the projectile is 5 a.u.

To calculate the BrS spectra for He + Xe collisions the formula (5.17) was used. The recoil effect in was neglected (see the discussion in Sect. 5.2.3). Thus, the function (5.9) for a diagonal transition was considered in the form $S(Q) = Z - F(Q)$. The simplified expression (5.21) was applied to electron- and α-Xe collisions. The form-factors of He and Xe atoms were calculated in the Hartree-Fock approximation, the generalized polarizability of Xe, $\alpha_{\mathrm{Xe}}(\omega, q)$—within the RPAE scheme as it was done in [12]. The generalized polarizability of He was approximated by $\alpha_{\mathrm{He}}(\omega, q) \approx -F_{\mathrm{He}}(q)/\omega^2$ (see (2.4)) since in the considered range of the photon energies (50 ... 140 eV) the strong inequality $\omega \gg I_{He} \approx 25$ eV is valid. Although $\left|\alpha_{\mathrm{He}}(\omega, q)\right|^2 \ll \left|\alpha_{\mathrm{Xe}}(\omega, q)\right|^2$ in the range $\omega \sim I_{4d}$, the two terms in the brackets in the integrand from (5.17) are of the same order of magnitude, and the interference between them is important when calculating the spectrum.

The most striking feature clearly seen in the figure is that in the whole spectral interval the magnitude of $d\sigma/d\omega$ for the collision α+Xe exceeds that formed in the collision e^-+Xe. The qualitative explanation of this effect is as follows. In the collision e^-+Xe both channels, the ordinary and the polarizational, contribute to the amplitude/spectrum. The amplitude of the process is described by (2.1) where one uses $m = 1$ and $e = 1$. In the case of α-particle scattering, $f_{\mathrm{ord}} \to 0$ because of a large mass of the projectile, but $f_{\mathrm{pol}} \propto 2\alpha_{\mathrm{Xe}}(\omega, q)$ is two times higher than for an electron. Therefore, the polarizational part of the spectrum for an α-particle is approximately four times larger than the PBrS of an electron. In the latter case, however, there are contributions of the terms $d\sigma_{\mathrm{ord}}$ and $d\sigma_{\mathrm{int}}$ which reduce the discrepancy.

Another important feature which is illustrated by the figure is that the intensity of the radiation in collision of a neutral and compact He atom with Xe is, in the order of magnitude, equal to that for α- and e^-–Xe collisions. This is totally due to the polarizational BrS which appears as a result of virtual polarization of Xe during the collision.

5.3 Effect of Coherence and Total Cross Section of BrS

Once the internal dynamic structure of a target is taken into account, the next logical step is to consider the radiative processes which are accompanied by the excitation or ionization of the target. Following [39, 56] we call BrS processes of this type 'inelastic' BrS contrary to the 'elastic' one, when the target remains in its ground state after the collision. The question which arises is: what is the role of inelastic BrS processes in the formation of the total BrS spectrum compared with the BrS process without change of the target's state. This question is not purely of theoretical interest since experimentally it is quite difficult to separate elastic and inelastic BrS. To do this it is necessary to observe the final state of the target with simultaneous detection of the emitted photon.

Within the framework the non-relativistic Born approximation the amplitude of the scattering process of a structureless particle (with the charge Z_0 and mass M) accompanied by the atomic transition from the initial ground state 0 to the final state m may be written as follows [100, 414]:

$$f_{\rm tot}^{(m)} = f_{\rm ord}^{(m)} + f_{\rm pol}^{(m)} = \frac{4\pi Z_0}{q^2\omega}\left[\mathbf{e}\cdot\mathbf{q}\,\frac{Z_0}{M}\,(Z\delta_{m0} - F_{m0}(\mathbf{q})) + \omega^2 A_{m0}(\mathbf{e},\omega,\mathbf{q})\right]. \tag{5.22}$$

Here $\mathbf{q} = \mathbf{p}_1 - \mathbf{p}_2$ is the momentum transfer,

$$F_{m0}(\mathbf{q}) = \langle m|\sum_{j=1}^{N}\exp(\mathrm{i}\mathbf{q}\cdot\mathbf{r}_j)|0\rangle \tag{5.23}$$

is a non-diagonal form-factor of the target electrons, and the function $A_{m0}(\mathbf{e},\omega,\mathbf{q})$ is defined as

$$A_{m0}(\mathbf{e},\omega,\mathbf{q}) = \sum_n\left\{\frac{\left\langle m|\mathbf{e}\cdot\sum_j\hat{\mathbf{p}}_j|n\right\rangle F_{n0}(\mathbf{q})}{\omega_{nm}-\omega-\mathrm{i}0} + \frac{F_{mn}(\mathbf{q})\left\langle n|\mathbf{e}\cdot\sum_j\hat{\mathbf{p}}_j|0\right\rangle}{\omega_{n0}+\omega}\right\}. \tag{5.24}$$

In the case of elastic BrS, $|m\rangle = |0\rangle$, this function reduces to $\mathrm{i}(\mathbf{e}\cdot\mathbf{q})\alpha(\omega, q)$ in accordance with (2.14).

As in (2.1) the two terms in (5.23) correspond to the ordinary and the polarizational parts, respectively.

It has been demonstrated that in a wide region of the photon frequencies, the elastic BrS dominates over the inelastic one in the total BrS spectrum for both heavy [33, 34, 37, 56, 98, 204, 205] and light [36, 56, 98, 456] projectiles scattered on many-electron target. Semi-quantitatively, the cross sections of the elastic BrS, of both the ordinary and the polarizational nature, exceed those of the inelastic BrS by a factor of N which is the number of the target electrons. The explanation is as follows [36, 37]. During the elastic BrS the contributions of each atomic electron to the polarizational part of the total amplitude (2.1) are coherent, as in Rayleigh scattering of light. Considering the case of a neutral atom and having in mind that the ordinary part of the elastic BrS spectrum is proportional, approximately, to the squared nuclear charge, one finds that the total elastic cross section is proportional to Z^2. In contrast, during the inelastic BrS the contributions of each electrons must be summed in the cross section rather than in the amplitude (5.22). Hence, the inelastic BrS cross section is proportional to Z and is parametrically small in the case of a many-electron target, when $Z \gg 1$. For low-Z targets the contribution of both types of process to the total BrS spectrum are of the same order of magnitude [98].

The region of photon frequencies, in which the above mentioned coherence effect plays an essential role, is estimated as [36, 37, 56]

$$I < \omega < \frac{v_1}{R_{\rm at}}, \tag{5.25}$$

where I is the ionization potential of the target and v_1 is the initial velocity of the projectile.

For a heavy projectile, i.e. in the case when OBrS is negligible, the nonelastic channels, as a rule, dominate beyond the region of coherence (5.25). An exception of this rule occurs in collisions of fast heavy charged particles ($Z \gg v_1 \gg 1$) with atoms/ions collisions (as well as in atom–atom, ion–atom and ion–ion) in the region of high photon frequencies. In this region the process of inelastic BrS has a threshold, which is equal to $\omega_{\max} \approx v_1^2/2$. However, the elastic BrS takes place at higher energies, up to $\omega \geq 2v_1^2$, dominating in this region in the total photon emission spectrum. Therefore, the photon energy range $v_1^2/2 \leq \omega \leq 2v_1^2$ is convenient for the observation of the elastic PBrS [34, 205]. We note that in this ω region there is a peculiar feature in the spectrum of PBrS [241] similar to that which occurs in inelastic scattering, where it is known as the Bethe ridge [279]. In more detail this phenomenon is discussed in Sect. 4.7.

The numerical comparison of the relative role of elastic and inelastic channels in proton-atom collisions was performed in [59, 241].

In electron/positron–many-electron atom scattering elastic BrS dominates parametrically over the inelastic one in the region (5.25). However, if one is interested in accurate data on the total BrS cross section it is necessary to include the inelastic channels into the computational scheme. Up to now, mostly due to technical difficulties, numerical investigations of the role of inelastic BrS have not been as extensive as in the elastic case. The achievements in this field include the model theoretical study carried out in [428, 456] in connection with the experimental data on the intensity of the BrS spectrum in $e^- + \mathrm{Xe}$ collision as a function of the incoming electron energy ε_1 [401, 428] (see also [157–160] for the experimental data in electron collision with Ar, Kr and Xe atoms).

For low-Z targets the absolute magnitudes of both $f^{\rm ord}$ and $f^{\rm pol}$ from (2.1), and the terms $f^{(m)}_{\rm ord}$ and $f^{(m)}_{\rm pol}$ from (5.22) are all of the same order. In this case, it is the charge of the projectile which introduces peculiarities in the total radiative spectrum. It can be shown that, both for high frequencies of the photon, $\omega \gg v_1/R_{\rm at}$ [37, 98] and for the low ones, $\omega \ll 1$ [37, 237] the role of inelastic channels is negligibly small compared with the elastic BrS in the case of electron scattering, while for the positron-atom collision the situation is the opposite. It occurs mainly because of the difference in the behaviour of the interference between the ordinary and polarizational amplitudes in inelastic BrS. The interference is negative in the case of the electron projectile and positive in the positron case. The simplest way to trace this effect is to consider the inelastic BrS amplitude (5.22) in the limit of high photon frequencies, as was done above for elastic BrS. For $\omega \gg I_{1s}$ the polarizational inelastic amplitude reduces to [414]:

$$f_{\text{pol}}^{(m)} = -\frac{4\pi}{q^2}\frac{\mathbf{e}\cdot\mathbf{q}}{\omega} Z_0 F_{m0}(\mathbf{q}). \tag{5.26}$$

Using this result in (5.22) one notices that for a projectile electron ($e = -1$) the ordinary and polarizational terms cancel each other out, while for a positron ($e = +1$) the effect is opposite. A more detailed quantitative analysis [237] demonstrates that the intensity of inelastic BrS during $e^+ + H$ collision is higher by an order of magnitude than the intensity of the electron inelastic BrS. In the case of an $e^+ + H$ collision the integrated contribution of all inelastic channels is about 4 times higher than that of the elastic channel.

5.4 Polarizational BrS of Inner Electron Shells

For sufficiently large photon energies the PBrS process involves the polarization of electrons from inner shells of the target atom or ion. In this case the many-electron correlations play less important role, therefore, the dynamic response of the inner shells can be treated within frameworks of comparatively simple single-electron approximations, like the HF approximation, or even the hydrogen-line approximation. The latter, as it was demonstrated in [256] (see also [257]), being quite reliable approximation, allows one to construct analytical expressions for the PBrS cross sections applicable for further effective numerical computations.

Let us discuss in more detail the approximate method of calculation of the PBrS cross section for the photon energies in the vicinity or beyond the K- and L- ionization thresholds. We restrict ourselves to non-relativistic collision velocities and photon energies. Hence, the approach described below is applicable to low- and intermediate-Z atoms. For the sake of simplicity, we consider the collision of a structureless charged projectile with a target atom or ion. The arguments presented can be generalized to the collisions of two complex systems.

The idea of the method is to calculate the generalized dynamic polarizability $\alpha(\omega, q)$, which defines the PBrS amplitude (see 2.1), using the Coulomb Green's function and the hydrogen-like wavefunctions for the inner-shell electrons (see Sect. 3.1). This approach is reasonable for the description of the dynamic response of the electrons from the K- and L-atomic shells or in the case of the hydrogen-like ions in the region of frequencies comparable to or above the K- and L-ionization potentials. In this frequency range the electrons of these shells provide the main contribution to the polarizational BrS cross section.

The main difficulty in the description of the many-electron system when using single-electron hydrogen-like wavefunctions consists in correctly accounting for the Pauli principle. In [256] it was demonstrated that the Pauli principle can be taken into account by using the subtraction procedure, which cancels the electron transitions from the ground state to the occupied atomic levels. Hence, within the framework of this approach the exact generalized polarizability $\alpha(\omega, q)$ of the inner subshell with quantum numbers n and l (the principal and the orbital quantum numbers) is approximated as follows:

$$\alpha_{nl}(\omega, q) \approx \tilde{\alpha}_{nl}(\omega, q) - \sum_{n'l'} \tilde{\alpha}_{nl\to n'l'}(\omega, q)\,. \tag{5.27}$$

Here $\tilde{\alpha}_{nl}(\omega, q)$ is the generalized dynamic polarizability of the nl-state but calculated in the point Coulomb field of the charge Z_{eff}. The latter can be chosen, for example, from the condition $I_{nl} = Z_{\text{eff}}^2/2n^2$, i.e. the ionization potential is approximated by the corresponding hydrogen-like value, or one may crudely put $Z_{\text{eff}} = Z$ in the case of the K-shell electrons. The quantities n' and l' on the right-hand side of (5.27) are the quantum numbers of the occupied subshells in the many-electron atom the transition to which is allowed by the dipole selection rules but is forbidden by the Pauli principle. Hence, each term $\tilde{\alpha}_{nl\to n'l'}(\omega, q)$ corresponds to the contribution to $\tilde{\alpha}_{nl\to n'l'}(\omega, q)$ due to the discrete dipole transition $(n, l) \to (n', l')$. To meet the Pauli principle all these terms must be subtracted from $\tilde{\alpha}_{nl}(\omega, q)$.

The total polarizability of the atom is the sum of all $\alpha_{nl}(\omega, q)$. The approximation (5.27) fails for the subshells which exhibit a strong correlated reaction to the action of an external field. For such subshells the model of independent electrons is inapplicable. In contrast, the dynamics of electrons in the inner shells (K- and/or L-shells) is mainly governed by the action of the field of nucleus which outpowers the inter-electron correlations. Therefore, the substitution of the exact polarizability $\alpha_{nl}(\omega, q)$ with its point Coulomb analogue is justified. In many-electron atoms the ionization potentials of the K- and L-shells are well-separated. Therefore, if one is interested in the magnitude of $\alpha(\omega, q)$ in the range, for example, $\omega \sim I_K$, then one can neglect the contribution of all the shells but K to the generalized polarizability and to approximate $\alpha(\omega, q) \approx \alpha_{1s}(\omega, q) \approx \tilde{\alpha}_{1s}(\omega, q) - \sum_{n'l'} \tilde{\alpha}_{1s\to n'l'}(\omega, q)$. The analytic formulae for $\tilde{\alpha}_{nl}(\omega, q)$ and $\tilde{\alpha}_{1s\to n'l'}(\omega, q)$ can be found in [253].

In the cited chapter the applicability of the new method was established by a comparison with the cross sections calculated in the HF approximation. A good agreement was demonstrated for several case studies (see Sect. 5.4.2). The important advantage of the new method consists in the fact that the hydrogen-like model provides rather simple analytic expressions for the PBrS cross sections. These expressions allow one to describe the scaling behaviour of the cross sections and to obtain reliable values of the cross sections of the polarizational BrS in the region of high frequencies for the non-relativistic atoms or ions.

5.4.1 Scaling Behaviour for the Inner-Shell PBrS Cross Section

To analyze the scaling behaviour of the BrS cross section let us consider the collision of a fast charged particle with a non-relativistic atom. In this case the cross section of the process can be described in the Born approximation. The result of this approach reads as (see, e.g., [260])

$$\omega \frac{\mathrm{d}\sigma}{\mathrm{d}\omega} = \frac{16}{3} \frac{\omega^4 Z_0^2}{c^3 \mu^2 v_1^2} \int_{q_{\min}}^{q_{\max}} \frac{dq}{q} \left| \frac{Z - F(q)}{\omega^2} - \mu\alpha(\omega, q) \right|^2 . \quad (5.28)$$

The atomic form-factor $F(q)$ and the generalized polarizability $\alpha(\omega, q)$ have been defined in Chap. 2, the notation μ stands for the reduced mass of the colliders. The upper and lower limits of integration over the transferred momentum q are given by

$$q_{\min} = \mu v_1 \left(1 - \sqrt{1 - 2\omega/\mu v_1^2}\right), \quad q_{\max} = \mu v_1 \left(1 + \sqrt{1 - 2\omega/\mu v_1^2}\right). \quad (5.29)$$

The first and the second terms under the modulus sign in (5.28) describe the ordinary and the polarizational BrS. For electron or positron impact ($\mu \approx 1$) both contributions are of the same order of magnitude. For a heavy projectile ($\mu \gg 1$) the second term dominates the cross section. In what follows we investigate the scaling behaviour of the polarizational contribution to the total BrS cross section.

Using the fact that in the hydrogen-like approximation distance and energy can be scaled by the factors $1/Z$ and Z^2, respectively, one can derive the following scaling law for the generalized polarizability:

$$\alpha(\omega, q) = \frac{1}{Z^4} \alpha^{\mathrm{red}} \left(\frac{\omega}{Z^2}, \frac{q}{Z} \right). \quad (5.30)$$

Here we have introduced the reduced polarizability α^{red}. For the hydrogen-like ion this polarizability coincides with the usual generalized polarizability calculated with the use of the Coulomb Green's function (see [253] and Sect. 3.1). For the inner shells of a many-electron atom the Pauli principle must be taken into account. This is important, because in the many-electron system all the transitions to the occupied electron levels are forbidden. The procedure accounting for the Pauli principle is described by (5.27).

Using the result (5.30) the scaling law for the PBrS cross section can be obtained:

$$\mathrm{d}\sigma_{\mathrm{pol}}(\omega, v_1) \equiv \omega \left(\frac{\mathrm{d}\sigma(\omega, v_1)}{\mathrm{d}\omega} \right)_{\mathrm{pol}} = \frac{N^2}{Z^2} \omega \left(\frac{\mathrm{d}\sigma(\omega/Z^2, v_1/Z)}{\mathrm{d}\omega} \right)_{\mathrm{pol}}^{\mathrm{red}}, \quad (5.31)$$

where N is the number of electrons in a shell.

The scaling law (5.31) shows that there are two independent parameters characterizing the PBrS cross section. The first one is the Coulomb parameter, v_1/Z, which characterizes the rapidity of the collision (fast or slow). The second scaling parameter determines the characteristic impact parameter of the process, v_1/ω, as compared to the size of the target $\sim 1/Z$.

In the high-frequency limit the polarizability is given by asymptotic formula (2.4). The simplification of the polarizability results in the reduction of the number of independent scaling parameters. The cross section in this case depends on the single parameter, $\omega/(v_1 Z)$. The scaling behaviour of the PBrS cross section in the high-frequency limit was studied in [206].

For the sake of completeness, let us briefly discuss the scaling behaviour for the OBrS cross section. The scaling law for the atomic form-factor in the hydrogen approximation is as follows:

$$F(q) = NF^{\rm red}\left(\frac{q}{Z}\right). \tag{5.32}$$

Introducing the parameter $\xi \equiv N/Z$, one derives the following scaling law:

$$\omega\left(\frac{{\rm d}\sigma(\omega, v_1)}{{\rm d}\omega}\right)_{\rm ord} = \omega\left(\frac{{\rm d}\sigma(\omega/Z^2, v_1/Z, \xi)}{{\rm d}\omega}\right)^{\rm red}_{\rm ord}. \tag{5.33}$$

Here

$$\omega\left(\frac{{\rm d}\sigma(\omega, v, \xi)}{{\rm d}\omega}\right)^{\rm red}_{\rm ord} = \frac{16}{3}\frac{Z_0^2}{c^3\mu^2 v^2}\int_{q_{\rm min}}^{q_{\rm max}}\frac{{\rm d}q}{q}\left|1 - \xi F^{\rm red}(q)\right|^2. \tag{5.34}$$

Note that the scaling law (5.33) contains three independent scaling parameters. The two of these are the same as in the case of PBrS. The third variable ξ is responsible for the screening effect, which is essential when calculating the OBrS cross section for an atom.

5.4.2 Numerical Results

The scaling formulae (5.31) and (5.33) allow one to obtain the cross sections for different atoms and ions, collision velocities, frequency ranges, using the results of the hydrogen approach. Indeed, one has to calculate the dependence of the cross sections on the scaling parameters only once, using the hydrogen-like wavefunctions. For the hydrogen-like systems, it is possible to derive a relatively simple expression for the generalized dynamic polarizability of the ground and excited states [124, 253].

The hydrogen-like polarizability cannot be used directly for the correct description of the dynamic properties of the many-electron system. Nevertheless, many-electron effects, such as screening, many-electron correlations, the Pauli principle, can be incorporated into the hydrogen-like model. Indeed, accounting for the screening effect and many-electron correlations can be done by the appropriate choice of the effective nuclear charge. In this way one can treat part of the correlation interaction responsible for the shift of the ionization potentials of the inner shells from their hydrogen values. Note that the hydrogen-like model is essentially single particle. Therefore it can be used only in those cases when many-electron correlations are not too strong, as it is in the K- or L-atomic shells.

To incorporate the Pauli principle into the hydrogen-like model the subtraction procedure (5.27) can be applied. For example, for the K-shell of the *Ar* atom one has to subtract the terms corresponding to electron excitations from the 1s to the 2p and 3p states. These terms read as follows [256]:

$$\alpha_{1s\to 2p}(\omega, q) = \frac{2^{17}}{9(9+4q^2)^3(9-64\omega^2)}, \tag{5.35}$$

$$\alpha_{1s\to 3p}(\omega, q) = \frac{2^2 3^8(16+27q^2)}{(16+9q^2)^4(16-81\omega^2)}. \tag{5.36}$$

For atoms with semi-filled 2p or 3p shells the terms $\alpha_{1s\to 2p}$ and $\alpha_{1s\to 3p}$ have to be taken with appropriate weights. The simplest choice of the weight is the number of electrons in the subshell divided by 6 (maximum number of electrons in the np-subshell). As a result, one has to substitute into expression (5.30) the following reduced polarizability:

$$\alpha_{red}(\omega, q) = \alpha_H(\omega, q) - w_{2p}\,\alpha_{1s\to 2p}(\omega, q) - w_{3p}\alpha_{1s\to 3p}(\omega, q), \tag{5.37}$$

where $w_{2p,3p}$ are the weights, which are determined by the electron configuration of the target atom.

In Fig. 5.3 we present the results of calculations of the cross section $d\sigma_{pol}(\omega, v_1)$ of a proton in collision with Ne, Na, P and Ar atoms in the photon energy range beyond the K-shell ionization potential. The values of the initial velocity v_1 of the proton, of the effective charge of nucleus Z_{eff} (see the explanation given below) and of the weights $w_{2p,3p}$ are summarized in Table 5.1. It is seen that each collision is characterized by the same value of the Coulomb parameter $v_1/Z = 4$.

Full curves represent the PBrS cross section obtained in the HF approximation. Dotted curves represent the cross section obtained using the scaling law (5.30) and (5.31) with the hydrogen polarizability $\alpha_{red} = \alpha_H$, which does not take into account the Pauli principle. The comparison of dotted and full curves shows that the cross sections obtained in the HF and the purely hydrogen-like approximations have the same order of magnitude but do not agree well enough.

The discrepancy between the two models becomes much less when one takes into account the Pauli principle. As explained above, this can be done by using the subtraction procedure (5.37). The results of the calculation of $d\sigma_{pol}(\omega, v_1)$ in this case are plotted as broken curves. It is clearly seen from the figures that the broken curves reproduce the behaviour of the HF cross sections reasonably well. The discrepancy between the two approaches is comparable with the accuracy of the HF method itself.

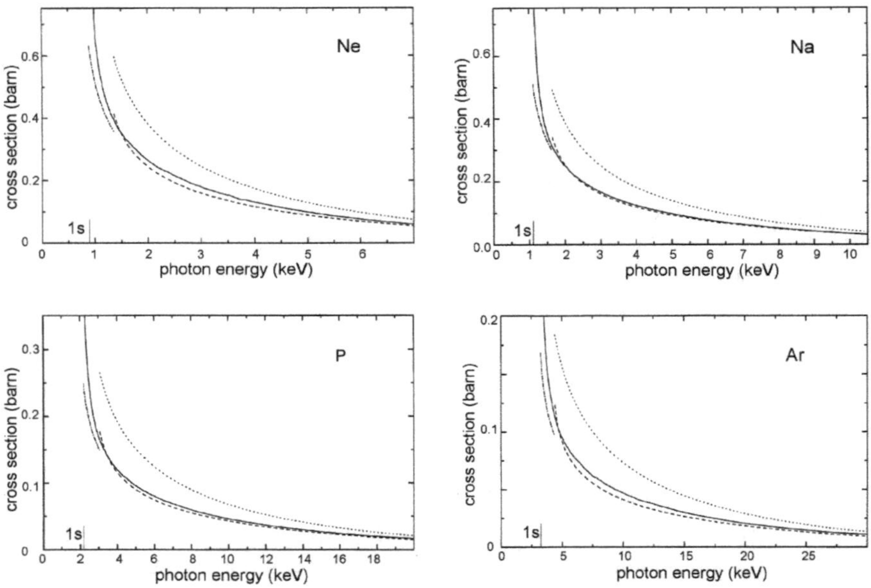

Fig. 5.3 The PBrS cross sections $d\sigma_{pol}(\omega, v_1)$ form in collisions of a proton with Ne, Na, P and Ar atoms for the photon energies above the K-shell ionization potential I_{1s} (marked by *vertical lines*) [256]. The collision velocities are listed in Table 5.1. *Full curve*, the cross section obtained in the Hartree-Fock approximation. *Dotted curve*, the cross section obtained in the hydrogen-like model (with $Z_{eff} = Z$) without the Pauli principle taken into account. *Broken curve*, the cross section obtained in the hydrogen-like model ($Z_{eff} = Z$) with the Pauli principle taken into account. *Chain curve*, the cross section obtained in the hydrogen-like model (with Z_{eff} calculated from $I_{1s} = Z_{eff}^2/2$, see Table 5.1 and explanation in the text) with the Pauli principle taken into account

Table 5.1 Nuclear charge Z, effective charge Z_{eff}, weights $w_{2p,3p}$ and initial velocity v_1 of the proton in collisions with Ne, Na, P and Ar atoms used to calculate the PBrS spectra in Fig. 5.3

	Z	Z_{eff}	v_1(a.u.)	w_{2p}	w_{3p}
Ne	10	8.1	40	1	0
Na	11	9	44	1	0
P	15	12.65	60	1	0.5
Ar	18	15.4	72	1	1

The deviation of the hydrogen approximation from the HF one becomes significant in the vicinity of the ionization potential of the K-shell. Shifting of the ionization potential of the K-shell calculated in the hydrogen-like approximation from its HF value occurs due to the many-electron correlation interaction. This phenomenon can be taken into account within the framework of the hydrogen-like model by the appropriate effective charge of the nucleus defined from $I_{1s} = Z_{eff}^2/2$ where I_{1s} can be either calculated within the HF scheme or taken from the experiment. The PBrS cross sections calculated with the appropriate effective nuclear charges in the frequency range between the HF and the hydrogen-like ionization potentials are shown in the figures by chain curves. It is seen that the continuation of the chain curves smoothly matches the broken curves in the vicinity of the hydrogen-like ionization potential. This means that the hydrogen-like approach provides a simple and effective method for the PBrS cross section calculations applicable over the whole range of photon frequencies above the ionization potentials of the inner atomic shells. Note that in the vicinity of the K-shell ionization potential the discrepancy between the hydrogen-like and the HF approximations increases. This can be a consequence of the fact that the HF approximation itself does not properly take the many-electron correlations into account.

Our method can be extended for heavy atomic systems, where relativistic effects are important. In this case a consistent many-body treatment of the polarizational BrS process is a rather difficult problem (see Sect. 6.5) and the hydrogen-like approximation seems to be even more appropriate compared to the non-relativistic case.

5.5 BrS in Slow Collisions of Atomic Particles

In this section we consider the BrS process in collisions of slow atomic particles. It is demonstrated that PBrS dominates over the well-known molecular orbital X-ray emission [173] in the region of large impact parameters and small photon energies. The contribution of the PBrS channel to the total radiation spectrum is significant in a broad range of ω.

5.5.1 BrS Amplitude

The overlapping electronic shells of two slowly moving atomic particles (atoms or ions) form the quasi molecule (QM) orbitals during the collision. The evolution of such system can be described within the framework of the adiabatic perturbation theory [320]. According to this formalism, the wavefunction of the system is sought the superposition of the orbitals:

$$\Psi_{\rm i}(\mathbf{r}, t) = \sum_n A_n(t)\, \Phi_n(\mathbf{r}, \mathbf{R}(t)) \exp\left(-{\rm i}\int_{-\infty}^{t} E_n(\tau)\, {\rm d}\tau\right), \tag{5.38}$$

where the QM orbitals $\Phi_n(\mathbf{r}, \mathbf{R}(t))$ are the solutions of the stationary two-center Schrödinger equation:

$$\hat{H}(\mathbf{r}, \mathbf{R})\, \Phi_n(\mathbf{r}, \mathbf{R}) = E_n(R)\, \Phi_n(\mathbf{r}, \mathbf{R}). \tag{5.39}$$

Here $\hat{H}(\mathbf{r};\mathbf{R})$ is the two-center Hamiltonian of QM; $\mathbf{r}$ denotes the coordinates of the electronic subsystem. The internuclear distance $\mathbf{R} = \mathbf{R}(t)$ is the slowly varying parameter of the adiabatic theory. The energy levels of $E_n(R)$ depend on the adiabatic parameter R.

At each instant t the solutions of the Schrödinger equation (5.39) define the wave-functions with corresponding to fixed $\mathbf{R}(t)$. The amplitudes $A_n(t)$, which enter (5.38), are the solutions of the following system of equations:

$$\dot{A}_n(t) = \sum_{n'} \langle n| \left(-\frac{\mathrm{d}}{\mathrm{d}t}\right) |n'\rangle \exp\left(\mathrm{i}\int_{-\infty}^{t} (E_n - E_{n'})\,\mathrm{d}\tau\right) A_{n'}, \tag{5.40}$$

where

$$\langle n| \left(-\frac{\mathrm{d}}{\mathrm{d}t}\right) |n'\rangle = -\int \mathrm{d}^3\mathbf{r}\, \Phi_n^*(\mathbf{r},\mathbf{R})\, \dot{\mathbf{R}} \cdot \frac{\partial \Phi_n(\mathbf{r},\mathbf{R})}{\partial \mathbf{R}}. \tag{5.41}$$

The wave functions (5.38) depend on the initial conditions of the system. These conditions are used for obtaining the correct amplitude $A_n(t)$. The subscript i in the notation $\Psi_{\mathrm{i}}(\mathbf{r}, t)$ indicates that the wavefunction is derived with the certain initial conditions.

The adiabatic perturbation theory if the collision velocity is small enough ensuring that the expression in the right-hand side in (5.40) is small and can be omitted in the lowest order of the perturbation theory. In this case, the wave function of the system is simplified:

$$\Psi_{\mathrm{i}}(\mathbf{r},\mathbf{R}) = \Phi_{n_{\mathrm{i}}}(\mathbf{r},\mathbf{R}(t)) \exp\left(-\mathrm{i}\int_{-\infty}^{t} E_{n_{\mathrm{i}}}(\tau)\,\mathrm{d}\tau\right). \tag{5.42}$$

The account of the initial and the final condition is important for the adiabatic calculations, because it has significant implications for the dynamic behaviour of the quasi-molecule during the collision. Very often, only two QM orbitals appear to be essential for the radiative process. Following [388], below in this section we discuss examples of this kind. Prior to doing this let us outline general formulae describing the radiation process within the framework of the two-orbital model.

Within this approach the wavefunction of the system can be written in the form

$$\Psi_{\mathrm{i}}(\mathbf{r},\mathbf{R}) = \frac{1}{\sqrt{2}}\left[\Phi_{\mathrm{g}}(\mathbf{r},\mathbf{R})\,\mathrm{e}^{-\mathrm{i}\int_{-\infty}^{t} E_{\mathrm{g}}(\tau)\mathrm{d}\tau} \pm \Phi_{\mathrm{u}}(\mathbf{r},\mathbf{R})\,\mathrm{e}^{-\mathrm{i}\int_{-\infty}^{t} E_{\mathrm{u}}(\tau)\mathrm{d}\tau}\right]. \tag{5.43}$$

The wavefunctions $\Phi_{\mathrm{g;u}}(\mathbf{r};\mathbf{R})$ are the superpositions In the case of the two-orbital model the superposition has the following form [320]:

$$\Phi_{\mathrm{g;u}} = \frac{1}{\sqrt{2}}(\Psi_1 \pm \Psi_2). \tag{5.44}$$

These functions are valid at least in the region of large internuclear distances $R(t)$. The wavefunction $\Psi_{\rm i}(\mathbf{r};\,\mathbf{R})$ with $\Phi_{\rm g;u}$ from (5.44) satisfies the initial conditions

$$\Psi_{\rm i}(\mathbf{r},\mathbf{R}) \overset{t\to-\infty}{\longrightarrow} \Psi_{1;2} \tag{5.45}$$

The choice of sign (+ or −) in (5.43) depends on which state, Ψ_1 or Ψ_2, is considered as the initial one.

The final state wavefunction $\Psi_{\rm f}(\mathbf{r},\mathbf{R})$, is obtained analogously:

$$\Psi_{\rm f}(\mathbf{r},\mathbf{R}) = \frac{1}{\sqrt{2}}\left[\Phi_{\rm g}(\mathbf{r},\mathbf{R})\,{\rm e}^{{\rm i}\int_t^{+\infty} E_{\rm g}(\tau){\rm d}\tau} \pm \Phi_{\rm u}(\mathbf{r},\mathbf{R})\,{\rm e}^{{\rm i}\int_t^{+\infty} E_{\rm u}(\tau){\rm d}\tau}\right]. \tag{5.46}$$

This wavefunction satisfies the condition

$$\Psi_{\rm f}(\mathbf{r},\mathbf{R}) \overset{t\to+\infty}{\longrightarrow} \Psi_{1;2}. \tag{5.47}$$

The radiation transition between i → f is described by the matrix element:

$$M = -\int_{-\infty}^{+\infty} {\rm d}t\,\left\langle \Psi_{\rm f}(\mathbf{r},\mathbf{R}) \left| \sum_{j=1}^{N} {\rm e}^{-{\rm i}\mathbf{k}\cdot\mathbf{r}_j}\mathbf{e}\cdot\hat{\mathbf{p}}_j \right| \Psi_{\rm i}(\mathbf{r},\mathbf{R}) \right\rangle {\rm e}^{{\rm i}\omega t}. \tag{5.48}$$

Here $\mathbf{r}_j$, $\hat{\mathbf{p}}_j$ are the coordinates and the momentum operators of electrons in the QM center-of-mass frame, and N is the number of electrons. In the dipole approximation ($kr_j \ll 1$) the right-hand side of (5.48) reduces to the matrix element of the dipole moment of the quasi-molecule. Below we consider the dipole-photon limit.

Supposing, for the sake of clarity, that the system was initially at the state Ψ_1, one substitutes the wavefunctions (5.43) and and (5.46) into (5.48), and, having performed the transformations, derives

$$M = -{\rm i}\omega\,\mathbf{e}\cdot\mathbf{I},$$

where

$$\mathbf{I} = \mathbf{I}_{\rm gg} \pm \mathbf{I}_{\rm uu} + \mathbf{I}_{\rm gu} \pm \mathbf{I}_{\rm ug}. \tag{5.49}$$

The vectors $\mathbf{I}_{\rm gg}$, $\mathbf{I}_{\rm gu}$ are defined as follows:

$$\mathbf{I}_{\rm gg} = \frac{1}{2}\exp\left(-{\rm i}\int_{-\infty}^{+\infty} E_{\rm g}{\rm d}\tau\right)\int_{-\infty}^{+\infty} {\rm d}t\,{\rm e}^{{\rm i}\omega t}\,\mathbf{d}_{\rm gg}(t), \tag{5.50}$$

$$\mathbf{I}_{\rm gu} = \frac{1}{2}\exp\left(-{\rm i}\int_{-\infty}^{+\infty} E_{\rm g}{\rm d}\tau\right)\int_{-\infty}^{+\infty} {\rm d}t\,{\rm e}^{{\rm i}\left[\omega t+\int_{-\infty}^{t}(E_{\rm g}-E_{\rm u}){\rm d}\tau\right]}\,\mathbf{d}_{\rm gu}(t). \tag{5.51}$$

Here $\mathbf{d}_{gg}(t)$ is the time-dependent dipole moment of the QM g-orbital, and $\mathbf{d}_{gu}$ is the time-dependent non-diagonal matrix element of $\mathbf{d}$. The vectors $\mathbf{I}_{uu}$ and $\mathbf{I}_{ug}$ in (5.49) are obtained from $\mathbf{I}_{gg}$ and $\mathbf{I}_{gu}$ by replacing the subscripts u and g. The upper sign in (5.49) corresponds to the radiative process without changing the initial state Ψ_1 of the system. The lower sign describes the radiative transition to the final state Ψ_2.

The four components of the amplitude (5.49) correspond to the four different radiative channels available in the two-orbital system. The first two terms describe the radiation emitted by the dipole moments of g and u orbitals, alternating during the collision. These two terms are associated with the PBrS mechanism. The third term in (5.49) describes the radiative de-excitation, and is associated with molecular orbital X-ray process. The last term is related to the radiative process accompanied by the excitation of QM. We note that all these terms contribute to the spectrum of photons generated in the collisions with large impact parameters.

5.5.2 BrS Cross Section

Using the relation of between the amplitude and the cross section, one derives the following expression for the triply differential cross section:

$$\frac{\mathrm{d}^3\sigma}{\mathrm{d}\omega \mathrm{d}\Omega_{\mathbf{k}} \mathrm{d}\rho} = \frac{\rho\omega^3}{2\pi c^3}\left(|\mathbf{I}|^2 - |\mathbf{n_k}\cdot\mathbf{I}|^2\right). \tag{5.52}$$

Here $\Omega_{\mathbf{k}}$ is the solid angle of the photon, $\mathbf{n_k} = \mathbf{k}/k$, and ρ is the impact parameter of the collision.

Let us consider the collision geometry as it is shown in Fig. 5.4. The scattering process takes place in the (xz) plane. The z-axis is directed along the initial collision velocity $\mathbf{v}_0$. The solid curve represents the trajectory $\mathbf{R}(t)$ of the nuclei in the center-of-mass frame. The vector $\mathbf{d}_{xz}$ is the projection of the QM dipole momentum $\mathbf{d}$ onto the scattering plane. Assuming that $\mathbf{d}$ has also the non-zero y-component, one writes the components of the vector $\mathbf{I}$ as follows:

$$I_x = \frac{1}{2}\sum_{jk} \mathrm{sign}_{jk}\, \mathrm{e}^{-\mathrm{i}\int_{-\infty}^{+\infty} E_j \mathrm{d}\tau} \int_{-\infty}^{+\infty} \mathrm{d}t\, \mathrm{e}^{\mathrm{i}(\omega t + \int_{-\infty}^{t} \omega_{jk}(\tau)\mathrm{d}\tau)} d_{xz}^{jk} \sin\Phi_{jk}, \tag{5.53}$$

$$I_z = \frac{1}{2}\sum_{jk} \mathrm{sign}_{jk}\, \mathrm{e}^{-\mathrm{i}\int_{-\infty}^{+\infty} E_j \mathrm{d}\tau} \int_{-\infty}^{+\infty} \mathrm{d}t\, \mathrm{e}^{\mathrm{i}(\omega t + \int_{-\infty}^{t} \omega_{jk}(\tau)\mathrm{d}\tau)} d_{xz}^{jk} \cos\Phi_{jk} \tag{5.54}$$

$$I_y = \frac{1}{2}\sum_{jk} \mathrm{sign}_{jk}\, \mathrm{e}^{-\mathrm{i}\int_{-\infty}^{+\infty} E_j \mathrm{d}\tau} \int_{-\infty}^{+\infty} \mathrm{d}t\, \mathrm{e}^{\mathrm{i}(\omega t + \int_{-\infty}^{t} \omega_{jk}(\tau)\mathrm{d}\tau)} d_{y}^{jk}. \tag{5.55}$$

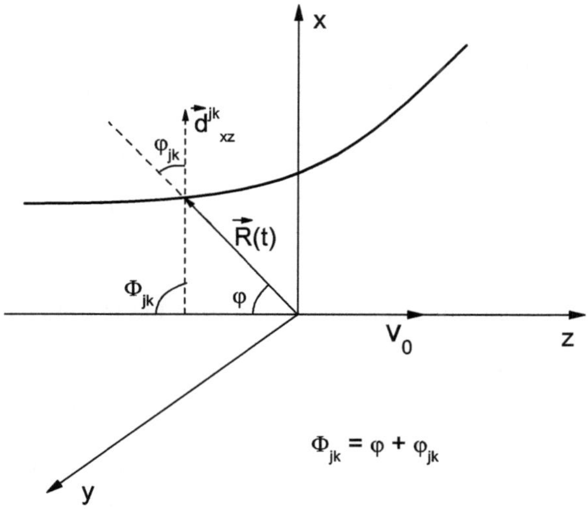

Fig. 5.4 Geometry of the collision

In these formulae the summation is performed over the available radiative channels $(jk) = \{gg, uu, gu, ug\}$. The signs ($\mathrm{sign}_{jk}$) are chosen in accordance with (5.49). The meaning of the angles Φ_{jk} is clear from Fig. 5.4: $\Phi_{jk} = \varphi + \varphi_{jk}$. Integrating (5.52) over the photon azimuth angle, one arrives at

$$\frac{\mathrm{d}^3\sigma}{\mathrm{d}\omega \mathrm{d}\cos\theta_{\mathbf{k}} \mathrm{d}\rho} = \frac{2\rho\omega^3}{3c^3} |I|^2 \left[1 + \left(1 - \frac{3|I_z|^2}{2|I|^2}\right)\left(3\cos^2\theta_{\mathbf{k}} - 1\right)\right]. \tag{5.56}$$

Here $|I|^2 = |I_x|^2 + |I_y|^2 + |I_z|^2$ and $\theta_{\mathbf{k}}$ is the polar angle of the photon emission. For further transformations, it is convenient to express $\sin\Phi_{jk}$ and $\cos\Phi_{jk}$ in (5.53) and (5.54) via the sums of exponents and instead of I_x and I_z to use the integrals I_1 and I_2, defined as

$$I_1 = \frac{1}{2}\sum_{jk} \mathrm{sign}_{jk}\, \mathrm{e}^{-\mathrm{i}\int_{-\infty}^{+\infty} E_j \mathrm{d}\tau} \int_{-\infty}^{+\infty} \mathrm{d}t \mathrm{e}^{\mathrm{i}(\omega t + \int_{-\infty}^{t} \omega_{jk}(\tau)\mathrm{d}\tau + \varphi(t))} \left[d_{xz}^{jk} e^{i\varphi_{jk}(t)}\right], \tag{5.57}$$

$$I_2 = \frac{1}{2}\sum_{jk} \mathrm{sign}_{jk}\, \mathrm{e}^{-\mathrm{i}\int_{-\infty}^{+\infty} E_j \mathrm{d}\tau} \int_{-\infty}^{+\infty} \mathrm{d}t \mathrm{e}^{\mathrm{i}(\omega t + \int_{-\infty}^{t} \omega_{jk}(\tau)\mathrm{d}\tau - \varphi(t))} \left[d_{xz}^{jk} e^{-i\varphi_{jk}(t)}\right], \tag{5.58}$$

$$I_3 = I_y. \tag{5.59}$$

For the sake of consistency the notation I_3 has been introduced for the integral I_y. The quantities $|I|^2$ and $|I_z|^2$ are expressed in terms of $I_{1,2,3}$ as follows:

$$|I|^2 = \frac{|I_1|^2 + |I_2|^2}{2} + |I_3|^2, \quad |I_z|^2 = \frac{|I_1|^2 + |I_2|^2 + 2\mathrm{Re}(I_1^* I_2)}{4}. \tag{5.60}$$

There are two reasons for altering the orientation of the dipole momentum **d** during the collision. First, the dipole moment changes due to the variation of the internuclear distance $R(t)$ in the proper frame of the QM. Second, **d** rotates with the proper frame during the collision. The time dependence of the angle $\varphi(t)$ (see Fig. 5.4) describing this rotation, can be obtained using the angular momentum conservation law:

$$\varphi(t) = v_0 \rho \int_{-\infty}^{t} \frac{\mathrm{d}t}{R^2(t)}. \tag{5.61}$$

This angle enters the phase factors in the integrands in (5.57)–(5.58). The angles $\varphi_{jk}(t)$ describe the orientation of the dipole moment with respect to the position vector $\mathbf{R}(t)$. They are either constants or slowly varying functions of time. This is why they are combined with the dipole matrix elements in (5.57)–(5.58). The extraction of the rapidly varying phase factors in the aforementioned integrals is necessary for their further evaluation by means of the stationary phase method.

Integrating the cross section (5.56) over $\theta_{\mathbf{k}}$, one derives:

$$\frac{\mathrm{d}^2\sigma}{\mathrm{d}\omega \mathrm{d}\rho} = \frac{4\rho\omega^3}{3c^3} |I|^2 . \tag{5.62}$$

The formulae presented above are written within the framework of the model of two coupled orbitals. They may be simply generalized to the case of many orbitals.

5.5.3 Hydrogen-Like System

Let us consider the radiative processes in a hydrogen-like system using the formalism developed in Sect. 5.5.3. The hydrogen-like system consists of the projectile nucleus of the charge Z_1 and the hydrogen-like ion as a target. The nuclear charge of the target is Z_2.

The wavefunctions and the energy levels of the hydrogen-like system are well-known (see, for example, [81]). For instance, in the case $Z_1 = Z_2 = Z$ the wavefunctions of the two lowest orbitals have the form of (5.44) in the region of large distances, where Ψ_1 and Ψ_2 are the atomic wave functions describing the electron in the vicinity of the projectile and the target nuclei respectively. The transition energy between g and u orbitals as a function of $R(t)$, according to [81], is

$$\omega_{\mathrm{gu}} = -\frac{4}{a_0^2} R(t)\, \mathrm{e}^{-R(t)/a_0 - 1}. \tag{5.63}$$

Here $a_0 = 1/Z$ is the radius of the ground state orbit in an isolated ion.

Taking the wavefunctions $\Phi_{g;u}$ as in (5.44) and substituting them in the expressions for the dipole matrix elements, one arrives at

$$\mathbf{d}_{\mathrm{gu}}^{(0)} = \mathbf{d}_{\mathrm{ug}}^{(0)} = -\frac{\mathbf{R}(t)}{2}, \qquad \mathbf{d}_{\mathrm{gg}}^{(0)} = \mathbf{d}_{\mathrm{uu}}^{(0)} = \frac{M_1 - M_2}{M_1 + M_2} \frac{\mathbf{R}(t)}{2}. \tag{5.64}$$

To derive these expressions we have assumed that the nuclei have different masses $M_{1;2}$ and that the functions $\Psi_{1;2}$ are the ground state functions in the isolated ion. In addition, in the second relation we have omitted the exponentially small corrections proportional to $\exp(-R/a_0)$. The corrections to the dipole moment which arise due to the ion-nucleus interaction are:

$$\mathbf{d}_{\mathrm{gg}}^{(1)} = \mathbf{d}_{\mathrm{uu}}^{(1)} = \frac{\alpha_1 \mathbf{E}_2 + \alpha_2 \mathbf{E}_1}{2}, \qquad \mathbf{d}_{\mathrm{gu}}^{(1)} = \mathbf{d}_{\mathrm{ug}}^{(1)} = \frac{\alpha_1 \mathbf{E}_2 - \alpha_2 \mathbf{E}_1}{2}. \tag{5.65}$$

Here $\alpha_{1;2} = 9a_{1;2}^3/2$ are the static dipole polarizabilities of the hydrogen-like ions with $Z_{1;2}$ [229], $a_{1;2}$ are the orbits radii. The electric field of the first (second) nucleus, acting on the ion partner, has been notated as $\mathbf{E}_{1;2} = \mp Z_{1;2}\mathbf{R}/R^3$. Let us note that the terms in (5.65) are not exponentially small but inversely proportional to the internuclear distance R.

Let us now evaluate the integrals or (5.57)–(5.59) by the stationary phase method [320]. The stationary phase equations read

$$\omega + \omega_{jk}(t) + \frac{v_0 \rho}{R(t)} = 0, \tag{5.66}$$

$$\omega + \omega_{jk}(t) - \frac{v_0 \rho}{R(t)} = 0, \tag{5.67}$$

$$\omega + \omega_{jk}(t) = 0. \tag{5.68}$$

The last term in (5.66)–(5.67) is the frequency of the QM dipole moment rotating during the collision (see (5.61)). When the rotating frequency is negligible, one arrives at the stationary phase equation known in the molecular orbital X-ray theory [173]. In this limit, the traditional formulae of molecular orbital X-ray theory follow from our considerations (see Sect. 5.5.4). In the opposite case, when the rotation frequency exceeds the transition frequency between levels of the QM, the second terms in (5.66)–(5.67) can be omitted. Then the radiation process differs from the molecular orbital X-ray one. This regime occurs in the impact parameter region

$$\frac{4\rho^2}{a_0^2} \mathrm{e}^{-\rho/a_0 - 1} \ll v_0, \tag{5.69}$$

as follows from (5.63) and (5.66)–(5.67).

Relation (5.69) shows that the rotation frequency of the QM dipole moment becomes larger than the transition frequency between levels in the region of large impact parameters.

Assuming the inequality (5.69) is met, let us derive the radiation spectrum. Substituting the matrix elements (5.64)–(5.65) into (5.53)–(5.55), one derives the result:

$$I_x = F_\pm \left(\alpha_1 - \frac{Z_1 - 1}{M_1\omega^2}\right)\frac{\mu\omega^2}{Z_1 - 1} X_\omega, \tag{5.70}$$

$$I_z = F_\pm \left(\alpha_1 - \frac{Z_1 - 1}{M_1\omega^2}\right)\frac{\mu\omega^2}{Z_1 - 1} Z_\omega, \tag{5.71}$$

$$I_y = 0. \tag{5.72}$$

Here X_ω and Z_ω are the Fourier transforms of x and z components of the vector $\mathbf{R}(t)$. The amplitudes $F_\pm$ are defined as

$$F_\pm = \frac{1}{2}\left(\mathrm{e}^{-\mathrm{i}\int_{-\infty}^{+\infty} E_\mathrm{g} \mathrm{d}\tau} \pm \mathrm{e}^{-\mathrm{i}\int_{-\infty}^{+\infty} E_\mathrm{u} \mathrm{d}\tau}\right). \tag{5.73}$$

The quantity F_- is the transition amplitude which accounts for the electron capture whereas F_+ is the transition amplitude without the capture (see, for example, [81]). The sum of modulus squares of these amplitudes is equal to one.

To derive the integrals (5.70)–(5.72) the nuclei equation of motion was accounted for:

$$\mu\ddot{\mathbf{R}} = (Z_1 - 1)Z_2 \frac{\mathbf{R}}{R^3}, \tag{5.74}$$

where $\mu = M_1M_2/(M_1 + M_2)$ is the reduced mass. The charge of the first nucleus is reduced by one due to the electron screening. Such an account of screening is correct at the distances exceeding greatly the ion radius.

Using the obtained expressions (5.70)–(5.72) and the well-known formulae for the Fourier transforms of the trajectory in a Coulomb field [279], one obtains

$$\begin{aligned}\frac{\mathrm{d}^2\sigma}{\mathrm{d}\omega\mathrm{d}\rho} &= \frac{4\pi^2 Z_2^2\omega^5}{3c^3 v_0^4}\left(\frac{Z_1 - 1}{\omega^2}\left(\frac{Z_1 - 1}{M_1} - \frac{Z_2}{M_2}\right) - \alpha_1\right)^2 \\ &\times \mathrm{e}^{2\pi\nu}\rho\left(\left|H_{\mathrm{i}\nu}^{(1)\prime}(\mathrm{i}\nu\varepsilon)\right|^2 - \frac{\varepsilon^2 - 1}{\varepsilon^2}\left|H_{\mathrm{i}\nu}^{(1)}(\mathrm{i}\nu\varepsilon)\right|^2\right)|F_\pm|^2. \end{aligned} \tag{5.75}$$

The contribution of the OBrS of the nuclei has been added in (5.75). The parameters ε and ν are equal to: $\varepsilon = \sqrt{1 + \rho^2/R_0^2}$ and $\nu = \omega R_0/v_0$, where $R_0 = 2Z(Z - 1)/\mu v_0^2$ is the turning point of the collision, $H_{i\nu}^{(1)}(i\nu\varepsilon)$ is the Hansel function of the first kind, $H_{i\nu}^{(1)\prime}(i\nu\varepsilon)$ is its derivative with respect to the argument. The factors $|F_+|^2$ and $|F_-|^2$ are equal to

$$|F_+|^2 = \cos^2\eta, \qquad |F_-|^2 = \sin^2\eta, \quad \text{with} \quad \eta = \frac{1}{2}\int_{-\infty}^{+\infty} \omega_{\mathrm{gu}}\,\mathrm{d}\tau\,. \tag{5.76}$$

The obtained cross section can be integrated over the impact parameter from ρ_0 up to infinity. The lower integration limit follows from (5.69) and is equal, in the order of magnitude, to the ionic radius. In addition to the integration the summation over the electron final states can be carried out. The result reads:

$$\frac{\mathrm{d}\sigma}{\mathrm{d}\omega} = \frac{4\pi^2 Z_2^2 \omega^3}{3c^3 v_0^2} \left(\frac{Z_1 - 1}{\omega^2} \left(\frac{Z_1 - 1}{M_1} - \frac{Z_2}{M_2} \right) - \alpha_1 \right)^2 \times \mathrm{e}^{2\pi \nu} \varepsilon \nu \left| H_{i\nu}^{(1)}(\mathrm{i}\nu\varepsilon) H_{\mathrm{i}\nu}^{(1)\prime}(\mathrm{i}\nu\varepsilon) \right|. \tag{5.77}$$

To derive (5.77) the known properties of the Hansel function and its derivative were used (see, for example, [171]). In the limiting case $\omega \ll v_0/R_0$ the parameter ν is small. In this case, the cross section can be simplified further. Within the logarithmic accuracy one writes

$$\frac{\mathrm{d}\sigma}{\mathrm{d}\omega} = \frac{16 Z_2^2 \omega^3}{3c^3 v_0^2} \left(\frac{Z_1 - 1}{\omega^2} \left(\frac{Z_1 - 1}{M_1} - \frac{Z_2}{M_2} \right) - \alpha_1 \right)^2 \ln \frac{2v_0}{\gamma \omega \sqrt{R_0^2 + \rho_0^2}}. \tag{5.78}$$

Here $\gamma = e^{\mathscr{C}}$ where $\mathscr{C}$ is the Euler constant. The logarithmic factor in (5.78) is a result of the integration over the region of large impact parameters. This is the only factor dependent on the cut-off radius ρ_0. Thus, the spectrum weakly (logarithmically) depends on the choice of the cut-off radius.

The second term is the contribution of PBrS which appears due to the polarization of the electron cloud in the field of the collision partner. The contribution of nuclear bremsstrahlung (the OBrS part) is negligible in comparison with that of PBrS in the region

$$\omega^2 \gg \frac{Z_1 - 1}{\alpha_1} \left(\frac{Z_1 - 1}{M_1} - \frac{Z_2}{M_2} \right). \tag{5.79}$$

The upper limit for the emitted photon frequency consistent with the discussed model can be derived from (5.69). Using there $\rho = v_0/\omega$ (these values of the impact parameter contribute most to the logarithm in (5.78)) one arrives at the condition:

$$\omega \ll v_0/\rho_0. \tag{5.80}$$

This inequality shows that (5.78) is valid when the argument of the logarithm is large. The inequalities (5.79) and (5.80), which can be fulfilled simultaneously, define the region of photon energies within which PBrS is the predominant radiation mechanism.

We have described the radiation process in which the electron was initially bound to the nucleus Z_1. If the electron was initially located in the vicinity of the nucleus Z_2 the corresponding formulae one obtains by interchanging the indices 1 and 2. For instance, (5.78) transforms into

$$\frac{d\sigma}{d\omega} = \frac{16Z_2^1\omega^3}{3c^3v_0^2}\left(\frac{Z_2-1}{\omega^2}\left(\frac{Z_2-1}{M_2}-\frac{Z_1}{M_1}\right)-\alpha_2\right)^2 \ln\frac{2v_0}{\gamma\omega\sqrt{R_0^2+\rho_0^2}}. \tag{5.81}$$

The formulae (5.78) and (5.81) show that the intensity of the PBrS radiation is proportional to the square of the strength of the Coulomb field of the nucleus (Z_1 or Z_2) and the polarizability of the ion (α_2 or α_1).

Equations (5.77)–(5.78) and (5.81) represent the total radiation cross section, summed over all final states of the electron. Let us discuss briefly the partial cross sections of the processes with and without electron capture. These cross sections are defined by the factors $|F_\pm|^2$ (see (5.75) and (5.76)). The integrals from (5.76) can be evaluated approximately. To do this one substitutes in the integral η the expression for ω_{gu} from (5.76) and performs the integration analogously to the standard radiative electron capture theory (see, for example, [279]). The result reads:

$$\eta \approx \frac{2\sqrt{2\pi}}{v_0}\left(\frac{\rho}{a_0}\right)^{3/2} e^{-\rho/a_0-1}. \tag{5.82}$$

This relation was derived assuming the projectile moves along the straight line. For the impact parameters $\rho \sim v_0/\omega$, which is the most important for the photon emission, the parameter η is much less than one. Therefore, the values of $|F_\pm|^2$ can be approximated as follows:

$$|F_+|^2 \approx 1 \gg |F_-|^2 \approx \eta^2. \tag{5.83}$$

Hence, the radiation process without capture is the predominant channel. The differential cross-section of the radiative capture contains the additional small factor η^2 which is the probability of the electron capture in the collision.

The presented formulae are applicable for the description of the adiabatic hydrogen-like system. Nevertheless, they can be easily generalized to the collisions involving many electron ions. Indeed, the expressions (5.78) and (5.81), resulting from the adiabatic consideration of the PBrS process, correspond to the results obtained within the framework of the Born approximation which is applicable to fast collisions (see [33, 34] and Sect. 5.2). In comparison with the results of the Born theory of BrS it follows that in the slow collisions all dynamic quantities, like dynamic polarizabilities of the colliding particles, are replaced with their static values. The atomic form-factors $Z - F(q)$ must be substituted with the net charges of the nucleus $Z - N$ (where N is the number of atomic/ionic electrons). Keeping these remarks in mind, one can start with (5.17), initially obtained in [33, 34], and derive:

$$\frac{d\sigma}{d\omega} = \frac{16\omega^3}{3c^3v_0^2}\left[\frac{(Z_1-N_1)(Z_2-N_2)}{\omega^2}\left(\frac{Z_2-N_2}{M_2}-\frac{Z_1-N_1}{M_1}\right)\right.$$
$$\left. + (Z_2-N_2)\alpha_1 - (Z_1-N_1)\alpha_2\right]^2 \ln\frac{2v_0}{\gamma\omega\sqrt{R_0^2+\rho_0^2}}. \tag{5.84}$$

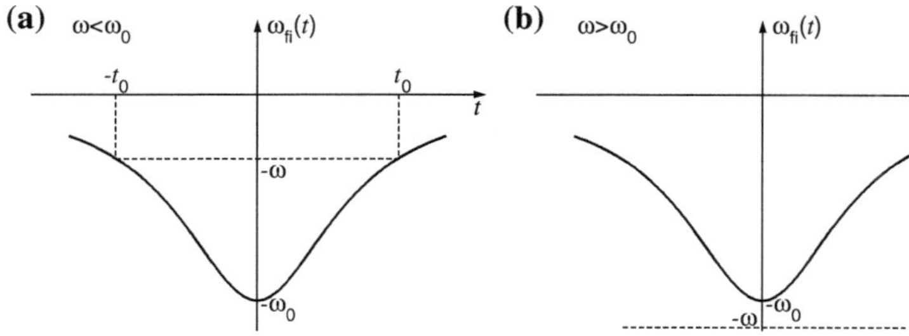

Fig. 5.5 Time dependence of the transition frequency $\omega_{fi}(t)$

Here $N_{1;2}$ are the number of electrons in the colliding ions/atoms. Note opposite signs of the terms containing the polarizabilities α_1 and α_2: thy indicate that the induced dipole moments of the colliding ions are oriented in the opposite direction (see Sect. 5.2.2 for more details).

Results of numerical calculations of the PBrS arising in collisions of a slow proton with the helium ion He^+ can be found in [388].

5.5.4 Molecular Orbital X-Rays

In this section we demonstrate that the well-known formulae of molecular orbital X-ray theory follow from our considerations. The rotation of the QM dipole moment turns out to be very important for the photon emission during distant collisions, as it was demonstrated in Sect. 5.5.3. Now let us consider the opposite case, when the rotation of the QM dipole moment is negligible. In this case the last terms in the stationary phase equations (5.66)–(5.67) can be omitted. This is correct for the photon energies satisfying the condition:

$$\omega \gg v_0/\rho_0. \tag{5.85}$$

This inequality is clear from the speculations in respect of (5.80). The stationary phase equations (5.66)–(5.67) reduce to

$$\omega + \omega_{jk}(t) = 0, \tag{5.86}$$

In a two-level system (5.86) has real roots only for the transition $g \leftarrow u$ with the transition frequency $\omega_{gu} < 0$. As it is seen from Fig. 5.5 The stationary phase equation may have either two real solutions (left panel), which can merge to a single solution for $\omega = \omega_0$ (not illustrated in the figure), or no real solutions (right panel).

The terms, which have stationary phase points in the integrands, most contribute to the sums on the right-hand sides in (5.57)–(5.59). Applying the stationary phase method one neglects the terms $\varphi(t)$ in the phase factors and, then, expands the factors up to the second order in the vicinity of the stationary phase points. Assuming the main contribution to the integral arise from the regions near the stationary phase points, one carries out the integration. This calculation leads to the following result for the cross section:

$$\frac{d\sigma}{d\omega d\rho} = \frac{32\pi\rho\omega^3}{3c^3} \frac{\left|\mathbf{d}_{gu}(\omega, t_0)\right|^2}{\left|\dot{\omega}_{gu}(t_0)\right|} \cos^2\left(\Delta + \frac{\pi}{4}\right). \tag{5.87}$$

Here the quantity Δ is defined as follows

$$\Delta = \omega t_0 + \int_0^{t_0} \omega_{gu}(\tau)\, d\tau.$$

The instants $\pm t_0$ are the two solutions of the phase equation (5.86) (see Fig. 5.5 left). This result is well known in molecular orbital X-ray theory. It is valid when the points $-t_0$ and t_0 are well separated.

Let us discuss briefly the case of two closely located (merging) solutions of the stationary phase equation (5.86). To carry out the integration in (5.57)–(5.59) one has to expand the phase factor into the power series of $(t \pm t_0)$ up to the third order and the matrix element up to the first order. Neglecting the rotation of QM during the collision, one obtains the cross section

$$\frac{d\sigma}{d\omega d\rho} = \frac{16\pi\rho\omega^3}{3c^3} \left|\mathbf{d}_{gu}(\omega, \mathbf{R}(0))\right|^2 \frac{\mathrm{Ai}^2(u)}{\Gamma^2}. \tag{5.88}$$

Here $\mathrm{Ai}(u)$ is the Airy function, the parameters u and Γ are

$$u = \frac{\omega + \omega_{gu}(0)}{\Gamma}, \qquad \Gamma = \left(\frac{\ddot{\omega}_{gu}(0)}{2}\right)^{1/3},$$

In these formulae the instant $t = 0$ is the turning point of the collision. The case of merging stationary phase points was previously discussed in [143].

Chapter 6
Relativistic Effects in the Polarizational BrS Process

6.1 Introductory Notes

In this chapter we describe peculiarities of the BrS process due to the relativistic effects. The discussion is focused on the polarizational part of the total BrS radiation. Taken alone, the PBrS mechanism defines the emission spectrum in collisions of heavy projectiles (structureless of structural, charged or neutral) with a many-electron target since the OBrS is suppressed due to the large reduced mass of the colliding particles. For a light projectile, both mechanisms must be taken into account, as it is done in Sects. 6.3 and 6.5.6, where the BrS process for relativistic electrons and positrons is considered. However, we do not discuss the properties of the OBrS of relativistic light particles. This can be found in the review articles and books, see, for example, [73, 189, 233, 327, 348, 350, 351, 358] and reference therein. The spectral and spectral angular distributions of the OBrS are tabulated [230, 352] for a wide range of electron energies and for a number of atomic targets (from helium up to uranium).

The relativistic effects in the PBrS process can be subdivided into the following categories. Firstly, there are those directly related to the relativistic velocities of the particles involved in the process. These are the velocity of a projectile and/or the velocity of the internal orbital motion of electrons inside the colliding particles. To account for these effects one has to rely on theoretical frameworks based on the Dirac equation rather than those based on the Schrödinger equation.

The relativistic effect of retardation, which modifies the interaction between a projectile and a target, can be attributed to another category. The retardation implies that the relativistic particle polarizes the target not only through the Coulomb field, but also via the field of the transverse virtual photons, or, in terms of classical electrodynamics, via the retarded vector potential [81, 278]. The effective radius of this field increases infinitely as the velocity v_1 of the projectile approaches the velocity of light c [2]. Therefore, in the ultra-relativistic limit, $v_1 \approx c$, the polarization mainly occurs via the exchange of the transverse virtual photons.

A. V. Korol and A. V. Solov'yov, *Polarization Bremsstrahlung*, Springer Series on Atomic, Optical, and Plasma Physics 80,
DOI: 10.1007/978-3-642-45224-6_6,

There are relativistic effects due to the multipole character of radiation emitted by the target electrons. The multipolarity of the PBrS radiation depends on the magnitude of kR_{at}, where $k = \omega/c$ is the photon momentum and R_{at} is the (mean) radius of the orbital motion of the radiating electron. For $kR_{at} \ll 1$ one can consider only the dipole-photon emission, whereas in the opposite limit, $kR_{at} \gtrsim 1$ the radiation into higher multipoles must be accounted for. In particular, the non-dipole effects will be pronounced in the PBrS emission by the electrons of the inner shells of high-Z elements but will be suppressed in other cases.

Finally, in the case when both colliders possess internal structure (for example, atom-atom, atom-ion, ion-ion pairs) it is necessary to account for the relativistic Doppler effect and the aberration of light. These are important factors which modify the PBrS of a relativistic complex projectile. The role of the above-mentioned relativistic effects was studied in [35, 38, 39, 43, 51, 56, 57, 61, 73, 168, 254, 255, 258, 259, 414] in various collisions of isolated atomic particles. In the first (historically) two papers [35, 57] on PBrS formed in relativistic collisions of charged structureless particles with isolated atoms (ion) the process was treated within the framework of the relativistic plane-wave BA for the projectile and in the non-relativistic approximation for the target electrons. The retardation effect was accounted for, whereas the emission of the photon via the polarizational mechanism was considered in the dipole-photon approximation. The following two important features were established in these papers. First, it was demonstrated that the PBrS cross section increases logarithmically with the energy ε_1 of a projectile. This feature is a result of the retardation in the projectile–atom interaction. Second, it was shown that the shape of the angular distribution of PBrS, in contrast to the ordinary BrS, is proportional to $(1+a\cos^2\theta_{\mathbf{k}})$ where the coefficient a weakly depends on ε_1. This feature reflects the nature of the PBrS process in which the radiation occurs due to the alteration of the induced dipole moment of the target during the collision.

It was demonstrated [35] that the PBrS amplitude can be expressed in terms of two polarizabilities, which define the dynamic atomic response to the joint actions of the fields of the projectile and of the dipole photon. In addition to the polarizability $\alpha(\omega, q)$, which accounts for the (non-retarded) Coulomb interaction, and thus defines the PBrS amplitude (2.16) for in the non-relativistic domain, another polarizability, $\beta(\omega, q)$, appears which describes the dynamic response to the retarded vector potential created by the projectile. In the limit $q \ll R_{at}^{-1}$ both polarizabilities are expressed via the dynamic dipole polarizability $\alpha(\omega)$ of the target. Small values of the transferred momenta q correspond to large distances between the projectile and the target, $\rho \gg R_{at}$, which are important for the PBrS mechanism [45, 454]. This fact, known from the non-relativistic theory of PBrS, was utilized in [35, 57] to derive the formulae for the PBrS spectrum and angular distribution within the large-distance approximation (another term is 'the logarithmic approximation' [45, 454]). In [61] the 'logarithmic' approximation was applied to study the total BrS process in combination with the use of the local density method for the computation of $\alpha(\omega)$.

The theory of PBrS formed in a collision of a charged particle with a many-electron atom/ion was developed further in the papers [254, 255, 258, 259] where the fully relativistic formalism and the results of numerical calculations were presented. The formalism accounts for the relativistic effects of all types mentioned above (with the exception for the Doppler shift and the aberration of light which are irrelevant in the case of a structureless projectile). To describe the scattering process the DPWA [258] and the BA [255] frameworks were utilized. The states of the target's electrons were described in terms of single-electron relativistic wavefunctions. The effect of retardation and the multipole expansion of the emitted photon wave were accounted for. It was demonstrated that the full relativistic PBrS amplitude is expressed in terms of the multipole generalized target polarizabilities of three different types corresponding to the allowed combinations of the types of virtual and emitted photons coupled in the amplitude. It was shown that the relativistic effects lead to noticeable changes in the spectral and angular characteristics of PBrS.

The PBrS mechanism defines the emission spectrum in collision of two complex particles, such as atoms or ions. The relativistic theory of such processes was developed in [38, 39, 43] where the details of the formalism were presented and the analysis of a number of physical phenomena and limiting cases was carried out. In these collisions the OBrS of the projectile is negligibly small because of the large mass. The PBrS of the target, which is at rest in a laboratory frame, bears the same features as in the particle-atom/ion collision. On the contrary, the spectrum and the angular distribution of PBrS emitted by a relativistic complex projectile is strongly influenced by the relativistic Doppler shift and the aberration of light. The degree of this influence depends on the magnitude of the photon energy measured in the rest frame of the projectile. For high values of ω the radiation, measured in the laboratory frame, is concentrated in the cone $\theta_{\mathbf{k}} \sim \gamma^{-1}$ (γ is the relativistic Lorenz factor of the projectile). For small values of ω the angular distribution is nearly isotropic. Therefore, as it was pointed out in [39, 43], for sufficiently high velocities of the collision there is a principal possibility to separate the radiation from the two colliders. Another important feature of the BrS process at relativistic velocities, which is also due to the Doppler and the aberration effects, is that the intensity of the dipole radiation formed in symmetric collisions does not vanish in contrast to the non-relativistic collisions.

The analysis of the contribution of "elastic" and "inelastic" BrS channels to the total emission spectrum was carried out in [38, 39] for collisions of two arbitrary atomic particles, and in [51, 56, 414] for collisions of structureless particles with atoms.

6.2 "Elastic" BrS in Atom–Atom Collisions

In this section the attention is devoted to the process of 'elastic' BrS in a relativistic collision of two atoms (ions). In such a process the colliding particles remain in their ground state after the collision. The theory of this process was developed in

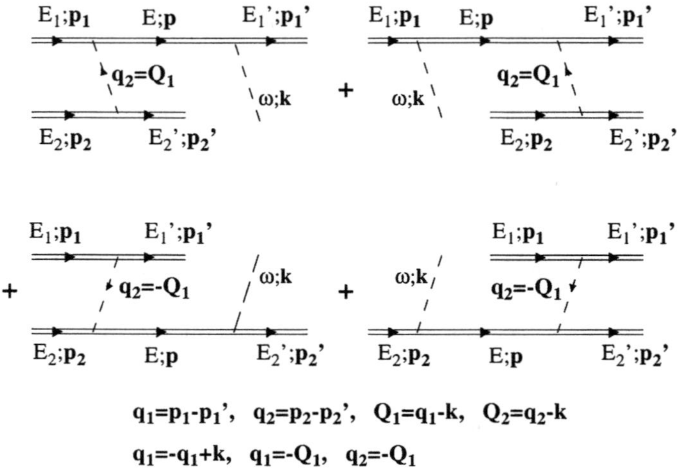

Fig. 6.1 Diagrammatic representation of the BrS amplitude in a relativistic atom-atom collision

[38, 39, 43]. The 'inelastic' BrS arisen in relativistic collisions is considered in Sect. 6.4 (see also [43]). It was demonstrated in [36, 57] that 'elastic' BrS dominates over the inelastic one in a wide range of photon frequencies, $\omega_{\rm at} \leq \omega \leq v_1 R_{\rm at}^{-1}$, where $\omega_{\rm at}$ stands for a 'characteristic' atomic frequency, which can be estimated as an ionization potential of the atomic electrons effectively participating in the process. The dominance is due to the coherent contribution of individual electrons to the amplitude of 'elastic' BrS, and, on the contrast, incoherent contributions in the case of 'inelastic' BrS. For non-relativistic collisions the coherence effect is discussed in Sect. 5.3. The generalization to the relativistic velocities of the colliding particles is given in Sect. 6.4.

In this section, the internal dynamics of the colliding particles is described in non-relativistic terms.

6.2.1 BrS Amplitude

In what follows we mark all the quantities referring to a projectile atom (or ion) with the subscript '1' (or, superscript '(1)'), while the subscript '2' and superscript '(1)' are designated for a target atom/ion which is at rest (in the laboratory frame) before the collision.

The Feynman diagrams describing the amplitude of the process are presented in Fig. 6.1. The upper double line in each diagram refers to the projectile, the lower line describes the target, the dashed line marked with $(\omega, \mathbf{k})$ stands for the emitted photon, and the inner dashed line describes the (retarded) interaction between the colliding atomic particles. The initial, intermediate and final states of each particle are characterized with the energy E (this includes the rest energy, the kinetic energy and the energy due to the internal dynamics) and the momentum $\mathbf{p}$ of the translational motion. The two upper diagrams describe the amplitude f_1 of the photon emission by a projectile which is virtually polarized by the target, the lower pair of diagrams represents f_2, which is the amplitude of the photon emission by the target.

The total BrS amplitude f is given by the sum

$$f = f_1 + f_2, \tag{6.1}$$

In contract to the non-relativistic limit (see Sect. 5.2 or the original papers [33, 34]), in which the analytic expression for f_1 can be obtained from that for f_2 by exchanging the characteristics of the atoms '1' and '2', in the relativistic collisions one should take care on the Lorenz transformations of various quantities which constitute the amplitudes. Therefore, to construct the amplitudes it is convenient, from the very beginning, to operate with the four-vectors and four-tensors which allow one to carry out all necessary transformation of the quantities using the standard rules of quantum electrodynamics [81]. Then, the amplitudes f_1 and f_2 can be written in the covariant form:

$$\begin{aligned} f_2 &= J^{(1)}_{\mu}(\mathbf{q}_1)\, D^{\mu\nu}(E_1 - E'_1, \mathbf{q}_1)\, T^{(2)}_{\lambda\nu}(\omega, \mathbf{k}, \mathbf{Q}_2)\, e^{\lambda}\,, \\ f_1 &= J^{(2)}_{\mu}(\mathbf{q}_2)\, D^{\mu\nu}(E_2 - E'_2, \mathbf{q}_2)\, T^{(1)}_{\lambda\nu}(\omega, \mathbf{k}, \mathbf{Q}_1)\, e^{\lambda}\,. \end{aligned} \tag{6.2}$$

Here e^{λ} is the 4-vector of the photon polarization, $J^{(j)}_{\mu}(\mathbf{q}_j)$ $(j = 1, 2)$ is the four-current of the jth atom due to the interaction with the emitted photon, $T^{(j)}_{\lambda\nu}(\omega, \mathbf{k}, \mathbf{Q}_j)$ is the four-tensor which describes the dynamic reaction of the particle to the action of the fields of the emitted and the virtual photons, and $D^{\mu\nu}(E_j - E'_j, \mathbf{q}_j)$ is the photon propagator. The momenta $\mathbf{q}_j$ and $\mathbf{Q}_j$ are explained in Fig. 6.1. The Greek indices mark the component of the four-vectors (tensors), the pair of the identical upper and lower indices implies the four-summation.

To calculate the quantities $J^{(j)}_{\mu}(\mathbf{q}_j)$ and $T^{(j)}_{\lambda\nu}(\omega, \mathbf{k}, \mathbf{Q}_j)$ one can first consider them in the rest frame of the jth atom, and then recalculate into the laboratory frame using the Lorenz transformations rules.

To present the details of the formalism we start with evaluating the amplitude f_2. The general expressions for the factors on the right-hand side of the upper equation in (6.2) read as follows [39]:

$$J^{(1)}_{\mu}(\mathbf{q}_1) = \frac{M_1 c^2}{E_1} \left\langle E'_1 \left| \sum_{a=1}^{N_1+1} e_a \gamma_{\mu}\, \mathrm{e}^{-\mathrm{i}\mathbf{q}_1 \cdot \mathbf{r}_a} \right| E_1 \right\rangle, \tag{6.3}$$

$$T^{(2)}_{\lambda\nu}(\omega, \mathbf{k}, \mathbf{Q}_2) = \left(\frac{M_2 c^2}{E_2}\right)^2 \times \sum_E \left\{ \frac{\left\langle E_2' \left| \sum_{j=a}^{N_2+1} e_a \gamma_\lambda \mathrm{e}^{-\mathrm{i}\mathbf{k}\cdot\mathbf{r}_a} \right| E \right\rangle \left\langle E \left| \sum_{a=1}^{N_2+1} \gamma_\nu e_a \mathrm{e}^{-\mathrm{i}\mathbf{Q}_2\cdot\mathbf{r}_a} \right| E_2 \right\rangle}{\omega - (E - E_2')} - \frac{\left\langle E_2' \left| \sum_{a=1}^{N_2+1} e_a \gamma_\nu \mathrm{e}^{-\mathrm{i}\mathbf{Q}_2\cdot\mathbf{r}_a} \right| E \right\rangle \left\langle E \left| \sum_{a=1}^{N_2+1} e_a \gamma_\lambda \mathrm{e}^{-\mathrm{i}\mathbf{k}\cdot\mathbf{r}_a} \right| E_2 \right\rangle}{\omega + (E - E_2)} \right\}. \tag{6.4}$$

In (6.2)–(6.4) the integrations over the center-of-mass position vectors of each atom have been carried out. To simplify the intermediate formulae, the factor $\delta(\mathbf{q}_2 + \mathbf{Q}_1)$, which ensures the momentum conservation law, as well as the normalizing factor in the photon wave function are omitted.

The following notations are used in these formulae. The Dirac matrices are γ_λ ($\gamma = 0, 1, 2, 3$). The quantities M_j ($j = 1, 2$) stand for the mass of the jth atom, $\mathbf{r}_a$ ($a = l, \ldots, N_j$) are the position vectors of the atomic electrons with respect to the center-off-mass, N_j is the total number of electrons in the atom, $\mathbf{r}_{N_j+1} \approx 0$ stands for the position vector of the nucleus (the recoil during the collision process is neglected), $e_a = -1$ with $a = 1, \ldots, N_j$) are the electron charges, $e_{N_j+1} \equiv Z_j$ is the charge of the nucleus. Summation $\sum_E$ is carried out over the complete spectrum of the positive- and negative-energy intermediate (virtual states of the target atom. The normalized factors $(M_j c^2/E_j)^{1/2} \approx (M_c^2/E_j')^{1/2}$, which enter the wave function of the translational motion of the jth atom, are accounted for when writing $J^{(1)}_\mu$ and $T^{(2)}_{\lambda\nu}$.

We took into account that in the kinematic domain, which will be of interest, the motion of the particles occurs almost along the straight line, so that $E_j - E_j' \ll E_j$, $|\mathbf{p}_j - \mathbf{p}_j'| \ll |\mathbf{p}_j|$.

The structure of the amplitude f_2 corresponds to the qualitative picture of the process. Indeed, $J^{(1)}_\mu$ describes the current of the first atom, and $T^{(2)}_{\lambda\nu}$ includes the dynamic response of the second atom, which leads to emission of a quantum. The interaction between the atoms is mediated by an exchange of the virtual photon represented by the Green function $D^{\mu\nu}(E_1 - E_1', \mathbf{q}_1)$.

In evaluating $J^{(1,2)}_\mu$ and $T^{(1,2)}_{\lambda\nu}$, it is important to have in mind that the characteristic transferred momenta in the collision are small: $|\mathbf{q}_{1,2}| \leq 1$. This estimate can be derived from simple physical considerations. For doing this let us distinguish two types of the collision.

In the collision of two neutral atoms the mutual polarization (which results in the photon emission) can effectively occur only when the electron clouds of the atoms overlap. Consequently an estimate for the impact parameter is $\rho \approx R_{\mathrm{at}}$. For the component of the transferred momenta perpendicular to the velocity $\mathbf{v}_1$ one writes

$$\left|\mathbf{q}^{\perp}_{1,2}\right| \approx \rho^{-1} \approx R_{\mathrm{at}}^{-1}. \tag{6.5}$$

The situation is different in ion–atom or ion–ion collisions. An ion strongly polarizes an atom (or another ion) even during the distant collisions, when $\rho \gg 1$ [35]. Therefore, the characteristic momentum transfers are small. As was shown in the cited paper, they can be estimated as follows:

$$\left|\mathbf{q}_{1,2}\right| \approx \rho^{-1} \sim \omega / v_1 \gamma \ll R_{\mathrm{at}}^{-1}, \tag{6.6}$$

where $\gamma = (1-\beta^2)^{-1/2}$ ($\beta = v_1/c$) is the Lorentz relativistic factor.

The parallel components of the momentum transfers, $q_j^{\parallel}$ in the BrS process are determined by the energy and momentum conservation laws. In the rest frame of the second atom, one obtains:

$$\left|\mathbf{q}_1^{\parallel}\right| = \omega/v_1, \qquad \left|\mathbf{q}_2^{\parallel}\right| = \left|\mathbf{Q}_1^{\parallel}\right| \approx (1-\beta\cos\theta_{\mathbf{k}})\omega/v_1,$$

where $\theta_{\mathbf{k}} = \widehat{\mathbf{k}\,\mathbf{v}_1}$ is the emission angle with respect to $\mathbf{v}_1$.

In what follows, we consider the BrS process in the ω-region where the polarizabilities of the colliding atomic particles are large and within dipole-photon approximation $\omega R_{\mathrm{at}}/c \ll 1$. Then, for the relativistic velocities v_1 the following relations are valid:

$$\left|\mathbf{q}_{1,2}^{\parallel}\right| \ll 1/R_{\mathrm{at}}. \tag{6.7}$$

From inequalities (6.5) and (6.7) follows, that in an atom–atom collision the parallel component of the transverse momentum is negligible, $q_{1,2}^{\parallel} \ll q_{1,2}^{\perp}$. Hence $\mathbf{q}_{1,2} \approx \mathbf{q}_{1,2}^{\perp}$.

In contrast, in an ion–ion (or, ion–atom) collision both components are important since $q_{1,2}^{\parallel} \sim q_{1,2}^{\perp} \sim q_{1,2}$.

The estimates (6.5)–(6.7) indicate that in the collision process the particles move almost without change of the velocity: $\delta v_{1,2} \approx q_{1,2}/M_{1,2} \ll 1$. This allows one to introduce the inertial rest frames for each of the colliding particles. In these frames the evaluation of $J_\mu^{(1,2)}$ and $T_{\lambda,\nu}^{(1;2)}$ is considerably simplified.[1]

Calculating the time and spatial components of the four-vector $J_\mu^{(1)}$ in the rest frame of the first atom, one writes:

$$\begin{cases} J_0^{(1)c} = Z_1 - \left\langle 0_1 \left| \sum\limits_{j=a}^{N_1} \exp(-i\mathbf{q}_1^{\mathrm{r}}\cdot\mathbf{r}_a) \right| 0_1 \right\rangle = Z_1 - F_1(\mathbf{q}_1^{\mathrm{r}}) \\ \mathbf{J}^{(1)\mathrm{r}} = -\left\langle 0_1 \left| \sum\limits_{a=1}^{N_1} \exp(-i\mathbf{q}_1^{\mathrm{r}}\cdot\mathbf{r}_a)\dfrac{\hat{\mathbf{p}}_a}{c} \right| 0_1 \right\rangle = -\dfrac{\mathbf{q}_1^{\mathrm{r}}}{2c} F_1(\mathbf{q}_1^{\mathrm{r}}) \end{cases}. \tag{6.8}$$

[1] Below in this section as well as in Sect. 6.4 the superscript 'r' indicates that the quantity is calculated in the corresponding rest frame.

Here F_1 is the electron form-factor of the first atom. In derivation of (6.8) the corrections of the order $q_1^r/c \ll 1$, $1/M_1 \ll 1$ are omitted.

Knowing $J_\mu^{(1)r}$ in the rest frame it is straightforward to obtain this quantity in the laboratory system, in which the particle moves with the velocity $\mathbf{v}_1$. Taking into account that $|\mathbf{J}^{(1)r}| \ll J_0^{(1)r}$, and that the combination $J_\mu^{(1)} E_1/M_1 c^2$ is a four-vector (see (6.3), one determines $J_\mu^{(1)}$ in the laboratory frame:

$$J_0^{(1)} = Z_1 - F_1(\mathbf{q}_1^r), \qquad \mathbf{J}^{(1)} = \left(Z_1 - F_1(\mathbf{q}_1^r)\right) \frac{\mathbf{v}_1}{c}, \tag{6.9}$$

where $\mathbf{q}_1^r = \gamma^{-1}\mathbf{q}_1^{\parallel} + \mathbf{q}_1^{\perp}$. In (6.9) and below in this section the indices $\parallel$ and $\perp$ denote the parallel and the perpendicular components of a vector in the laboratory frame.

The formulae obtained have a simple meaning. The time component describes the atomic form-factor deformed as a consequence of the Lorentz contraction. The spatial component $\mathbf{J}^{(1)}$ describes the current of the atomic charges due to their translational motion with velocity $\mathbf{v}_1$.

The four-tensor $T_{\lambda\nu}^{(2)}(\omega, \mathbf{k}, \mathbf{Q}_2)$ (see (6.3) is calculated similarly. In the rest frame of the second atom its components are as follows:

$$\begin{aligned} T_{\lambda\nu}^{(2)r} &= \begin{pmatrix} T_{00}^{(2)}(\omega^r, \mathbf{k}^r, \mathbf{Q}_2^r) & T_{0j}^{(2)}(\omega^r, \mathbf{k}^r, \mathbf{Q}_2^r) \\ T_{i0}^{(2)}(\omega^r, \mathbf{k}^r, \mathbf{Q}_2^r) & T_{ij}^{(2)}(\omega^r, \mathbf{k}^r, \mathbf{Q}_2^r) \end{pmatrix} \\ &\equiv \begin{pmatrix} \mathbf{k}^r \cdot \mathbf{Q}_2^r\, \alpha^{(2)}(\omega^r, \mathbf{Q}_2^r) & -(\omega^r/c)\, k_i^r \beta_{ij}^{(2)}(\omega^r, \mathbf{Q}_2^r) \\ (\omega^r/c)\, Q_{2i}^r \alpha^{(2)}(\omega^r, \mathbf{Q}_2^r) & -(\omega^r/c)^2 \beta_{ij}^{(2)}(\omega^r, \mathbf{Q}_2^r) \end{pmatrix}. \end{aligned} \tag{6.10}$$

Here the subscripts $i, j = (x, y, z)$ denote the spatial components. The generalized polarizability $\alpha^{(2)}(\omega^r, \mathbf{Q}_2^r)$ and the tensor polarizability $\beta_{ij}^{(2)}(\omega^r, \mathbf{Q}_2^r)$ are introduced in accordance with the definitions:

$$\mathbf{Q}_2^r \alpha^{(2)}(\omega^r, \mathbf{Q}_2^r) = \mathrm{i} \sum_{n_2} \frac{2\omega_{n_2 0_2}^r}{(\omega^r)^2 - \left(\omega_{n_2 0_2}^r\right)^2} \mathbf{D}_{0_2 n_2}^{(2)} F_{n_2 0_2}^{(2)}(\mathbf{Q}_2^r), \tag{6.11}$$

$$\mathrm{i}\frac{\omega^r}{c} \beta_{ij}^{(2)}(\omega^r, \mathbf{Q}_2^r) = \sum_{n_2} \left[\frac{\left(\mathbf{D}_{0_2 n_2}^{(2)}\right)_i \left(\mathbf{J}_{n_2 0_2}^{(2)}(\mathbf{Q}_2^r)\right)_j}{\omega^r - \omega_{n_2 0_2}^r} - \frac{\left(\mathbf{J}_{0_2 n_2}^{(2)}(\mathbf{Q}_2^r)\right)_j \left(\mathbf{D}_{n_2 0_2}^{(2)}\right)_i}{\omega^r + \omega_{n_2 0_2}^r} \right]. \tag{6.12}$$

Here $(\mathbf{D}_{n_2 0_2}^{(2)})_j$ is the jth spatial component of the matrix element of the atomic dipole moment and

$$\begin{cases} F^{(2)}_{n_2 0_2}(\mathbf{Q}^{\mathrm{r}}_2) = \left\langle n_2 \left| \sum_{a=1}^{N_2} \exp(-\mathrm{i}\mathbf{Q}^{\mathrm{r}}_2 \cdot \mathbf{r}_a) \right| 0_2 \right\rangle , \\ \mathbf{J}^{(2)}_{n_2 0_2}(\mathbf{Q}^{\mathrm{r}}_2) = \left\langle n_2 \left| \sum_{j=1}^{N_2} \exp(-\mathrm{i}\mathbf{Q}^{\mathrm{r}}_2 \cdot \mathbf{r}_j) \frac{\hat{\mathbf{p}}_j}{c} \right| 0_2 \right\rangle . \end{cases} \tag{6.13}$$

In the region of small transferred momentum, $Q^{\mathrm{r}}_2 R_{\mathrm{at}} \ll 1$, the polarizabilities (6.11) and (6.12) can be expressed in terms of the dynamic dipole polarizability $\alpha^{(2)}(\omega^{\mathrm{r}})$ of the second atom:

$$\alpha^{(2)}(\omega^{\mathrm{r}}, \mathbf{Q}^{\mathrm{r}}_2) \approx \alpha^{(2)}(\omega^{\mathrm{r}}), \qquad \beta^{(2)}_{ij}(\omega^{\mathrm{r}}, \mathbf{Q}^{\mathrm{r}}_2) \approx \delta_{ij}\alpha^{(2)}(\omega^{\mathrm{r}}).$$

The expression (6.13) can be considerably simplified in the kinematic regions defined by (6.5)–(6.7).

In the case of neutral atom–atom collision, (6.5) combined with the dipole photon condition results in the estimate $|\mathbf{Q}^{\mathrm{r}}_2| \gg \omega^{\mathrm{r}}/c$. Using this, one finds

$$T^{(2)\mathrm{r}}_{\lambda\nu} \approx \alpha^{(2)}(\omega^{\mathrm{r}}, \mathbf{Q}^{\mathrm{r}}_2)\, \frac{\omega^{\mathrm{r}}}{c} \begin{pmatrix} \mathbf{n}^{\mathrm{r}} \cdot \mathbf{Q}^{\mathrm{r}}_2 & \mathbf{0}_j \\ \mathbf{Q}^{\mathrm{r}}_{2i} & 0_{ij} \end{pmatrix} . \tag{6.14}$$

The symbols $\mathbf{0}_j$ and 0_{ij} denote the components of the zero vector and tensor, and $\mathbf{n}^{\mathrm{r}} = \mathbf{k}^{\mathrm{r}}/|\mathbf{k}^{\mathrm{r}}|$ is the unit vector along $\mathbf{k}^{\mathrm{r}}$.

In the collisions which involve at least one ion, the condition (6.6) leads to

$$T^{(2)\mathrm{r}}_{\lambda\nu} \approx \alpha^{(2)}(\omega^{\mathrm{r}})\, \frac{\omega^{\mathrm{r}}}{c} \begin{pmatrix} \mathbf{n}^{\mathrm{r}} \cdot \mathbf{Q}^{\mathrm{r}}_2 & -\mathbf{k}^{\mathrm{r}}_j \\ \mathbf{Q}^{\mathrm{r}}_{2i} & -(\omega^{\mathrm{r}}/c)\,\delta_{ij} \end{pmatrix} . \tag{6.15}$$

Similar to the transformation of the four-vector $J^{(1)}_\mu$, the components of the tensor $T^{(2)}_{\lambda\nu}$ in the laboratory frame can be expressed via $T^{(2)\mathrm{r}}_{\lambda\nu}$ by means of the Lorentz transformation.

Having constructed $J^{(j)}_\mu$ and $T^{(j)}_{\lambda\nu}$ $(j = 1, 2)$ both in the rest and the laboratory frames, one evaluates the amplitudes $f_{1,2}$, (6.2). Omitting the intermediate algebra, we reproduce the final results [39].

For an atom–atom collision the amplitudes read:

$$f_2 = \frac{4\pi}{c}\, \frac{Z_1 - F^{(1)}(\mathbf{q}^{\perp}_1)}{(q^{\perp}_1)^2}\, (\mathbf{e} \cdot \mathbf{q}^{\perp}_1)\, \omega\, \alpha^{(2)}(\omega, \mathbf{q}^{\perp}_1) , \tag{6.16}$$

$$f_1 = \frac{4\pi}{c}\, \frac{Z_2 - F^{(2)}(\mathbf{q}^{\perp}_1)}{(q^{\perp}_1)^2} \left[\omega^{\mathrm{r}}\, \mathbf{e} \cdot \mathbf{q}^{\perp}_1 + \gamma(\mathbf{e} \cdot \mathbf{v}_1) \left(\mathbf{k} \cdot \mathbf{q}^{\perp}_1\right) \right] \alpha^{(1)}(\omega^{\mathrm{r}}, \mathbf{q}^{\perp}_1) . \tag{6.17}$$

For an ion–atom (or, ion–ion) collision, the amplitudes are equal to

$$f_2 = \frac{4\pi}{c} \frac{Z_1 - N_1}{q_1^2 - \omega^2/c^2} \left[\mathbf{e} \cdot \mathbf{q}_1 - \frac{\omega}{c^2} \mathbf{e} \cdot \mathbf{v}_1 \right] \omega \alpha^{(2)}(\omega), \tag{6.18}$$

$$f_1 = \frac{4\pi}{c} \frac{Z_2 - N_2}{q_2^2} \left[\omega^{\mathrm{r}} \mathbf{e} \cdot \mathbf{q}_2^{\perp} + \gamma (\mathbf{e} \cdot \mathbf{v}_1)(\mathbf{k} \cdot \mathbf{q}_2^{\perp}) - \frac{\omega^{\mathrm{r}}}{\gamma^2} \mathbf{e} \cdot \mathbf{q}_1^{\parallel} \right] \alpha^{(1)}(\omega^{\mathrm{r}}). \tag{6.19}$$

The photon energy ω^{r} calculated in the rest frame of the particle '1' is expressed in terms of the photon energy and the emission angle in the laboratory frame (the rest frame of particle '2'):

$$\omega^{\mathrm{r}} = \omega \gamma \left(1 - \frac{v_1}{c} \cos \theta_{\mathbf{k}} \right).$$

For the collision velocities $v_1 \ll c$, (6.17) and (6.19) reduce to the corresponding non-relativistic formulae (see Sect. 5.2.1 and [33, 34]).

The amplitude f_2 from (6.16) coincides with its non-relativistic limit. The coincidence is not accidental. Indeed, in a collision of two neutral atoms, the radiation occurs at small distances $\rho \sim R_{\mathrm{at}}$, where the retardation effects in the interaction of the atoms are not pronounced. The momentum transfer $\mathbf{q}$ during the Coulomb part is equal, basically, to its transverse component $\mathbf{q}_1^{\perp}$. Therefore, the polarization of the target atom in the collision process is transverse and does not depend on the choice of the reference frame. It follows from this that the amplitude f_1, (6.17), can be derived from f_2 by transformation to the reference frame moving with the velocity $\mathbf{v}_1$ and the subsequent interchange of the atomic indices. In this transformation only the photon energy and polarization vector are re-calculated, whereas the atomic polarizabilities and form-factors are not affected.

The bremsstrahlung process in an ion-ion collision occurs somewhat differently. In this case, the collision occurs at large distances, $\rho \gg R_{\mathrm{at}}$, so that the retardation in the interaction between the ions and its transverse nature become important. The polarization of each of the ions at large distances is due mainly to the field of the net charge $(Z_j - N_j)$ of the partner ion. This explains the agreement between the amplitude f_2 from (6.18) with the amplitude of PBrS of the relativistic structureless charged particle, which was obtained in [35, 37] (see (6.48) the right-hand side of which should be multiplied by the bet charge $(Z_1 - N_1)$ to explicitly match that of (6.18)). The radiative amplitude by the projectile ion, f_1 from (6.19), can be derived from (6.18) by the transformation to the reference frame moving with $\mathbf{v}_1$ and the subsequent interchange of the ionic indices. Here, in contrast to the case of a neutral atom–atom collision, it is necessary to take into account also the transformation of the longitudinal component of the momentum transfer $\mathbf{q}^{\parallel}$, which induces a longitudinal polarization of the ion (see the third term in the square brackets in (6.19)). The magnitude of the longitudinal polarization of the first ion in the laboratory frame is γ^2 times smaller than that of the transverse polarization. This circumstance must be taken into account in the conversion from (6.18) to (6.19).

6.2.2 BrS Cross Section

Using the well-known rules [81] relating the amplitude of a radiative process to its differential cross section, one writes

$$d^4\sigma \equiv \frac{d^4\sigma}{d\omega d\Omega_{\mathbf{k}} dq_1 d\varphi_{\mathbf{q}_1}} = \frac{\omega q_1}{(2\pi)^4 c v_1^2} \, |f_1 + f_2|^2 = d^4\sigma_1 + d^4\sigma_2 + d^4\sigma_{12} \,. \quad (6.20)$$

Here $\varphi_{\mathbf{q}_1}$ is the azimuthal angle of the vector $\mathbf{q}_1$ with respect to coordinate system with the z-axis along $\mathbf{v}_1$. The right-hand side of (6.20) explicitly stresses that the total cross section is a sum of three terms: the BrS cross sections of the projectile, $d^4\sigma_1$, and of the target, $d^4\sigma_2$, and the interference term $d^4\sigma_{12}$.

Substituting into (6.20) the amplitudes derived in Sect. 6.2.1, one calculates the BrS cross section for the different collisions.

6.2.2.1 Atom–Atom Collisions

In atom–atom collision, the cross section (6.20), integrated over $dq_1 d\varphi_{\mathbf{q}_1}$ and summed over the photon polarizations, has the following form [39]:

$$\frac{d^2\sigma_2}{d\omega d\Omega_{\mathbf{k}}} = \frac{\omega^3(1+\cos^2\theta_{\mathbf{k}})}{\pi v_1^2 c^3} \int_0^\infty \frac{dq_1^\perp}{q_1^\perp} \left(Z_1 - F_1(q_1^\perp)\right)^2 \left|\alpha^{(2)}(\omega, q_1^\perp)\right|^2, \quad (6.21)$$

$$\frac{d^2\sigma_1}{d\omega d\Omega_{\mathbf{k}}} = \frac{\omega^3(1+\cos^2\theta_{\mathbf{k}}^{\mathrm{r}})}{\pi v_1^2 c^3} \left(\frac{\omega^{\mathrm{r}}}{\omega}\right)^2 \int_0^\infty \frac{dq_1^\perp}{q_1^\perp} \left(Z_2 - F_2(q_1^\perp)\right)^2 \left|\alpha^{(1)}(\omega^{\mathrm{r}}, q_1^\perp)\right|^2, \quad (6.22)$$

$$\frac{d^2\sigma_{12}}{d\omega d\Omega_{\mathbf{k}}} = -\frac{2\omega^3(1+\cos\theta_{\mathbf{k}}\cos\theta_{\mathbf{k}}^{\mathrm{r}})}{\pi v_1^2 c^3} \frac{\omega^{\mathrm{r}}}{\omega} \int_0^\infty \frac{dq_1^\perp}{q_1^\perp} \left(Z_1 - F_1(q_1^\perp)\right)\left(Z_2 - F_2(q_1^\perp)\right)$$
$$\times \mathrm{Re}\left[\alpha^{(1)}(\omega^{\mathrm{r}}, q_1^\perp)\alpha^{(2)*}(\omega, q_1^\perp)\right]. \quad (6.23)$$

Here

$$\omega^{\mathrm{r}} = \omega\gamma(1-\beta\cos\theta_{\mathbf{k}}), \qquad \cos\theta_{\mathbf{k}}^{\mathrm{r}} = \frac{\cos\theta_{\mathbf{k}} - \beta}{1-\beta\cos\theta_{\mathbf{k}}}. \quad (6.24)$$

In derivation of these equations it was assumed that each atoms has a spherically symmetric charge distribution in the rest frame. Equations (6.21)–(6.23) show that the main contribution to the BrS spectrum formed in a collision of two neutral atoms comes from the momentum transfer region $q_1^\perp R_{\mathrm{at}} \sim 1$, thereby validating the estimate (6.5). Outside this region either $Z_j - F_j(q_1^\perp) \ll 1$ or $|\alpha^{(j)}(\omega, q_1^\perp)| \ll 1$.

Comparing the terms $\mathrm{d}^2\sigma_2/\mathrm{d}\omega\mathrm{d}\Omega_{\mathbf{k}}$ and $\mathrm{d}^2\sigma_1/\mathrm{d}\omega\mathrm{d}\Omega_{\mathbf{k}}$, one notices that in the latter the effect of aberration modifies the angular distribution of radiation, $(1+\cos^2\theta_{\mathbf{k}}) \to (1+\cos^2\theta^{\mathrm{r}}_{\mathbf{k}})$, whereas the Doppler effect modifies the photon frequency, $\omega \to \omega^{\mathrm{r}}$.

The angular distribution of the radiation emitted by the projectile and that of the interference term $\mathrm{d}^2\sigma_{12}/\mathrm{d}\omega\mathrm{d}\Omega_{\mathbf{k}}$, depends strongly on the polarizability of the projectile.

In the limit of small photon energies $\omega \ll \omega_{\mathrm{at}}/\gamma$ such that for all angles $0 \le \theta_{\mathbf{k}} \le \pi$ the condition $\omega^{\mathrm{r}} \ll \omega_{\mathrm{at}}$ is met, the dependence of $\alpha^{(1)}(\omega^{\mathrm{r}}, q_1^{\perp})$ ω and $\theta_{\mathbf{k}}$ can be neglected, so that the polarizability can be substituted with its static value. In this limit, the angular distribution of radiation emitted by a relativistic projectile has the following shape:

$$\mathrm{d}^2\sigma_2/\mathrm{d}\omega\mathrm{d}\Omega_{\mathbf{k}} \sim \left(1 - \frac{v_1}{c}\cos\theta_{\mathbf{k}}\right)^2 + \left(\cos\theta_{\mathbf{k}} - \frac{v_1}{c}\right)^2 .$$

Thus, $\mathrm{d}^2\sigma_2/\mathrm{d}\omega\mathrm{d}\Omega_{\mathbf{k}}$ does not exhibit a well-known feature of the angular distribution of the OBrS radiation of a relativistic charge [81], when the photons are predominantly emitted in the cone with the opening angle γ^{-1} along the vector $\mathbf{v}_1$. Comparing of the cross sections (6.21) and (6.22) one finds, that the ratio $\xi = \mathrm{d}^2\sigma_2/\mathrm{d}^2\sigma_1$ is of the order of magnitude γ^2 for the emission angles $\theta_{\mathbf{k}} \le \gamma^{-1}$, but $\xi \sim \gamma^{-2}$ for $\theta_{\mathbf{k}} \gg \gamma^{-1}$. Hence, the BrS of the target dominates at small angles $\theta_{\mathbf{k}} \le \gamma^{-1}$, whereas the BrS of the projectile is more intense for the larger angles.

The above-mentioned feature in the angular distribution of the radiation emitted by the projectile appears for large photon energies $\omega \gg \omega_{\mathrm{at}}\gamma$. In this case, the condition $\omega^{\mathrm{r}} \gg \omega_{\mathrm{at}}$ is met nearly in the whole range of $\theta_{\mathbf{k}}$, so that $\alpha^{(1)}(\omega^{\mathrm{r}}, q_1^{\perp}) \sim (\omega^{\mathrm{r}})^{-2}$. This leads to the peculiarity $(d^2\sigma_1/\mathrm{d}\omega\mathrm{d}\Omega_{\mathbf{k}}) \sim \gamma^{-2}(1+\beta\cos\theta_{\mathbf{k}})^{-2}$. The ratio of the cross sections (6.21) and (6.22) equals to $\xi \sim \gamma^{-2}$ for $\theta_{\mathbf{k}} \le \gamma^{-1}$ and to $\xi \sim \gamma^2$ for $\theta_{\mathbf{k}} \gg \gamma^{-1}$. Hence, the BrS of the target dominates at large angles $\theta_{\mathbf{k}} \gg \gamma^{-1}$, while the emission by the projectile forms the radiation spectrum at $\theta_{\mathbf{k}} \le \gamma^{-1}$. We note here, that the angular distribution of the BrS radiation formed in a collision of a relativistic electron with an atom has similar behaviour (see Sect. 6.3.2 and [35, 57]). This agreement is due to the fact that in the limit $\omega \gg \omega_{\mathrm{at}}$, the electrons in the projectile atom can be considered as (quasi-)free ones, so that the BrS process of the projectile reduces to the coherent radiation by these electrons in the field of the target.

In general it can be stated that if the photon energy and the emission angle satisfy the conditions

$$\left(\frac{\omega^{\mathrm{r}}}{c}\right)^2 \left|\alpha^{(1)}(\omega^{\mathrm{r}}, q_1^{\perp})\right|^2 \gg \left|\alpha^{(2)}(\omega, q_1^{\perp})\right|^2, \qquad q_1^{\perp} R_{\mathrm{at}} \sim 1, \tag{6.25}$$

then the BrS of a projectile dominates over that of a target. If the sign of the inequality is changed, then the BrS of the target becomes predominant. Therefore we can conclude that at sufficiently high collision velocities, there is a principal possibility to distinguish the radiation emitted by a projectile from the BrS of the target.

Integrating the double differential cross sections (6.21)–(6.23) over the emission angles, one finds the formulae for the spectral distribution of the radiation:

$$\frac{d\sigma_2}{d\omega} = \frac{16\omega^3}{3v_1^2c^3}\int_0^\infty \frac{dq_1^\perp}{q_1^\perp}\left(Z_1 - F_1(q_1^\perp)\right)^2 \left|\alpha^{(2)}(\omega, q_1^\perp)\right|^2, \tag{6.26}$$

$$\frac{d\sigma_1}{d\omega} = \frac{2\omega^3}{v_1^2c^3}\int_0^\infty \frac{dq_1^\perp}{q_1^\perp}\int_{\gamma(1-\beta)}^{\gamma(1+\beta)} dx \left[\frac{x^2}{\beta\gamma} + \frac{(1-\gamma x)^2}{(\beta\gamma)^3}\right] \times \left(Z_2 - F_2(q_1^\perp)\right)^2 \left|\alpha^{(1)}(\omega x, q_1^\perp)\right|^2, \tag{6.27}$$

$$\frac{d\sigma_{12}}{d\omega} = -\frac{4\omega^3}{v_1^2c^3}\int_0^\infty \frac{dq_1^\perp}{q_1^\perp}\int_{\gamma(1-\beta)}^{\gamma(1+\beta)} dx \left[\frac{x}{\beta\gamma} + \frac{(1-\gamma x)(\gamma - x)}{(\beta\gamma)^3}\right] \times \left(Z_1 - F_1(q_1^\perp)\right)\left(Z_2 - F_2(q_1^\perp)\right) \mathrm{Re}\left[\alpha^{(1)}(\omega x, q_1^\perp)\alpha^{(2)*}(\omega, q_1^\perp)\right]. \tag{6.28}$$

Here $x = \gamma(1 - \beta\cos\theta_{\mathbf{k}})$.

An important feature of these expressions, is that the BrS spectrum of a projectile and the interference term at a given frequency ω_0 depend on the behaviour of the dynamic polarizability of the projectile atom $\alpha^{(1)}(\omega, q_1^\perp)$ over a broad range of the first argument, from $\omega = \omega_0/\gamma(1+\beta)$ up to $\omega = \omega_0\gamma(1+\beta)$. This leads, in particular, to the result that narrow spectral lines, which can be observed in the spectrum of the atom '1' when it is at rest, become broadened over the interval $\omega_{\mathrm{at}}\gamma^{-1} \le \omega \le \omega_{\mathrm{at}}\gamma$ in the BrS spectrum of the projectile.

Equations (6.27) and (6.28) can be considerably simplified for small photon energies $\omega \ll \omega_{\mathrm{at}}\gamma^{-1}$, where the polarizability $\alpha^{(1)}(\omega x, q_1^\perp)$ can be approximated with its static value $\alpha^{(1)}(0, q_1^\perp)$. Then, the integration over x is carried out explicitly resulting in:

$$\frac{d\sigma_1}{d\omega} = \frac{16\omega^3(1+\beta^2)\gamma^2}{3v_1^2c^3}\int_0^\infty \frac{dq_1^\perp}{q_1^\perp}\left(Z_2 - F_2(q_1^\perp)\right)^2 \left(\alpha^{(1)}(0, q_1^\perp)\right)^2, \tag{6.29}$$

$$\frac{d\sigma_{12}}{d\omega} = -\frac{32\omega^3\gamma}{3v_1^2c^3}\int_0^\infty \frac{dq_1^\perp}{q_1^\perp}\left(Z_1 - F_1(q_1^\perp)\right)\left(Z_2 - F_2(q_1^\perp)\right) \times\alpha^{(1)}(0, q_1^\perp)\alpha^{(2)}(0, q_1^\perp). \tag{6.30}$$

The term $d\sigma_2/d\omega$ is still given by (6.26). Comparing (6.29) with (6.26) and (6.30), one finds that the BrS of the projectile is γ^2 times more intensive that of the target and by a factor γ exceeds the interference term.

In the region of high photon frequencies $\omega \gg \omega_{\rm at}\gamma$, the polarizability of the projectile atom can be approximated as $\alpha^{(1)}(\omega x; q_1^{\perp}) \approx -F_1(q_1^{\perp})/\omega^2 x^2$ for all x within the integration interval in (6.27)– (6.28). This approximation also allows one to simplify the expressions for $\mathrm{d}\sigma_1$ and $\mathrm{d}\sigma_{12}$:

$$\frac{\mathrm{d}\sigma_1}{\mathrm{d}\omega} = \frac{16}{3\omega v_1^2 c^3} \int\limits_0^{\infty} \frac{\mathrm{d}q_1^{\perp}}{q_1^{\perp}} \left[F_1(q_1^{\perp}) \left(Z_2 - F_2(q_1^{\perp})\right)\right\}^2 , \tag{6.31}$$

$$\frac{\mathrm{d}\sigma_{12}}{\mathrm{d}\omega} = -\frac{16}{\omega v_1^2 c^3} \left[\frac{1}{\beta^2\gamma} - \frac{\ln\gamma(1+\beta)}{(\beta\gamma)^3}\right] \int\limits_0^{\infty} \frac{\mathrm{d}q_1^{\perp}}{q_1^{\perp}} F_1(q_1^{\perp}) F_2(q_1^{\perp}) \times \left(Z_1 - F_1(q_1^{\perp})\right) \left(Z_2 - F_2(q_1^{\perp})\right) . \tag{6.32}$$

Equations (6.29)–(6.32) generalize the corresponding formulae of the non-relativistic theory, which one obtains by carrying the limit $v_1/c \to 0$ (see also Sect. 5.2.2). Since $\alpha^{(j)}(0; q_1^{\perp})$ and $F_j(q_1^{\perp})$ are real quantities, the signs for complex conjugations in these formulae are omitted.

Summing (6.26) with (6.29) and (6.30) or/and with (6.31) and (6.32), one obtains the expressions for the total BrS spectrum in relativistic atom–atom collisions in different regions of the emitted photon energy. The resulting formulae produce non-zero results for the collision of two identical atoms (see Sect. 6.2.2.3 for more details). This is a direct consequence of the Doppler effect and the aberration of light in the BrS process of the projectile. For $\omega \ll \omega_{\rm at}\gamma^{-1}$ and for a sufficiently large Lorentz factor γ, the total BrS spectrum is dominated by the emission of the projectile. In the region $\omega \gg \omega_{\rm at}\gamma$, the spectra $\mathrm{d}\sigma_1/\mathrm{d}\omega$ and $\mathrm{d}\sigma_2/\mathrm{d}\omega$ are of the same order of magnitude and exceed the interference term by a factor of γ. The small magnitude of $\mathrm{d}\sigma_{12}/\mathrm{d}\omega$ is due to essentially different character of the angular distributions: the radiation emitted by the projectile is concentrated within the narrow cone $\theta_{\mathbf{k}} \leq \gamma^{-1}$ whereas the radiation by the target, being weakly anisotropic, is emitted into all angles.

6.2.2.2 Ion–Ion Collisions

Substituting the amplitudes (6.18) and (6.19) into the general formula (6.20), and carrying out the integration over $\mathrm{d}\varphi_{\mathbf{q}_1}$ and summation over photon polarizations, one derives the triply differential cross sections [39]:

$$\frac{\mathrm{d}^3\sigma_2}{\mathrm{d}\omega\mathrm{d}\Omega_{\mathbf{k}}\mathrm{d}q_1} = \frac{\mathscr{Z}_1^2}{\pi c^3 v_1^2}\,\omega^3 |\alpha^{(2)}(\omega)|^2$$

$$\times q_1 \frac{(q_1^{\perp})^2(1+\cos^2\theta_{\mathbf{k}}) + 2\sin^2\theta_{\mathbf{k}}\left(q_1^{\parallel} - \beta\frac{\omega}{c}\right)^2}{(q_1^2 - \omega^2/c^2)^2}, \tag{6.33}$$

$$\frac{d^3\sigma_1}{d\omega d\Omega_{\mathbf{k}} dq_2} = \frac{\mathscr{Z}_2^2}{\pi c^3 v_1^2}\left(\frac{\omega^{\mathrm{r}}}{\omega}\right)^2 \omega^3 |\alpha^{(1)}(\omega^{\mathrm{r}})|^2$$
$$\times \frac{(q_2^{\perp})^2(1+\cos^2\theta_{\mathbf{k}}^{\mathrm{r}}) + 2\gamma^{-2}(q_2^{\parallel})^2\sin^2\theta_{\mathbf{k}}^{\mathrm{r}}}{q_2^3}. \tag{6.34}$$

Here

$$\sin\theta_{\mathbf{k}}^{\mathrm{r}} = \gamma^{-1}\frac{\sin\theta_{\mathbf{k}}}{1-\beta\cos\theta_{\mathbf{k}}}.$$

The notation $\mathscr{Z}_j = Z_j - N_j$ $(j = 1, 2)$ stands for the net ionic charge.

Equations (6.33) and (6.34) are valid in the region $q_{1,2}R_{\mathrm{at}} \ll 1$. The cross section $d^3\sigma_2$ is the same as in the collision of a structureless relativistic particles of the charge $\mathscr{Z}_1$ with atomic/ionic target (see Sect. 6.3). This is a quite obvious result for the region of large impact parameters (small transferred momenta), when the projectile ion acts as a net charge. The structure of (6.34) is also clear. The term proportional to $(q_2^{\perp})^2$ is due to the transverse polarization of the projectile ion by the static Coulomb field of the target, while the term $\propto (q_2^{\parallel})^2$ results from the longitudinal polarization by the field of the virtual photon. Instead of the direct evaluation, the cross section (6.34) could be obtained from the (6.33) considering the latter in the rest frame of the projectile and interchanging the indices of colliding particles. When doing this, one takes into account that the longitudinal polarization of a relativistic projectile is γ^2 times smaller than its transverse polarization, since the latter does not depend on the choice of the reference frame.

Integrating (6.33) and (6.34) over q_1 and q_2 from the lower values

$$q_1^{\min} = \omega/v_1, \quad q_2^{\min} = (1-\beta\cos\theta_{\mathbf{k}})\omega/v_1$$

up to $q_{\max} \sim R_{\mathrm{at}}^{-1}$, and keeping only logarithmically large terms, one arrives at:

$$\frac{d^2\sigma_2}{d\omega d\Omega_{\mathbf{k}}} = \frac{\mathscr{Z}_1^2}{\pi c^3 v_1^2}\omega^3\left|\alpha^{(2)}(\omega)\right|^2 (1+\cos^2\theta_{\mathbf{k}})\ln\frac{v_1\gamma}{\omega R_{\mathrm{at}}}, \tag{6.35}$$

$$\frac{d^2\sigma_1}{d\omega d\Omega_{\mathbf{k}}} = \frac{\mathscr{Z}_2^2}{\pi c^3 v_1^2}\left(\frac{\omega^{\mathrm{r}}}{\omega}\right)^2\omega^3\left|\alpha_d^{(1)}(\omega^{\mathrm{r}})\right|^2(1+\cos^2\theta_{\mathbf{k}}^{\mathrm{r}})\ln\frac{\gamma v_1}{\omega^{\mathrm{r}}R_{\mathrm{at}}}, \tag{6.36}$$

$$\frac{d^2\sigma_{12}}{d\omega d\Omega_{\mathbf{k}}} = -\frac{2\omega^3\mathscr{Z}_1\mathscr{Z}_2}{\pi c^3 v_1^2}\frac{\omega^{\mathrm{r}}}{\omega}(1+\cos\theta_{\mathbf{k}}\cos\theta_{\mathbf{k}}^{\mathrm{r}})\mathrm{Re}\left[\alpha^{(1)}(\omega^{\mathrm{r}})\alpha^{(2)*}(\omega)\right]\ln\frac{v_1\gamma}{\omega^{\mathrm{r}}R_{\mathrm{at}}}. \tag{6.37}$$

The structures of the terms on the right-hand sides of (6.35) and (6.37) are similar to those of (6.21)–(6.23). Therefore the analysis of the behavior of the spectral-angular dependence in ion–ion collisions is identical to that which was carried out for atom–atom collisions. The presence of long-range forces due to the net charges in a system of colliding ions leads to a logarithmic growth of the cross sections with the energy of the incident particle.

The formulae obtained, are applicable to the ion–atom collision as well.

6.2.2.3 Collisions of Identical Atomic Particles

As discussed above (see Sect. 5.2) In the non-relativistic limit, $v_1/c \to 0$, the total BrS cross section is identically zero for a symmetric collision. Physical reason for this is in the fact, that the induced moments of the colliding particles, being equal in magnitude, are oppositely-directed. Therefore, the total dipole moment is zero at any instant leading to the absence of the dipole radiation.

Allowance for the Doppler effect and the aberration of light lifts the forbidenness on the dipole radiation. To demonstrate it, let us analyze the formulae obtained in Sect. 6.2.2.1 and in the case of the symmetric collision occurring at sufficiently high but non-relativistic velocity, $1 \ll v_1 \ll c$. In what follows, we keep the terms in cross sections linear in $\beta = v_1/c$ but neglect the higher powers of this ratio.

For the atom–atom symmetric collision, (6.21)–(6.23), omitting the atomic indices '1' and '2', one derives the following formulae for the double differential cross section $\mathrm{d}^2\sigma = \mathrm{d}^2\sigma_1 + \mathrm{d}^2\sigma_2 + \mathrm{d}^2\sigma_3$ of the total BrS:

$$\frac{\mathrm{d}^2\sigma}{\mathrm{d}\omega \mathrm{d}\Omega_{\mathbf{k}}} = \frac{\omega^3}{\pi c^5} \int_0^\infty \frac{\mathrm{d}q}{q} (Z - F(q))^2 \left\{ \cos^2\theta_{\mathbf{k}} \left| \frac{\mathrm{d}}{\mathrm{d}\omega} (\omega\alpha(\omega, q)) \right|^2 \right.$$
$$\left. + \left| \alpha(\omega, q) + \omega \cos^2\theta_{\mathbf{k}} \frac{\mathrm{d}}{\mathrm{d}\omega} \alpha(\omega, q) \right|^2 \right\}, \qquad \frac{\omega R_{\mathrm{at}}}{v_1} \ll 1. \tag{6.38}$$

The cross section (6.38) does not depend on the collision velocity and is β^2 times smaller than the characteristic values of the BrS cross section in collision of non-identical atoms. Equation (6.38) takes into account all relativistic corrections to the amplitude, which are proportional to β. In addition to these, the amplitude contains the terms corresponding to the quadrupole-photon emission. The quadrupole terms give a correction to the dipole-photon amplitude of the order $R_{\mathrm{at}}\omega/c \ll 1$. This correction, however, is β^2 in the order of magnitude and, thus, can be neglected within the range of applicability of formula (6.38), which is $R_{\mathrm{at}}\omega/v_1 \ll 1$.

Integrating (6.38) over the emission angles, one derives the spectral distribution of BrS for a symmetric collision:

$$\frac{\mathrm{d}\sigma}{\mathrm{d}\omega} = \frac{16\omega^3}{3c^5} \int_0^\infty \frac{\mathrm{d}q}{q} \left(Z - F(q)\right)$$
$$\times \left[|\alpha(\omega, q)|^2 + \omega \mathrm{Re}\left(\alpha^*(\omega, q) \frac{\mathrm{d}\,\alpha(\omega, q)}{\mathrm{d}\omega} \right) + \frac{2\omega^2}{5} \left| \frac{\mathrm{d}\,\alpha(\omega, q)}{\mathrm{d}\omega} \right|^2 \right]. \tag{6.39}$$

For the sake of consistency, let us present the formulae for $\mathrm{d}^2\sigma/\mathrm{d}\omega\mathrm{d}\Omega_{\mathbf{k}}$ and $\mathrm{d}\sigma/\mathrm{d}\omega$ for the ion–ion symmetric collision calculated within the same approximation as (6.38) and (6.39). Using (6.35)–(6.37), one derives:

$$\frac{\mathrm{d}^2\sigma}{\mathrm{d}\omega\mathrm{d}\Omega_{\mathbf{k}}} = \frac{\mathscr{Z}^2\omega^3}{\pi c^5} \left[\cos^2\theta_{\mathbf{k}} \left| \frac{\mathrm{d}\left(\omega\alpha(\omega)\right)}{\mathrm{d}\omega} \right|^2 + \left| \alpha(\omega) + \omega\cos^2\theta_{\mathbf{k}} \frac{\mathrm{d}\alpha(\omega)}{\mathrm{d}\omega} \right|^2 \right]$$
$$\times \ln\frac{v_1}{\omega R_{\mathrm{at}}}, \tag{6.40}$$

$$\frac{\mathrm{d}\sigma}{\mathrm{d}\omega} = \frac{16\mathscr{Z}^2\omega^3}{3c^5} \left[|\alpha(\omega)|^2 + \frac{\omega}{2}\frac{\mathrm{d}\,|\alpha(\omega)|^2}{\mathrm{d}\omega} + \frac{2\omega^2}{5} \left| \frac{\mathrm{d}\alpha(\omega)}{\mathrm{d}\omega} \right|^2 \right] \ln\frac{v}{\omega R_{\mathrm{at}}}. \tag{6.41}$$

In contrast to (6.38) and (6.39), cross sections (6.40) and (6.41) weakly (logarithmically) depend on the collision velocity. This is due to the presence of the long-range Coulomb forces.

6.3 Electron-Atom Collisions

In this section we use the relativistic system of units, $\hbar = c = 1$. The notation m stands for the electron mass.

6.3.1 BrS Amplitude in a Relativistic Collision

Consider the collision of an electron with an energy $\varepsilon_1 = (\mathbf{p}_1^2 + m^2)^{1/2}$ with an atom in its ground state. After the collision the electron energy $\varepsilon_2 = (\mathbf{p}_2^2 + m^2)^{1/2}$ is decreased due to the emission of a photon of energy ω, whereas the atom is left in the ground state. We assume that the electron and photon energies satisfy the inequalities $\varepsilon_{1,2} \gg \omega$.

The collision process will be treated within the Born approximation which implies that the following condition is met: $Ze^2/v_{1,2} \ll 1$ where Z is the nucleus charge. In this case, to construct the amplitude of the process it is sufficient to consider the lowest orders of the perturbation theory in the interaction between the projectile electron and the atom as well as in the interaction of the colliding particles with electromagnetic field.

We describe the electron before and after the collision by the Dirac plane waves, while the motion of atomic electrons is described by the non-relativistic wave functions $\Psi_n(\mathbf{r}_1;\ldots;\mathbf{r}_z)$ (the subscript n denotes a set of quantum numbers) which are the solutions of the Schrödinger equation.[2]

The OBrS amplitude, evaluated within the framework of relativistic Born approximation, is well-known (see, for example, [81]):

$$f_{\text{ord}} = -e^3 \frac{Z - F(\mathbf{q}_1)}{q_1^2} \times \overline{u}_{\mu_2}(p_2) \left[\mathbf{e}\cdot\boldsymbol{\gamma}\, \frac{\not{p}_2 + \not{k} + m}{(p_2+k)^2 - m^2}\gamma^0 + \gamma^0 \frac{\not{p}_1 - \not{k} + m}{(p_1-k)^2 - m^2}\,\mathbf{e}\cdot\boldsymbol{\gamma} \right] u_{\mu_1}(p_1). \tag{6.42}$$

Here $F(\mathbf{q}_1)$ is the form-factor of atomic electrons, $\mathbf{q}_1 = \mathbf{p}_1 - \mathbf{p}_2 - \mathbf{k}$ is the momentum transferred to the atom, the following notations are used for the momenta four-vectors of the photon and the projectile: $k^\nu = (\omega, \mathbf{k})$, $p_{1,2}^\nu = (\varepsilon_{1,2}, \mathbf{p}_{1,2})$, the subscripts $\mu_{1,2}$ denote the initial and final spin states of the electron, $u_{\mu_{1,2}}(p_{1,2})$ is the bi-spinor amplitude of the free electron. The bar over u denotes the Dirac conjugation: $\overline{u}_{\mu_2}(p_2) = u_{\mu_2}^\dagger(p_2)$ with the superscript '†' standing for the Hermitian conjugation. The so-called Feynman slash notation, $\not{a}$, is used for a scalar product $\not{a} = \sum_{\nu=0}^{3} a_\nu\gamma^\nu \equiv a_\nu\gamma^\nu$ of the four-vectors $a^\nu = (a^0, \mathbf{a})$ and $\gamma^\nu = (\gamma^0, \boldsymbol{\gamma})$ (here γ^0 and $\boldsymbol{\gamma}$ are the Dirac matrices).

To derive the PBrS amplitude we assume that the photon momentum is small enough to ensure the validity of the dipole-photon approximation: $\mathbf{k}\cdot\mathbf{r}_a \sim \omega R_{\text{at}} \ll 1$ where $\mathbf{r}_a$ is the position vector of an ath atomic electron and R_{at} denotes the (mean) atomic radius. Choosing the photon Green's function the Coulomb gauge (see, for example, [81]), one can write the PBrS amplitude as the sum of three terms [35]:

$$f_{\text{pol}} = f_{\text{pol}}^{(1)} + f_{\text{pol}}^{(2)} + f_{\text{pol}}^{(3)}. \tag{6.43}$$

Here

$$f_{\text{pol}}^{(1)} = \frac{e^3}{m}\sum_n \left[\frac{b_0}{q^2}\frac{\langle 0|\sum_a \mathbf{e}\cdot\hat{\mathbf{p}}_a|n\rangle\langle n|\sum_a e^{i\mathbf{q}\mathbf{r}_a}|0\rangle}{\omega-\omega_{n0}} - \frac{1}{m(\omega^2-q^2)}\frac{\langle 0|\sum_a \mathbf{e}\cdot\hat{\mathbf{p}}_a|n\rangle\langle n|\sum_a e^{i\mathbf{q}\cdot\mathbf{r}_a}\left(\mathbf{b}\cdot\hat{\mathbf{p}}_a - (\mathbf{b}\cdot\mathbf{q})(\mathbf{q}\cdot\hat{\mathbf{p}}_a)/q^2\right)|0\rangle}{\omega-\omega_{n0}} \right], \tag{6.44}$$

$$f_{\text{pol}}^{(2)} = \frac{e^3}{m}\sum_n \left[\frac{b_0}{q^2}\frac{\langle 0|\sum_a e^{i\mathbf{q}\cdot\mathbf{r}_a}|n\rangle\langle n|\sum_a \mathbf{e}\cdot\hat{\mathbf{p}}_a|0\rangle}{\omega+\omega_{n0}} \right.$$

[2] The role of relativistic effects in the atomic electrons dynamics in forming PBrS is discussed in Sect. 6.5.

$$- \frac{1}{m(\omega^2 - q^2)} \frac{\langle 0 \left| \sum_a e^{i\mathbf{q}\cdot\mathbf{r}_a} \left(\mathbf{b}\hat{\mathbf{p}}_a - (\mathbf{b}\cdot\mathbf{q})(\mathbf{q}\cdot\hat{\mathbf{p}}_a)/q^2 \right) \right| n \rangle \langle n \left| \sum_a \mathbf{e}\cdot\hat{\mathbf{p}}_a \right| 0 \rangle}{\omega + \omega_{n0}} \Bigg], \tag{6.45}$$

$$f^{(3)}_{\rm pol} = \frac{e^3}{m} \left(\mathbf{e}\cdot\mathbf{b} - \frac{(\mathbf{e}\cdot\mathbf{q})(\mathbf{q}\cdot\mathbf{b})}{q^2} \right) \left\langle 0 \left| \sum_a e^{i\mathbf{q}\cdot\mathbf{r}_a} \right| 0 \right\rangle. \tag{6.46}$$

In (6.44)–(6.46), $\mathbf{q} = \mathbf{p}_1 - \mathbf{p}_2$ is the momentum transferred to atom, $\hat{\mathbf{p}}_a$ is the momentum operator of the ath electron. The four-vector $b^\nu = (b^0, \mathbf{b})$ is defined as follows: $b^\nu = \overline{u}_{\mu_2}(p_2)\gamma^\nu u_{\mu_1}(p_1)$. The summation over n also means integration over the continuous spectrum of atomic excitations.

Equations (6.44)–(6.46) can be simplified allowing one to express the amplitude (6.43) in terms of the dynamic dipole polarizability $\alpha(\omega)$ of the atom. To do this, let us determine the range of the momentum transfer q. In (6.44)–(6.46), the vector $\mathbf{q}$ enters the atomic matrix elements $\langle 0 | e^{i\mathbf{q}\cdot\mathbf{r}_a} | n \rangle$. For values of q such that $\mathbf{q}\cdot\mathbf{r}_q \sim q R_{\rm at} \gg 1$, this matrix elements contain rapidly oscillating function and, thus, are small. The minimum value of q is related to the energies $\varepsilon_{1,2}$, ω through

$$q_{\rm min} = |\mathbf{p}_1| - |\mathbf{p}_2| = |(\varepsilon_1^2 - m^2)^{1/2} - (\varepsilon_2^2 - m^2)^{1/2} \approx \omega/v_1. \tag{6.47}$$

For values of ω satisfying the condition $\omega \sim \omega_{\rm at}$ (where $\omega_{\rm at} \sim v_{\rm at}/R_{\rm at}$ is the characteristic atomic frequency), the inequality $q_{\rm min} R_{\rm at} \sim v_{\rm at}/v_1 \ll 1$ is met (recall that the motion of atomic electrons is treated non-relativistically). Thus, there is a region of the transferred momenta $q \ge q_{\rm min}$ within which the expansion $\exp(i\mathbf{q}\cdot\mathbf{r}_j) = 1 + i\mathbf{q}\cdot\mathbf{r}_j$ is valid. We will show below that, with the logarithmic accuracy, this region dominates the PBrS cross section so that large distances between the incident electron and the atom are most important. Using the aforementioned expansion the PBrS amplitude can be reduced to the following formula:

$$f_{\rm pol} = e^3 \left[b^0 \frac{(\mathbf{e}\cdot\mathbf{q})}{q^2} + \left(\mathbf{e}\cdot\mathbf{b} - \frac{(\mathbf{b}\cdot\mathbf{q})(\mathbf{e}\cdot\mathbf{q})}{q^2} \right) \frac{\omega}{\omega^2 - q^2} \right] \omega\alpha(\omega), \tag{6.48}$$

6.3.2 Characteristics of PBrS

Taking into account the ordinary and the polarization terms the total amplitude, we writes the BrS cross section as the sum of three terms:

$$d\sigma = d\sigma_{\rm ord} + d\sigma_{\rm pol} + d\sigma_{\rm int}. \tag{6.49}$$

The ordinary part $d\sigma_{\rm ord}$ of the cross section has been thoroughly studied (see, for example, [233]). Therefore, below the focus is made on the polarizational and interference terms.

6.3.2.1 Spectrum of PBrS

To derive an expression for the spectral distribution of photons, one first notes that the interference term, $\mathrm{d}\sigma_{\mathrm{int}}$, is negligibly small in comparison with the two other terms. The physical reason for its small value is that the PBrS is dominated by the region of large distances, $R \gg R_{\mathrm{at}}$, between the incident electron and the atom , while the OBrS is instead dominated by the distances $R \lesssim R_{\mathrm{at}}$. Therefore, the interference term can therefore be ignored in the total cross section.

The PBrS cross section, differential in the photon energy ω and the solid angle of the emission $\Omega_{\mathbf{k}}$, as well as in the solid angle $\Omega_{\mathbf{p}_2}$ of the scattered electron (its element can be expressed through the transferred momentum q as follows: $\mathrm{d}\Omega_{\mathbf{p}_2} = \mathrm{d}\varphi_{\mathbf{q}} q \mathrm{d}q / p_1 p_2$) where $\varphi_{\mathbf{q}}$ is the azimuthal angle of $\mathbf{q}$) is given by

$$\frac{\mathrm{d}^4\sigma_{\mathrm{pol}}}{\mathrm{d}\omega \mathrm{d}\Omega_{\mathbf{k}} \mathrm{d}\varphi_{\mathbf{q}} \mathrm{d}q} = \frac{q}{8\pi^2 v_1^2} \frac{\omega}{\varepsilon_1 \varepsilon_2} \sum_{\mu_1 \mu_2 \lambda} |f_{\mathrm{pol}}|^2. \tag{6.50}$$

Here the sum $\sum_\lambda$ is carried out over the photon polarizations.

Substituting expression (6.48) into (6.50) one carries out summations over the polarizational states of the particles, integrates over the angle variable and over the transferred momentum from $q = q_{\min}$ up to the cutoff value $q_{\max} \sim 1/R_{\mathrm{at}}$. In view of the uncertainty in the cutoff value, the resulting expression for $\mathrm{d}\,\sigma_{\mathrm{pol}}\,/\mathrm{d}\omega$ is presented in the 'logarithmic approximation', i.e. neglecting all non-logarithmic terms [35]:

$$\frac{\mathrm{d}\sigma_{\mathrm{pol}}}{\mathrm{d}\omega} = \frac{16 e^2}{3 v_1^2} \omega^3 \, |\alpha(\omega)|^2 \ln \frac{v_1 \varepsilon_1}{m \omega R_{\mathrm{at}}}. \tag{6.51}$$

The terms of the order $\omega/\varepsilon_1 \ll 1$ are omitted on the right-hand side of (6.51). The logarithm of the large quantity $v_1\varepsilon_1/m\omega R_{\mathrm{at}} \gg 1$ appears due to integration over the region $q \gtrsim R_{\mathrm{at}}^{-1}$.

Expression (6.51) generalizes the result of non-relativistic theory. Indeed, putting $\varepsilon_1 = m$ on the right-hand side one obtains the expression derived in [45, 454], where it was called the 'logarithmic approximation' for PBrS. Let us note here that the OBrS cross section of a relativistic projectile in collision with a neutral target is almost independent on ε_1 in the domain of high velocities $v_1 \lesssim 1$. Therefore, the total cross section also increases logarithmically with ε_1.

The Coulomb gauge for the photon propagator, used to derive the amplitude f_{pol}, allows one to explicitly determine the contributions to cross section (6.51) from the scalar (Coulomb) interaction and the vector interaction between the incident particle and the target atom. To this end, the logarithmic factor would be written as follows

$$\ln \frac{v_1 \varepsilon_1}{m \omega R_{\mathrm{at}}} = \ln \frac{v_1}{\omega R_{\mathrm{at}}} + \ln \frac{\varepsilon_1}{m}. \tag{6.52}$$

The Coulomb interaction results in the first term on the right-hand side of (6.52), while the exchange with the transverse virtual photon (the vector part of the interaction) leads to the second term. With increasing energy, the second term greatly exceeds the first one.

6.3.2.2 Spectral-Angular Distribution of PBrS

Integrating (6.50) over the angular variables of the scattered electron, one derives the two-fold cross section which characterized the spectral-angular distribution of the emitted radiation:

$$\frac{d^2\sigma_{\text{pol}}}{d\omega d\Omega_{\mathbf{k}}} = \frac{e^2\omega^3}{\pi v_1^2}|\alpha(\omega)|^2(1+\cos^2\theta_{\mathbf{k}})\ln\frac{v_1\varepsilon_1}{m\omega R_{\text{at}}}. \tag{6.53}$$

The angular profile of the PBrS radiation, which is characterized by the factor $(1+\cos^2\theta_{\mathbf{k}})$, is that of radiation emitted by a rotating electric dipole [278]. This weakly anisotropic angular distribution differs substantially from that of the OBrS. The latter is primarily into the cone along the vector $\mathbf{p}_1$ and the opening angle which decreases rapidly as the velocity v_1 increases. In the ultra-relativistic case, $v_1 \to 1$, all radiation emitted via the OBrS mechanism is concentrated in the cone $\theta_{\mathbf{k}} \sim 1/\gamma \ll 1$ where $\gamma = E_1/m$ is the relativistic Lorentz factor. Hence, the angular distribution of the total BrS radiation, $d^2\sigma_{\text{ord}} + d^2\sigma_{\text{pol}}$, is characterized by a smooth dependence for $\theta_{\mathbf{k}} > 1/\gamma$ and a sharp anisotropy as $\theta_{\mathbf{k}} < 1/\gamma$. A similar photon angular distribution was found in [197] for the bremsstrahlung arising in electron–nucleus collision with account for dynamic polarizability of the nucleus.

6.3.2.3 Polarization of PBrS

To find the angular distribution of the polarized photons in the collision of unpolarized electrons by an atomic system, we choose the vectors $\mathbf{e}^{(1)} = \mathbf{p}_1 \times \mathbf{k}/\left|\mathbf{p}\times\mathbf{k}\right|$ and $\mathbf{e}^{(2)} = \mathbf{k}\times\mathbf{e}^{(1)}/\omega$, as the polarization unit vectors. Then one finds

$$\frac{d^2\sigma_{\text{pol}}(\xi_i)}{d\omega d\Omega_{\mathbf{k}}} = \frac{e^2\omega^3}{2\pi v_1^2}|\alpha(\omega)|^2\left(1+\cos^2\theta_{\mathbf{k}}+\xi\sin^2\theta_{\mathbf{k}}\right)\ln\frac{v_1\varepsilon_1}{m\omega R_{\text{at}}}. \tag{6.54}$$

Here ξ is the Stoke parameter averaged over the angles of the scattered electron. This quantity is a measure of the degree of linear polarization of the photon along the vectors $\mathbf{e}^{(1)}$ and $\mathbf{e}^{(2)}$. Namely, the probability of the photon polarization along $\mathbf{e}^{(1)}$ is equal to $(1+\xi)/2$, and along $\mathbf{e}^{(2)}$ is $(1-\xi)/2$. Hence, the PBrS radiation from unpolarized electrons is linearly polarized as well as the OBrS radiation [81].

6.3.2.4 Angular Distribution of Scattered Electron

As the velocity of the incident electron increases a peculiar feature appears in the angular distribution of scattered electrons. To analyze it one begins with writing the double differential cross section of PBrS as a function of the momentum transfer q:

$$\frac{d^2\sigma_{\rm pol}}{d\omega dq} = \frac{16e^2}{3v_1^2}|\alpha(\omega)|^2\frac{q}{q^2-\omega^2}. \tag{6.55}$$

The denominator $q^2 - \omega^2$ becomes very small as $q \to q_{\rm min} \approx \omega/v_1$. In the ultra-relativistic limit $q_{\rm min}^2 - \omega^2 \propto \gamma^{-2} \ll 1$. Thus, the contribution of the small scattering angles $\theta_{\mathbf{p}_2}$ region to the cross section increases with the incident energy. In this region $d^2\sigma_{\rm pol}$ becomes especially large in comparison with $d^2\sigma_{\rm ord}$.

6.3.3 *Total BrS Spectrum*

Neglecting the interference term in (6.49), one writes the total BrS spectrum as

$$\frac{d\sigma}{d\omega} \approx \frac{d\sigma_{\rm pol}}{d\omega} + \frac{d\sigma_{\rm ord}}{d\omega}. \tag{6.56}$$

A cumbersome formula, which represents the OBrS spectrum in general relativistic case, is simplified considerably in the ultra-relativistic limit $E_{1,2} \gg m$. For a neutral atom the spectrum $d\sigma_{\rm ord}/d\omega$ reads (see, for example, [81]):

$$\left.\frac{d\sigma_{\rm ord}}{d\omega}\right|_{v_1\to 1} = \frac{16Z^2e^2}{3}\frac{r_e^2}{\omega}\ln(mR_{\rm at}). \tag{6.57}$$

Here $r_{\rm e}$ is the classical radius of electron. The logarithm $\ln(mR_{\rm at})$ is due to the screening effect of atomic electrons.

Let us demonstrate, that in the ultra-relativistic limit the total cross section is fundamentally different from that in the non-relativistic case.

The ultra-relativistic limit of $d\sigma_{\rm pol}/d\omega$ one obtains setting $v_1 = 1$ in (6.51):

$$\left.\frac{d\sigma_{\rm pol}}{d\omega}\right|_{v_1\to 1} = \frac{16e^2}{3}\,\omega^3\,|\alpha(\omega)|^2 \ln\frac{\varepsilon_1}{m\omega R_{\rm at}}. \tag{6.58}$$

Let us consider the spectrum for photon energies much larger than energies of the atomic electrons, i.e. the limit $\omega \gg I_{1s}$, where I_{1s} is the K-shell ionization potential. For such values of ω, the non-relativistic dynamic polarizability of a neutral atom with nucleus charge Z reduces to $\alpha(\omega) \approx -Z/m\omega^2$ (we remind, that the relativistic system of units is used in this section). Combining (6.57) and (6.58) one derives the following formula for the total BrS spectrum [35]:

$$\left.\frac{d\sigma_{tot}}{d\omega}\right|_{\substack{v_1\to 1\\ \omega>I_{1s}}} = \frac{16Z^2e^2}{3}\frac{r_e^2}{\omega}\ln\frac{\varepsilon_1}{\omega}. \tag{6.59}$$

Apart from the factor Z^2 the right-hand side of (6.59) reproduces the formula for the cross section of BrS emitted by a slow free electron in the collision with an ultra-relativistic electron or positron [81]. This coincidence is not accidental and has clear qualitative explanation [35]. For $\omega \gg I_{1s}$, the atomic electrons can be treated as free ones. If $Z\alpha \ll 1$ (with $\alpha \approx 1/137$ being the fine structure constant) then the velocities of all atomic electrons are small compared to that of the projectile. Then, the total BrS amplitude can be written as the sum of three terms $f_{tot} = Z\mathscr{F}_1 + Z\mathscr{F}_2 + \mathscr{F}_3$. Here $\mathscr{F}_1$ denotes the amplitude of the photon emission by the projectile electron/positron interacting with a free atomic electron, $\mathscr{F}_2$ is the amplitude of the emission by the atomic electron during this interaction, and $\mathscr{F}_3$ is the amplitude due to the interaction of the projectile with the nucleus. As known (see, e.g., [81]), the BrS amplitude of an ultra-relativistic electron/positron scattered from a free slow particle depends on the particle's charge but does not depend on its mass. Therefore, the sum of two terms, $Z\mathscr{F}_1 + \mathscr{F}_3$ is identically equal to zero, and f_{tot} reduces to $Z\mathscr{F}_1$, i.e. only Z atomic electrons radiate in this process.

In a way, this result is opposite to the 'stripping' effect in the non-relativistic BrS, when $d\sigma_{tot}(\omega)$ in electron-atom collision reduces to that on the bare nucleus if $\omega \gg I_{1s}$ (see [12] and (4.21)) with $F_{in} \equiv 0$). The qualitative explanation of this difference between the non-relativistic and the ultra-relativistic cases is as follows. In the former case, the range of distances between the projectile and the target important in the PBrS process can be estimated as $R_{eff} \sim 1/q_{min} \approx v_1/\omega$. This value is much smaller than the photon wavelength $\lambda = 2\pi/\omega$. Therefore, the dipole approximation can be applied for the system 'the projectile + the target'. The only allowed radiation in this system in the range $\omega \gg I_{1s}$ (the limit of quasi-free atomic electrons) is that by a projectile on the nucleus. As was already mentioned, in the ultra-relativistic case $R_{eff} \sim 1/q_{min}^{\perp} \approx \gamma/\omega$. As $v_1 \to 1$ and $\gamma \to \infty$ these distances increase unrestrictedly and are much greater than λ. The retardation effects in the interaction between the projectile and the atomic electron become important, so that the dipole approximation is inapplicable to this system. This leads to the difference of (6.59) from its non-relativistic analogue (4.21).

Returning to expressions (6.57) and (6.58), we note that PBrS is determined entirely by a dynamic polarizability of the atom and does not depend on the mass of a projectile. In contrast, $d\sigma_{ord}$ is inversely proportional to the mass squared. Therefore, for massive projectiles one can neglect OBrS completely and observe PBrS in its pure form.

The total BrS amplitude for a relativistic positron on an atom equals to the sum of (6.42) and (6.48) but taken with the opposite sign (as the sign of a positron is opposite to that of an electron). Equations (6.51)–(6.59) for the cross sections of polarizational and total BrS are thus also applicable to the case of a relativistic positron.

6.4 Inelastic Collisions. Coherence Effect

In this section we analyze the role of PBrS in formation of the total radiation spectrum in the collision of two relativistic atomic particles. For doing this we evaluate the formulae describing spectral-angular and spectral distributions of the emitted radiation [43]. From theoretical viewpoint, it is not a simple task to take into account all possible radiative processes which contribute to the total spectrum. The complexity is due, on the one hand, to the fact that in collision of two (complex) atomic particles there occur, along with 'pure' PBrS, a number of other radiative processes accompanied by the atom-projectile or/and atom-target excitation or ionization. On the other hand, the relativistic collision velocities make the problem more complicated than in the non-relativistic domain (see Sect. 5.2).

In what follows, apart from presenting the general formalism, we pay special attention to evaluating the relative contributions of the elastic and non-elastic BrS channels (see Sect. 5.3) to the total spectrum. It is shown that there exist broad ranges of photon energy ω and emission angle $\theta_{\mathbf{k}}$ within which the elastic PBrS dominates, thus making possible its experimental study. At the same time, there exist quite substantial region of the parameters where the situation is the opposite [43].

6.4.1 *Amplitude of "Inelastic" PBrS*

Let us consider the 'inelastic' radiative collision of a pair of relativistic atomic particles (called, for clarity, atoms), in which the emission of a photon of momentum **k** and polarization **e** is accompanied by excitation or ionization of the atoms. The case of 'elastic' BrS, when after the collision the atoms remain in the initial (ground) states is considered in Sect. 6.2.

In the lowest order of perturbation theory in the atom–atom interaction and in the photon–atom interaction, the amplitude of the process is described by the four Feynman diagrams presented in Fig. 6.1. The description of the diagrams, presented in Sect. 6.2.1, is applicable to the 'inelastic' BrS as well with the only difference, that the final electron states of the colliding atoms differ from their initial states.

The notations used below are the same as in Sect. 6.2: the lower/upper indices '1'/'(1)' refer to the projectile, the lower/upper indices '2'/'(2)'—to the target atom.

To calculate the amplitude of inelastic PBrS we apply the approach, described in Sect. 6.2.1. Namely, the amplitude f_1, which stands for the photon emission by the projectile (the fist two diagrams in Fig. 6.1), is proportional to the product of the four-current $J_{\mu}^{(2)}$ of the target and the polarizability four-tensor of the $T_{\mu\nu}^{(1)}$ projectile. Two other diagrams, which describe the amplitude f_2 of the photon emission by the target, one derives from f_1 by interchanging the indices 1 and 2. The calculation of the four-currents $J_{\mu}^{(j)}$ and the polarizability four-tensors $T_{\mu\nu}^{(j)}$ $(j = 1, 2)$ is simplified in the rest frames of the projectile and the target. The applicability of the concept of an inertial rest frame for either one of the colliding atoms is justified, since the momenta transferred in the collision are small enough (it is demonstrated below) thus allowing one to approximate the trajectories of the particles with straight lines. Once the quantities $J_\mu^{(j)}$ and $T_{\mu\nu}^{(j)}$ are calculated in the corresponding rest frame they can be re-calculated into the laboratory reference frame with the help of the Lorentz transformations.

Omitting the intermediate algebra, which is quite similar to the one outlined in evaluating the formulae for the elastic BrS (see [38, 39] and Sect. 6.2.1) we present the final formulae for the amplitudes $f_{1,2}$ which constitute the total amplitude $f = f_1 + f_2$ of the inelastic BrS. Additionally, below the amplitudes are considered in the limit $\omega^{\mathrm{r}} \gg \omega_{\mathrm{at}}$ for each atom (with ω_{at} standing for the characteristic atomic frequency) where the expressions are simplified considerably [43]:

$$f_2 = \frac{4\pi}{q_1^2 - (\omega + \omega_{m_2 0_2})^2 \big/ c^2} \frac{\omega}{c} \left[\mathbf{e} \cdot \mathbf{q}_1 - \frac{\omega}{c^2} \mathbf{e} \cdot \mathbf{v}_1 \right] \times S_{m_1 0_1}^{(1)}(\mathbf{q}_1^{\mathrm{r}})\, \alpha_{m_2 0_2}^{(2)}(\omega, \mathbf{Q}_2)\,, \tag{6.60}$$

$$f_1 = \frac{4\pi}{q_2^2 - \omega_{m_2 0_2}^2 / c^2} \frac{1}{c} \left[\omega^{\mathrm{r}} \mathbf{e} \cdot \mathbf{q}_2^{\perp} + \gamma (\mathbf{e} \cdot \mathbf{v}_1)(\mathbf{k} \cdot \mathbf{q}_2^{\perp}) - \frac{\omega}{\gamma^2} \mathbf{e} \cdot \mathbf{q}_1^{\parallel} \right] \times S_{m_2 0_2}^{(2)}(\mathbf{q}_2^{\mathrm{r}})\, \alpha_{m_1 0_1}^{(1)}(\omega^{\mathrm{r}}, \mathbf{Q}_1^{\mathrm{r}})\,. \tag{6.61}$$

Here 0_j and m_j ($j = 1, 2$) denote the initial and the final electronic states of the atoms, $\omega_{m_j 0_j} = \varepsilon_{m_j} - \varepsilon_{0_j}$ is the transition energy, the energy ω^{r} is defined as in (6.24). The superscript 'r' indicates that the quantity is calculated in the corresponding rest frame (see also Sect. 6.2). The quantity $S_{m_j 0_j}^{(j)}(\mathbf{q})$ is equal to

$$S_{m_j 0_j}^{(j)}(\mathbf{q}) = Z_j \delta_{m_j 0_j} - F_{m_j 0_j}^{(j)}(\mathbf{q}) \tag{6.62}$$

where $F_{m_j 0_j}^{(j)}(\mathbf{q})$, defined in (6.13), is a non-diagonal form-factor of atomic electrons.

In the limit of large photon energies the expressions for non-diagonal generalized polarizabilities $\alpha_{m_j 0_j}^{(j)}(\omega, \mathbf{Q}_j)$ ($j = 1, 2$) are quite simple. Here we present the formula for the polarizability of the projectile in its rest frame:

$$\alpha_{m_1 0_1}^{(1)}(\omega^{\mathrm{r}}, \mathbf{Q}_1^{\mathrm{r}}) \approx -\frac{1}{(\omega^{\mathrm{r}})^2} \left[\frac{(Z_1 - N_1)\, S_{m_1 0_1}^{(1)}(\mathbf{Q}_1^{\mathrm{r}})}{M_1} + F_{m_1 0_1}^{(1)}(\mathbf{Q}_1^{\mathrm{r}}) \right]. \tag{6.63}$$

The transferred momenta (also calculated in the rest frame) are given by

$$\begin{cases} \mathbf{q}_1^{\mathrm{r}} = \gamma^{-1} \mathbf{q}_1^{\parallel} + \dfrac{\omega_{m_1 0_1}}{c} \dfrac{\mathbf{v}_1}{c} + \mathbf{q}_1^{\perp} \approx \gamma^{-1} \mathbf{q}_1^{\parallel} + \mathbf{q}_1^{\perp}, \\ \mathbf{Q}_1^{\mathrm{r}} = \dfrac{\omega^{\mathrm{r}}}{v_1} \dfrac{\omega_{m_1 0_1}}{c} \dfrac{\mathbf{v}_1}{c} + \mathbf{q}_1^{\perp} - \mathbf{k}^{\perp} \approx \dfrac{\omega^{\mathrm{r}}}{v_1^2} \mathbf{v}_1 + \mathbf{q}_1^{\perp} - \mathbf{k}^{\perp}\,. \end{cases} \tag{6.64}$$

In deriving (6.60) and (6.61) the condition $q_1^{\mathrm{r}} \gg \omega_{m_1 0_1}/c$, valid for $\omega^{\mathrm{r}} \gg \omega_{\mathrm{at}}^{\mathrm{r}}$, was used.

The two terms in the right-hand side of (6.63) are of different nature (analogous arguments are applicable to the two terms in the polarizability $\alpha_{m_2 0_2}^{(2)}(\omega^{\mathrm{r}}, \mathbf{Q}_2^{\mathrm{r}})$ which are not written explicitly here). The first one appears due to the radiation of the first atom as it radiates as a whole in the field of the second atom. The second term results from taking into account the polarization of the atom in the collision process. For a neutral ($Z_1 = N_1$) or heavy ($M_1 \gg 1$) atoms the first term can be neglected. In the case of a structureless particle (electron, positron) only the first term remains.

In the non-relativistic domain of the collision velocities, the amplitude $f = f_1 + f_2$ reduces to its non-relativistic limit

$$f = \frac{4\pi}{\omega c} \frac{\mathbf{e} \cdot \mathbf{q}_1}{q_1^2} \left[\left(\frac{Z_1 - N_1}{M_1} - \frac{Z_2 - N_2}{M_2} \right) S_{m_1 0_1}^{(1)}(\mathbf{q}_1) S_{m_2 0_2}^{(2)}(-\mathbf{q}_1) \right.$$
$$\left. - \left(Z_1 \delta_{m_1 0_1} F_{m_2 0_2}^{(2)}(-\mathbf{q}_1) - Z_2 \delta_{m_2 0_2} F_{m_2 0_2}^{(1)}(\mathbf{q}_1) \right) \right]. \qquad (6.65)$$

This expression is a generalization of the formula (5.14) for the amplitude of 'elastic' BrS in the collision of two non-relativistic atomic particles to the case of arbitrary final states of the atoms.

Using formulae (6.60)–(6.64) one can derive the BrS cross sections in neutral atom–atom collisions as well as in collisions involving relativistic ions.

6.4.2 Spectral-Angular Distribution of PBrS

The differential cross section of BrS is related to the total amplitude $f = f_1 + f_2$ via

$$\frac{\mathrm{d}^4\sigma}{\mathrm{d}\omega \mathrm{d}\Omega_{\mathbf{k}} \mathrm{d}q_1 \mathrm{d}\varphi_{\mathbf{q}_1}} = \frac{1}{(2\pi)^4} \frac{\omega q_1}{c v_1^2} |f_1 + f_2|^2 . \qquad (6.66)$$

Substituting (6.60) and (6.61) into (6.66), one evaluates the sums over the photon polarizations and the final states of $m_{1,2}$ of the atoms, and integrates over the azimuthal angle $\varphi_{\mathbf{q}_1}$ of the vector $\mathbf{q}_1$. As a result, the double differential cross sections characterizing the spectral-angular distributions of the radiation emitted by the projectile and the target (the terms $\mathrm{d}^2\sigma_1$ and $\mathrm{d}^2\sigma_2$)[3] can be written as follows:

[3] The interference term $\mathrm{d}^2\sigma_{12}$ is not written down in the intermediate transformations but will appear in the final formulae.

$$\frac{\mathrm{d}^2\sigma_2}{\mathrm{d}\omega \mathrm{d}\Omega_{\mathbf{k}}} = \sum_{m_1,m_2} \int\limits_{q_{1,\min}^{(m_1,m_2)}}^{q_{1,\max}^{(m_1,m_2)}} \frac{\mathrm{d}^3\sigma_2(m_1,m_2)}{d\omega \mathrm{d}q_1 \mathrm{d}\Omega_{\mathbf{k}}}\, \mathrm{d}q_1,$$

$$\frac{\mathrm{d}^2\sigma_1}{\mathrm{d}\omega \mathrm{d}\Omega_{\mathbf{k}}} = \sum_{m_1,m_2} \int\limits_{q_{2,\min}^{(m_1,m_2)}}^{q_{2,\max}^{(m_1,m_2)}} \frac{\mathrm{d}^3\sigma_1(m_1,m_2)}{d\omega \mathrm{d}q_1 \mathrm{d}\Omega_{\mathbf{k}}}\, \mathrm{d}q_1, \tag{6.67}$$

where

$$\frac{\mathrm{d}^3\sigma_2(m_1,m_2)}{d\omega \mathrm{d}\Omega_{\mathbf{k}} \mathrm{d}q_1} = \frac{q_1}{\pi c^3 v_1^2 \omega} \frac{\left(q_1^{\perp}\right)^2 (1+\cos^2\theta_{\mathbf{k}}) + 2\sin^2\theta_{\mathbf{k}} \left(q_1^{\parallel} - \beta\omega/c\right)^2}{\left(q_1^2 - (\omega+\omega_{m_2 0_2})^2/c^2\right)^2}$$
$$\times \left|\left(Z_1\delta_{m_1 0_1} - W^{(1)}_{m_1 0_1}(\mathbf{q}_1^{\mathrm{r}})\right) W^{(2)}_{m_2 0_2}(-\mathbf{q}_1)\right|^2, \tag{6.68}$$

$$\frac{\mathrm{d}^3\sigma_1(m_1,m_2)}{d\omega \mathrm{d}\Omega_{\mathbf{k}} \mathrm{d}q_2} = \frac{q_2}{\pi c^3 v_1^2 \omega} \left(\frac{\omega}{\omega^{\mathrm{r}}}\right)^2 \frac{\left(q_1^{\perp}\right)^2 (1+\cos^2\theta_{\mathbf{k}}^{\mathrm{r}}) + 2\left(q_2^{\parallel}\right)^2 \sin^2\theta_{\mathbf{k}}^{\mathrm{r}}}{\left(q_2^2 - \omega_{m_2 0_2}^2/c^2\right)^2}$$
$$\times \left|\left(Z_2\delta_{m_2 0_2} - W^{(2)}_{m_2 0_2}(\mathbf{q}_2)\right) W^{(1)}_{m_1 0_1}(\mathbf{Q}_1^{\mathrm{r}})\right|^2, \tag{6.69}$$

The vectors $\mathbf{q}_1^{\mathrm{r}}$ and $\mathbf{Q}_1^{\mathrm{r}}$ are defined in (6.64).

The upper and lower limits of integration in (6.67)–(6.68) one derives from the energy conservation law:

$$\begin{cases} q_{1,\min}^{m_1,m_2} = \dfrac{\omega + \omega_{m_1 0_1} + \omega_{m_2 0_2}}{v_1} \\ q_{2,\min}^{m_1,m_2} = q_{1,\min}^{m_1,m_2} - \dfrac{\omega}{c}\cos\theta_{\mathbf{k}} \end{cases}, \qquad \begin{cases} q_{1,\max}^{m_1,m_2} = 2p_1 - q_{1,\min}^{m_1,m_2} \\ q_{2,\max}^{m_1,m_2} = 2p_1 + k - q_{2,\min}^{m_1,m_2} \end{cases} \tag{6.70}$$

Below, it is shown that the main contribution to the spectrum comes from the non-relativistic region of transferred momentum. Therefore, the upper limits can be set to infinity.

To carry out the summation over the final states of the atoms one notices the following. In the region of the momenta transfer $q_j^2 \sim 2\omega_{m_j 0_j}$ $(j = 1, 2)$, which is of the most importance for the integrals in (6.67) and (6.68), the quantities $(\omega_{m_j 0_j}/c)^2 \sim q_j^4/c^2 \ll q_j^2$ and $(\omega_{m_j 0_j}/c)(\omega/c) \sim q_j^2\omega/c^2 \ll q_j^2$ can be omitted in the integrands. As a result, the term $\mathrm{d}^2\sigma_2/\mathrm{d}\omega\mathrm{d}\Omega_{\mathbf{k}}$ can be written as follows:

$$\frac{d^2\sigma_2}{d\omega d\Omega_{\mathbf{k}}} = \frac{1}{\pi c^3 v_1^2 \omega}\left\{ \int\limits_{\omega/v_1}^{\infty} \frac{q_1\,dq_1}{q_1^2-\omega^2/c^2}\Big(Z_1 - F_1(q_1)\Big)^2 F_2^2(q_1)\,\mathscr{P}_2(q_1,\theta_{\mathbf{k}})\right.$$
$$+ \int\limits_{\omega/v_1}^{\infty} \frac{q_1\,dq_1}{q_1^2-\omega^2/c^2}\Big(N_1 + W_1(q_1) - F_1^2(q_1)\Big) F_2^2(q_1)\,\mathscr{P}_2(q_1,\theta_{\mathbf{k}})$$
$$+ \int\limits_{\omega/v_1}^{2v_1} \frac{q_1\,dq_1}{q_1^2-\omega^2/c^2}\Big(Z_1 - F_1(q_1)\Big)^2 \Big(N_2 + W_2(q_1) - F_2^2(q_1)\Big)\,\mathscr{P}_2(q_1,\theta_{\mathbf{k}})$$
$$+ \int\limits_{\omega/v_1}^{v_1} \frac{q_1\,dq_1}{q_1^2-\omega^2/c^2}\Big(N_1 + W_1(q_1) - F_1^2(q_1)\Big)$$
$$\left. \times \Big(N_2 + W_2(q_1) - F_2^2(q_1)\Big)\,\mathscr{P}_2(q_1,\theta_{\mathbf{k}})\right\}. \tag{6.71}$$

Here $F_{1,2}(q_1)$ are the ground-state electron form-factors of the colliding atoms, and the following notations have been introduced:

$$\mathscr{P}_2(q_1,\theta_{\mathbf{k}}) = 1 + \cos^2\theta_{\mathbf{k}} - \frac{v_1^2/\omega^2\gamma^2}{q_1^2-\omega^2/c^2}\left[2 - \left(3 - 2\frac{v_1^2}{c^2}\right)\sin^2\theta_{\mathbf{k}}\right],$$
$$F_{1,2}(q_1) = \left\langle 0_{1,2}\left|\sum_{a\neq b}\exp\Big[\mathrm{i}\mathbf{q}_1\cdot(\mathbf{r}_a - \mathbf{r}_b)\Big]\right|0_{1,2}\right\rangle. \tag{6.72}$$

In writing (6.72) the argument in the functions $F_1(q_1^{\mathrm{r}})$ and $W_1(q_1^{\mathrm{r}})$ have been replaced with q_1, which is correct in the region $\omega R_{\mathrm{at}}/v_1\gamma \sim \omega R_{\mathrm{at}}/c\gamma \ll 1$, or if the condition $q_1^{\perp} \gg q_1^{\parallel}$ is met, when $\exp(\mathrm{i}\mathbf{q}_1^{\parallel}\cdot\mathbf{r}_j/\gamma) \approx 1$.

Expression (6.72) contains four terms each of different nature. The first integral accounts for the elastic BrS of the target without exciting the projectile. The second term describes the elastic BrS of the target but accompanied by the projectile excitation (ionization). The upper limit of integration in this term can be made infinite since the main contribution comes from the region $q_1 \sim R_{\mathrm{at}}^{-1}$. Outside this region either $F_2(q_1) \to 0$ (for $q_1 R_{\mathrm{at}} \gg 1$), $N_1 + W_1(q_1) - F_2^2(q_1) \to 0$ (for $q_1 R_{\mathrm{at}} \ll 1$). The third term accounts for all processes of the target excitation (ionization) but without change in the state of the projectile. Finally, the last term presents the sum over all inelastic radiative channels for the target and the sum over all non-radiative inelastic processes for the projectile. Different upper limits of integration in the last two terms is due to the following reason. The thirds term accounts for the radiation emitted by the target electrons in the collision with a heavy projectile whereas in the forth term the target electrons radiate in collision with the with the projectile ones. The difference in the reduced masses in these two cases lead to different values of the maximum momentum transfer: $q_{1,\max} = 2v_1$ and $q_{1,\max} = v_1$.

The similar calculation scheme applied to $\mathrm{d}^2\sigma_1/\mathrm{d}\omega\,\mathrm{d}\Omega_{\mathbf{k}}$ (see (6.67) and (6.69)), lead to the result:

$$\frac{\mathrm{d}^2\sigma_1}{\mathrm{d}\omega \mathrm{d}\Omega_{\mathbf{k}}} = \frac{1}{\pi c^3 v_1^2 \omega}\left(\frac{\omega}{\omega^{\mathrm{r}}}\right)^2 \left\{ \int_{\omega^{\mathrm{r}}/\gamma v_1}^{\infty} \frac{\mathrm{d}q_2}{q_2} |Z_2 - F_2(q_2)|^2 F_1^2(q_2) \mathscr{P}_1(q_2, \theta_{\mathbf{k}}^{\mathrm{r}}) \right.$$
$$+ \int_{\omega^{\mathrm{r}}/\gamma v_1}^{\infty} \frac{\mathrm{d}q_2}{q_2} \left(N_2 + W_2(q_2) - F_2^2(q_2)\right) F_1^2(q_2)\, \mathscr{P}_1(q_2, \theta_{\mathbf{k}}^{\mathrm{r}})$$
$$+ \int_{\omega^{\mathrm{r}}/\gamma v_1}^{2v_1} \frac{\mathrm{d}q_2}{q_2} (Z_2 - F_2(q_2))^2 \left(N_1 + W_1(q_2) - F_1^2(q_2)\right) \mathscr{P}_1(q_2, \theta_{\mathbf{k}}^{\mathrm{r}})$$
$$+ \int_{\omega^{\mathrm{r}}/\gamma v_1}^{v_1} \frac{\mathrm{d}q_2}{q_2} \left(N_2 + W_2(q_2) - F_2^2(q_2)\right)$$
$$\left. \times \left(N_1 + W_1(q_2) - F_1^2(q_2)\right) \mathscr{P}_1(q_2, \theta_{\mathbf{k}}^{\mathrm{r}}) \right\}. \tag{6.73}$$

where

$$\mathscr{P}_1(q_2, \theta_{\mathbf{k}}^{\mathrm{r}}) = 1 + \cos^2\theta_{\mathbf{k}}^{\mathrm{r}} + (1 - 3\cos^2\theta_{\mathbf{k}}^{\mathrm{r}}) \left(\frac{q_2^{\parallel}}{q_2}\right)^2 . \tag{6.74}$$

Here, as in (6.71), we have ignored the difference between Q_1^{r} and q_2 in the arguments of the functions $W_1(Q_1^{\mathrm{r}})$ and $F_1(Q_1^{\mathrm{r}})$, which is valid either for $\omega^{\mathrm{r}} R_{\mathrm{at}}/c \ll 1$ or when $q_2^{\perp} \gg q_2^{\parallel}$.

The first two terms in (6.73) accounts for the elastic BrS of the projectile without exciting the target (the first term) or accompanied by the target excitation/ionization (the second term). The third term accounts for all processes of the projectile excitation (ionization) but without change in the state of the target. Finally, the last term presents the sum over all inelastic radiative channels for the projectile and the sum over all non-radiative inelastic processes for the target.

The expressions obtained for $\mathrm{d}^2\sigma_{1,2}/\mathrm{d}\omega\,\mathrm{d}\Omega_{\mathbf{k}}$ can be simplified further noticing, that the terms proportional to $\omega^2/\left(v_1^2\gamma(q_1^2 - \omega^2/c^2)\right)$ in the integrands in (6.71) and to $(q_2^{\parallel})^2/q_2^2$ in the integrands in (6.73), are of the same order of magnitude with the other terms only in the region $q_j \sim q_{j,\min}$ ($j = 1, 2$). In the rest part of the integration intervals, i.e. for $q_{j,\max} > q_j \gg q_{j,\min}$, the aforementioned terms are negligibly small and do not contribute to the cross sections. Thus, ignoring these terms, one arrives at [43]:

$$\frac{d^2\sigma_2}{d\omega d\Omega_{\mathbf{k}}} = \frac{1+\cos^2\theta_{\mathbf{k}}}{\pi c^3 v_1^2 \omega} \left\{ \int\limits_{\omega/v_1}^{\infty} \frac{q_1\, dq_1}{q_1^2 - \omega^2/c^2} \Big(Z_1 - F_1(q_1)\Big)^2 F_2^2(q_1) \right.$$
$$+ \int\limits_{\omega/v_1}^{\infty} \frac{q_1\, dq_1}{q_1^2 - \omega^2/c^2} \Big(N_1 + W_1(q_1) - F_1^2(q_1)\Big) F_2^2(q_1)$$
$$+ \int\limits_{\omega/v_1}^{2v_1} \frac{q_1\, dq_1}{q_1^2 - \omega^2/c^2} \Big(Z_1 - F_1(q_1)\Big)^2 \left(N_2 + W_2(q_1) - F_2^2(q_1)\right)$$
$$\left. + \int\limits_{\omega/v_1}^{v_1} \frac{q_1\, dq_1}{q_1^2 - \omega^2/c^2} \Big(N_1 + W_1(q_1) - F_1^2(q_1)\Big)\Big(N_2 + W_2(q_1) - F_2^2(q_1)\Big) \right\}, \tag{6.75}$$

$$\frac{d^2\sigma_1}{d\omega d\Omega_{\mathbf{k}}} = \frac{1+\cos^2\theta_{\mathbf{k}}^{\mathrm{r}}}{\pi c^3 v_1^2 \omega} \left(\frac{\omega}{\omega^{\mathrm{r}}}\right)^2 \left\{ \int\limits_{\omega^{\mathrm{r}}/\gamma v_1}^{\infty} \frac{dq_2}{q_2} \Big(Z_2 - F_2(q_2)\Big)^2 F_1^2(q_2) \right.$$
$$+ \int\limits_{\omega^{\mathrm{r}}/\gamma v_1}^{\infty} \frac{dq_2}{q_2} \Big(N_2 + W_2(q_2) - F_2^2(q_2)\Big) F_1^2(q_2)$$
$$+ \int\limits_{\omega^{\mathrm{r}}/\gamma v_1}^{2v_1} \frac{dq_2}{q_2} \Big(Z_2 - F_2(q_2)\Big)^2 \Big(N_1 + W_1(q_2) - F_1^2(q_2)\Big)$$
$$\left. + \int\limits_{\omega^{\mathrm{r}}/\gamma v_1}^{v_1} \frac{dq_2}{q_2} \Big(N_2 + W_2(q_2) - F_2^2(q_2)\Big)\Big(N_1 + W_1(q_2) - F_1^2(q_2)\Big) \right\}. \tag{6.76}$$

Within the same approximation the interference term in the cross section (6.66) reads [43]:

$$\frac{d^2\sigma_{12}}{d\omega d\Omega_{\mathbf{k}}} = -\frac{2\omega}{\omega^{\mathrm{r}}} \frac{1+\cos\theta_{\mathbf{k}}\cos\theta_{\mathbf{k}}^{\mathrm{r}}}{\pi c^3 v_1^2 \omega}$$
$$\times \left\{ \int\limits_{\omega^{\mathrm{r}}/\gamma v_1}^{\infty} \frac{dq_2}{q_2} \Big(Z_1 - F_1(q_2)\Big)\Big(Z_2 - F_2(q_2)\Big) F_1(q_2) F_2(q_2) \right.$$

$$
\begin{aligned}
& - \int\limits_{\omega^{\mathrm{r}}/\gamma v_1}^{\infty} \frac{\mathrm{d}q_2}{q_2} \Big(Z_1 - F_1(q_2)\Big) F_1(q_2) \Big(N_2 + W_2(q_2) - F_2^2(q_2)\Big) \\
& - \int\limits_{\omega^{\mathrm{r}}/\gamma v_1}^{2v_1} \frac{\mathrm{d}q_2}{q_2} \Big(Z_2 - F_2(q_2)\Big) F_2(q_2) \Big(N_1 + W_1(q_2) - F_1^2(q_2)\Big) \\
& + \int\limits_{\omega^{\mathrm{r}}/\gamma v_1}^{v_1} \frac{\mathrm{d}q_2}{q_2} \Big(N_1 + W_1(q_2) - F_1^2(q_2)\Big) \Big(N_2 + W_2(q_2) - F_2^2(q_2)\Big) \Bigg\} .
\end{aligned}
\tag{6.77}
$$

Expressions (6.75)–(6.77) demonstrate that the angular distributions of the radiation emitted by the projectile and by the target are quite different in the region of sufficiently high velocities of collision. For the target, its profile $\mathrm{d}^2\sigma_2/d\omega\, d\Omega_{\mathbf{k}} \propto (1+\cos^2\theta_{\mathbf{k}})$, is weakly anisotropic. On the contrary, the radiation of the projectile, being proportional to

$$
\frac{\mathrm{d}^2\sigma_1}{\mathrm{d}\omega \mathrm{d}\Omega_{\mathbf{k}}} \sim \left. \frac{1+\cos^2\theta_{\mathbf{k}}}{\gamma^2(1-\beta\cos^2\theta_{\mathbf{k}})^2} \right|_{\beta = v_1/c}, \tag{6.78}
$$

is emitted in a narrow cone $\theta_{\mathbf{k}} \le \gamma^{-1}$ along the direction of motion for $v_1 \lesssim c$.

Comparing (6.75) and (6.77) one finds, that for large emission angles $\theta_{\mathbf{k}} \gg \gamma^{-1}$ (and for the photon frequencies $\omega^{\mathrm{r}} \gg \omega_{\mathrm{at}}$) the radiation intensity of the target exceeds that of the projectile by the factor γ^2. For small emission angles, $\theta_{\mathbf{k}} \ll \gamma^{-1}$, the situation is inverse: it is the radiation intensity of the projectile which exceeds that of the target by the same factor. Thus, the emission of high-energy photons by each of the atoms is concentrated in its own angular region. The interference term becomes important only in the intermediate region $\theta_{\mathbf{k}} \sim \gamma^{-1}$.

Let us demonstrate the role of the coherent radiation of the electrons in the process of elastic BrS. For clarity, below we consider the ion–ion collision. To this end, one replaces $F_{1,2}(q) = N_{1,2}$ and $N_{1,2} + W_{1,2}(q) - F_{1,2}^2(q)$ with zero in the right-hand sides of (6.75)–(6.77) when integrating over the region of small momentum transfer, $q_{1,2} \le R_{\mathrm{at}}^{-1}$. The integration over the region $q_{1,2} \gg R_{\mathrm{at}}^{-1}$ is carried out assuming $F_{1,2}(q) \to 0$, $W_{1,2}(q) \to 0$.

In the region $(q_{1,2})_{\min} \ll R_{\mathrm{at}}^{-1}$, carrying out the integrals and keeping only logarithmically large terms, one derives the following formulae for the cross sections:

$$
\frac{\mathrm{d}^2\sigma_2}{\mathrm{d}\omega \mathrm{d}\Omega_{\mathbf{k}}} = \frac{1+\cos^2\theta_{\mathbf{k}}}{\pi c^3 v_1^2 \omega} N_2 \left[\mathscr{Z}_1^2 N_2 \ln \frac{v_1\gamma}{\omega R_{\mathrm{at}}} + \Big(Z_1^2 + N_1\Big) \ln(v_1 R_{\mathrm{at}}) \right], \tag{6.79}
$$

$$
\frac{\mathrm{d}^2\sigma_1}{\mathrm{d}\omega \mathrm{d}\Omega_{\mathbf{k}}} = \frac{1+\cos^2\theta_{\mathbf{k}}}{\pi c^3 v_1^2 \omega} \left(\frac{\omega}{\omega^{\mathrm{r}}}\right)^2 N_1 \left[\mathscr{Z}_2^2 N_1 \ln \frac{v_1\gamma}{\omega^{\mathrm{r}} R_{\mathrm{at}}} + \Big(Z_2^2 + N_2\Big) \ln(v_1 R_{\mathrm{at}}) \right], \tag{6.80}
$$

$$\frac{d^2\sigma_{12}}{d\omega d\Omega_{\mathbf{k}}} = -\frac{2\omega}{\omega^{\mathrm{r}}} \frac{1+\cos\theta_{\mathbf{k}}\cos\theta_{\mathbf{k}}^{\mathrm{r}}}{\pi c^3 v_1^2 \omega} N_1 N_2 \left[\mathscr{Z}_1 \mathscr{Z}_2 \ln \frac{v_1 \gamma}{\omega^{\mathrm{r}} R_{\mathrm{at}}} + \ln(v_1 R_{\mathrm{at}}) \right]. \tag{6.81}$$

where $\mathscr{Z}_j = Z_j - N_j$ $(j = 1, 2)$.

The first term in brackets on the right-hand side of (6.79) describes the elastic BrS of the target without exciting the projectile. The second and the third terms stand for the inelastic BrS of the target without and with exciting the projectile, respectively. The meaning of the terms on the right-hand side of (6.80) is similar but with account for the interchange 'target ↔ projectile'.

In the region $(q_{1,2})_{\min} \gg R_{\mathrm{at}}^{-1}$, the cross sections (6.75)–(6.77) reduce to simpler formulae:

$$\frac{d^2\sigma_2}{d\omega d\Omega_{\mathbf{k}}} = \frac{1+\cos^2\theta_{\mathbf{k}}}{\pi c^3 v_1^2 \omega} N_2 \left(Z_1^2 + N_1 \right) \ln \frac{v_1^2 \gamma}{\omega}, \tag{6.82}$$

$$\frac{d^2\sigma_1}{d\omega d\Omega_{\mathbf{k}}} = \frac{1+\cos^2\theta_{\mathbf{k}}^{\mathrm{r}}}{\pi c^3 v_1^2 \omega} \left(\frac{\omega}{\omega^{\mathrm{r}}} \right)^2 N_1 \left(Z_2^2 + N_2 \right) \ln \frac{v_1^2 \gamma}{\omega^{\mathrm{r}}}, \tag{6.83}$$

$$\frac{d^2\sigma_{12}}{d\omega d\Omega_{\mathbf{k}}} = \frac{2\omega}{\omega^{\mathrm{r}}} \frac{1+\cos\theta_{\mathbf{k}}\cos\theta_{\mathbf{k}}^{\mathrm{r}}}{\pi c^3 v_1^2 \omega} N_1 N_2 \ln \frac{v_1^2 \gamma}{\omega^{\mathrm{r}}} \tag{6.84}$$

Let us elucidate the relative contribution of the elastic and inelastic channels to the PBrS cross section. Using (6.79) and (6.80), one obtains for the ratios $\xi_{1,2}$ of the elastic to inelastic BrS cross sections for the target and the projectile, respectively:

$$\xi_2 = \frac{\mathscr{Z}_1^2 N_2 \ln(v_1\gamma/\omega R_{\mathrm{at}})}{(Z_1^2 + N_1)\ln(v_1 R_{\mathrm{at}})} \le \frac{\mathscr{Z}_1^2 N_2 \ln(v_1\gamma/\omega R_{\mathrm{at}})}{Z_1^2 \ln(v_1 R_{\mathrm{at}})} \tag{6.85}$$

$$\xi_1 = \frac{\mathscr{Z}_2^2 N_1 \ln(v_1\gamma/\omega^{\mathrm{r}} R_{\mathrm{at}}}{(Z_2^2 + N_2)\ln(v_1 R_{\mathrm{at}})} \le \frac{\mathscr{Z}_2^2 N_1 \ln(v_1\gamma/\omega^{\mathrm{r}} R_{\mathrm{at}})}{Z_2^2 \ln(v_1 R_{\mathrm{at}})}. \tag{6.86}$$

The quantity ξ_2 characterizes the relative contribution to the total radiation spectrum in the angular region $\theta_{\mathbf{k}} \gg \gamma^{-1}$, where the radiation by the target dominates. For small emission angles $\theta_{\mathbf{k}} \ll \gamma^{-1}$ the radiation by the projectile is more intensive, so that the relative contribution is found from (6.86).

The ratios (6.85) and (6.86) demonstrate the increased role of the coherent radiation of the electrons in the BrS process at relativistic velocities. For example, in the collision of two identical or nearly identical multi-charged ions ($N_1 \approx N_2 \gg 1$, $Z_{1,2} \gg N_{1,2}$) the elastic channels dominate practically in the whole range of the emission angles, since $\xi_{1,2} \ge N_{1,2} \gg 1$. It is instructive to recall, that in non-relativistic collisions of the identical ions the elastic BrS is suppressed due to destructing interference of the radiation by the projectile and by the target (see Sects. 5.2.2 and 5.2.3).

Analyzing (6.85) and (6.86) one can establish the cases when the elastic channels dominates only in one of the ratios, i.e., either for small or large emission angles. For example, considering $\mathscr{Z}_1 \sim 1$, $\mathscr{Z}_2 \sim 1$, $Z_1 \gg Z_2$, one derives $\xi_2 \sim N_2/Z_1^2 \ll 1$

but $\xi_1 \sim N_1/Z_2^2 \gg 1$. Thus, in this case the small-angle emission of the projectile is dominated by the elastic channels whereas the inelastic BrS dominates for $\theta_{\mathbf{k}} \gg \gamma^{-1}$.

Equations (6.85) and (6.86) are inapplicable in the photon energy range $\omega \leq \omega_{\rm at}$. In this limit, the contribution of the elastic BrS to the total spectrum must be described by the formulae derived in Sect. 6.2. The contribution of the inelastic channels, calculated within the logarithmic accuracy, can be still described by the formulae of this section, since the inelastic BrS occurs mainly with large momentum transfer, so that the atomic electrons can be treated as quasi-free ones. In the region $\omega \sim \omega_{\rm at}$ the elastic BrS can be additionally intensified by the increase of the atomic polarizabilities.

Finally, we note similar analysis of the relative roles of elastic and inelastic channels can be carried out for neutral atom–atom and atom–ion collisions.

6.4.3 Spectral Distribution of PBrS

Integrating the double differential cross sections (6.75)–(6.84) over the solid angle $\Omega_{\mathbf{k}}$ of the emission, one finds the spectral distribution of the emitted radiation.

For ion–ion collisions and in the photon energy range $\omega_{\rm at}\gamma \ll \omega \ll v_1 R_{\rm at}^{-1}$, one derives the following formulae written in the logarithmic approximation:

$$\frac{{\rm d}\sigma_2}{{\rm d}\omega} = \frac{16}{3c^3 v_1^2 \omega} N_2 \left[\mathscr{Z}_1^2 N_2 \ln \frac{v_1 \gamma}{\omega R_{\rm at}} + (Z_1^2 + N_1) \ln(v_1 R_{\rm at}) \right], \tag{6.87}$$

$$\frac{{\rm d}\sigma_1}{{\rm d}\omega} = \frac{16}{3c^3 v_1^2 \omega} N_1 \left[\mathscr{Z}_2^2 N_1 \ln \frac{v_1 \gamma^2 (1+\beta)}{\omega R_{\rm at}} + (Z_2^2 + N_2) \ln(v_1 R_{\rm at}) \right], \tag{6.88}$$

$$\frac{{\rm d}\sigma_{12}}{{\rm d}\omega} = -\frac{16}{3c^3 v_1^2 \omega} \left(1 - \frac{\ln \gamma(1+\beta)}{\beta\gamma^2}\right) \frac{N_1 N_2}{\beta^2 \gamma} \left[\mathscr{Z}_1 \mathscr{Z}_2 \ln \frac{v_1 \gamma}{\omega^{\rm r} R_{\rm at}} + \ln(v_1 R_{\rm at}) \right], \tag{6.89}$$

where $\beta = v_1/c$.

The spectral dependencies corresponding to (6.82)–(6.84) have the following form (for $\gamma^2 v_1 R_{\rm at}^{-1} \ll \omega \ll c^2$):

$$\frac{{\rm d}\sigma_2}{{\rm d}\omega} = \frac{16}{3c^3 v_1^2 \omega} N_2 \left[Z_1^2 \ln \frac{2 v_1^2 \gamma}{\omega} + N_1 \ln \frac{v_1^2 \gamma}{\omega} \right] \tag{6.90}$$

$$\frac{{\rm d}\sigma_1}{{\rm d}\omega} = \frac{16}{3c^3 v_1^2 \omega} N_1 \left[Z_2^2 \ln \frac{2(1+\beta) v_1^2 \gamma^2}{\omega} + N_2 \ln \frac{(1+\beta) v_1^2 \gamma^2}{\omega} \right] \tag{6.91}$$

$$\frac{{\rm d}\sigma_{12}}{{\rm d}\omega} = -\frac{16}{3c^3 v_1^2 \omega} \frac{1}{\beta^2 \gamma} \left(1 - \frac{\ln \gamma(1+\beta)}{\beta\gamma^2}\right) N_1 N_2 \ln \frac{(1+\beta) v_1^2 \gamma^2}{\omega}. \tag{6.92}$$

The structure of the spectra (6.87)–(6.92) is similar to that of the double differential cross sections (6.79)–(6.84). The interference term is smaller by a factor γ than the contributions of the projectile and target radiation, which is quite natural in view of what has been pointed out for the angular distribution of the radiation. The argument of one of the logarithms in $\mathrm{d}\sigma_1/\mathrm{d}\omega$ has attained an additional large factor $\gamma(1+\beta)$ in comparison with the corresponding term from (6.80). This modification favours enhancement of the elastic BrS of the projectile. The relative role of elastic and inelastic channels in the total spectrum of radiation is determined by the expressions analogous to (6.85) and (6.86). By making use of (6.79)–(6.84) written in the ultra-relativistic limit $\gamma \ll 1$, one can derive the cross sections $\mathrm{d}\sigma_{1,2}/\mathrm{d}\omega$ and $\mathrm{d}\sigma_{12}/\mathrm{d}\omega$ in the region of intermediate photon energies $v_1 R_{\mathrm{at}}^{-1} \ll \omega \ll \gamma^2 v_1 R_{\mathrm{at}}^{-1}$.

The quantity R_{at}, characterizing the atom size, has completely disappeared from the spectra (6.90)–(6.92). This is not accidental. Indeed in the region $\omega \gg \gamma^2 v_1 R_{\mathrm{at}}$ the minimum momentum transfers $q_{1,\min}$ and $q_{2,\min}$ exceed greatly the characteristic atomic momentum $q_{\mathrm{at}} \sim R_{\mathrm{at}}^{-1}$. The energy of the emitted photon is also much larger than ω_{at}. Thus, the total spectrum is predominantly formed due to inelastic radiative processes of quasi-free electrons of each of the colliding particles in the field of the partner. The radiation spectra of an ultra-relativistic/recoil electrons in collision with heavy particle at rest/ultra-relativistic are known, as well as that of an ultra-relativistic and a recoil electrons in the electron–electron collision [81]. In the collision of two ions the total radiation spectrum of the projectile and the target in the region $\gamma^2 v_1 R_{\mathrm{at}}^{-1}$ is represented by the sum of the above-mentioned spectra multiplied by the factors determined by the numbers $N_{1,2}$ of electrons in the ions and by the nucleus charges $Z_{1,2}$. These arguments are explicitly applicable to explain the structure of the terms in is exactly the structure of the terms in (6.90)–(6.92). Thus, e.g., the quantity $\mathrm{d}\sigma_2/\mathrm{d}\omega$ represents the sum of (a) N_2 radiative cross sections of a recoil electron in the collision with an ultra-relativistic massive particle with the charge Z_1, and (b) of $N_1 N_2$ radiative cross sections of a recoil electron in the collision with an ultra-relativistic electron.

The structure of expressions (6.90) and (6.91) suggests a simple way in which they can be refined and extended into the relativistic domain of the emitted photon energies $\omega \sim c^2$. To this end, instead of the radiative cross sections for an ultra-relativistic electron and for a recoil electrons calculated in the logarithmic approximation, one can use exact formulae for these cross sections in the whole ω-range (see, e.g., [75]) in the right-hand sides of (6.90) and (6.91). The atomic properties will be still determined by the numbers $N_{1,2}$ and $Z_{1,2}$.

The simple relation established between the BrS cross section in atomic collisions with that in the ultra-relativistic electron–electron collision indicates, that the radiation intensity in the collision of two heavy relativistic atomic particles can greatly exceed that of an ultra-relativistic electron, or even a heavy particle of charge Z, in collision with an electron.

The expressions for the total spectral-angular and spectral distributions of PBrS generalize the corresponding formulae derived within the framework of the non-relativistic theory (see Sect. 5.3 and [33, 34, 37]) and reduce to the latter in the limit $v_1 \ll c$.

6.5 Relativistic Effects Due to Internal Structure of Particles

In this section, following the approach developed in [255, 258, 259], the full relativistic description of the elastic PBrS process in a charged-particle–atom collision is presented.

Specific features of PBrS due to the relativistic effects can be subdivided into the following categories. First, there are effects directly related to the relativistic velocities of the particles involved in the process: the velocity of a projectile and the velocity of the electron orbital motion in the target. To account for these effects one has to describe the dynamics of the colliding particles using the Dirac equation rather than the Schrödinger equation. In contrast to Sect. 6.3, where the target atom was considered within the framework of the non-relativistic theory, below in this section we apply the relativistic formalism to describe the dynamics of the target electrons.

The relativistic effect of *retardation*, which modifies the interaction between a projectile and a target, can be attributed to another category. Retardation implies that the projectile particle polarizes the target not only through the Coulomb field, but also via the field of the transverse virtual photons (see, e.g., [81]), or, in terms of classical electrodynamics, via the retarded vector potential [278]. The effective radius of this field increases infinitely as the speed v_1 of the projectile approaches the speed of light c. Therefore, the ultra-relativistic charge, $v_1 \approx c$, polarizes the target mainly via the exchange of the transverse virtual photons.

There are relativistic effects due to the *multipole character* of radiation emitted by the target electrons. The multipolarity of the PBrS radiation depends solely on the magnitude of kR_{at}, where k is the photon momentum and R_{at} is the (mean) radius of the orbit of the electron who radiates. If this parameter is small, then the dipole-photon approximation is applicable, if otherwise, then it is necessary to take into account the radiation in higher multipoles.

In [258] the formalism was developed which allowed one to derive the DPWA and the multipole series of the BrS amplitude and the cross section. It was shown, that if all the mentioned relativistic effects are taken into account, the characteristics of PBrS are expressed through generalized multipole polarizabilities of three types corresponding to the allowed combinations of the types (longitudinal, electric and magnetic) of the virtual and emitted photons in the PBrS amplitude. Later on [255] the PBrS process was described within the relativistic plane-wave BA but with the retardation effect and the radiation into higher multipoles taken into account.

In what follows we will mainly consider collisions of a heavy projectile thus assuming that the relativistic plane-wave BA is applicable to describe its motion.[4] The results of numerical calculations, presented in Sect. 6.5.4, of the spectral and angular distributions of the PBrS radiation illustrate the contribution of the inner-shells electrons. Such a study allows us to discuss all specific features of PBrS in the relativistic domain by means of comparatively simple frameworks (such as the hydrogen-like and the Hartree-Fock-Dirac approximations) which are applicable

[4] For the sake of completeness, in the Supplementary Sect. 6.5.5 the relativistic DPWA formalism is outlined [258].

to describe the dynamics of the inner-shell electrons for which the many-electron correlations are not of the prime importance.

We will show that emission into higher multipoles results in a substantial asymmetry of the angular distribution as compared to the case of the dipole-photon radiation. The important feature of the logarithmic growth of the PBrS intensity with the energy of the incident particle, being related to the retardation effect, is clearly demonstrated.

Such calculations are of importance for comparison with the experimental data on the BrS radiation in collisions of 10 ... 100 keV electrons with various solid-state, thin-film and gaseous targets, such as Al, Cu, Ni, Ag, Au, Ne, Ar, Kr, Xe, [4, 5, 345, 346, 355–357, 362, 436]. Some of the experimental data [345, 346, 356, 357] are indicative of the PBrS contribution to the total radiation spectrum. In addition, the PBrS emitted in relativistic heavy-ion collisions are of interest in connection with the experimental attempts [291]. In the cited paper the BrS spectrum was measured in collisions of helium-like uranium ions with a 223 MeV/nucleon energy with Ne and Ar gas targets.

In this section we use the relativistic system of units $\hbar = m_e = c = 1$. The 4-metrics with the signature $(+---)$ is adopted, $g_{\mu\nu}$ is the metric tensor, δ_{ij} is the Kronecker symbol. The four-product of two four-vectors $a^\nu = (a_0, \mathbf{a})$ and $b^\nu = (b_0, \mathbf{b})$ is equal to $a_0 b_0 - \mathbf{a}\cdot\mathbf{b} \equiv a_\nu b^\nu$. The Dirac matrices are $\gamma^\nu = (\gamma^0, \boldsymbol{\gamma})$.

6.5.1 Amplitude of PBrS

The process of elastic PBrS (see the diagram in Fig. 6.2) in a collision between a relativistic projectile of a charge $Z_0\, e$ and mass m_0 with a target (atom or ion) is a transition of the projectile from the initial state $(\mathbf{p}_1, \mu_1)$ (the energy $\varepsilon_1 = (\mathbf{p}_1^2 + m_0^2)^{1/2}$) to the final state $(\mathbf{p}_2, \mu_2)$ $(\varepsilon_2 = (\mathbf{p}_2^2 + m_0^2)^{1/2})$ accompanied by the emission of a photon. The initial and the final states of the target coincide, therefore, the energy conservation law reads as $\varepsilon_1 = \varepsilon_2 + \omega$.

The two-fold differential cross section $\mathrm{d}^2\sigma_{\mathrm{pol}}/\mathrm{d}\omega \mathrm{d}\Omega_{\mathbf{k}}$, which describes the spectral-angular distribution of the radiation, is obtained from the differential cross section by integrating over the directions of the scattered particle, summing over the photon polarizations λ and the polarizations of the scattered electron μ_2 and carrying out the averaging over the initial-state polarizations μ_1 of the projectile:

$$\frac{\mathrm{d}^2\sigma_{\mathrm{pol}}}{\mathrm{d}\omega \mathrm{d}\Omega_{\mathbf{k}}} = \frac{p_2}{8 p_1} \frac{\omega}{(2\pi)^4} \sum_{\lambda} \sum_{\mu_1,\mu_2} \int_{(4\pi)} \mathrm{d}\Omega_{\mathbf{p}_2} \left| f_{\mathrm{pol}} \right|^2 . \tag{6.93}$$

Here $\mathrm{d}\Omega_{\mathbf{k}}$ and $\mathrm{d}\Omega_{\mathbf{p}_2}$ are is the elements of solid angle in the direction of the photon emission and the scattered particles, respectively.

The amplitude of elastic PBrS, described by two Feynman diagrams (Fig. 6.2), represents by itself two second-order matrix elements corresponding to the atom's transition $0 \to n \to 0$ under the action of two fields: the field of the emitted

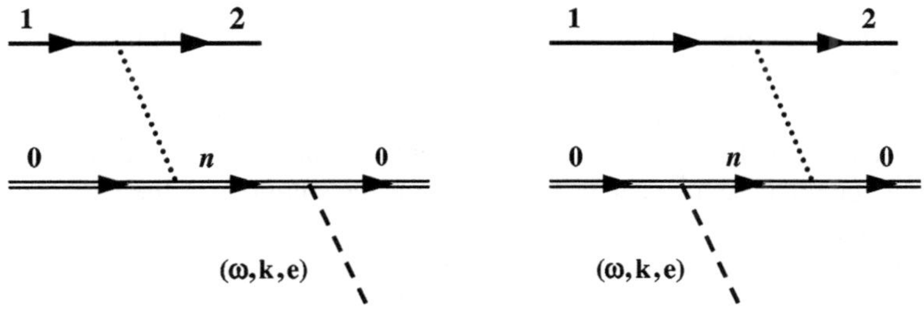

Fig. 6.2 Diagrammatic representation of the PBrS process for a relativistic structureless charged projectile scattered by a many-electron atom. *Thick solid lines* correspond to the projectile which moves in a field of the target. The initial ('1') and the final ('2') states of the projectile are characterized by the asymptotic momenta $\mathbf{p}_{1,2}$ and the polarizations $\mu_{1,2}$. The *double lines* denote the target. Index '0' marks the initial and the final states, index 'n' corresponds to the intermediate virtual state. The *dashed lines* designate the emitted (real) photon of energy ω, momentum $\mathbf{k}$ and the polarizational vector $\mathbf{e}$. The *dotted lines* stand for the virtual photon, the energy of which is also ω but the momentum $\mathbf{q}$ is not fixed by kinematic relations

photon, $\mathbf{A}_{(\gamma)}$, and that of the virtual photon, which is characterized by the 4-potential, $A^{\nu} = (\Phi, \mathbf{A})$:

$$f_{\rm pol} = e^2 \sum_n \left\{ \frac{\langle 0|\mathbf{A}_{(\gamma)} \cdot \boldsymbol{\gamma}|n\rangle\langle n|\gamma^{\nu} A_{\nu}|0\rangle}{\varepsilon_n(1-\mathrm{i}0) - \varepsilon_0 - \omega} + \frac{\langle 0|\gamma^{\nu} A_{\nu}|n\rangle\langle n|\mathbf{A}_{(\gamma)} \cdot \boldsymbol{\gamma}|0\rangle}{\varepsilon_n(1-\mathrm{i}0) - \varepsilon_0 + \omega} \right\} . \tag{6.94}$$

The sum is carried out over the quantum numbers of the complete spectrum of the (virtual) excited atomic states and contains the contributions of both the positive-energy, $\varepsilon_n > 0$, and the negative-energy, $\varepsilon_n < 0$, states.

The vector field of the emitted (real) photon is [5]

$$\mathbf{A}_{(\gamma)} = \sum_{a=1}^{N} \mathbf{e}\, \mathrm{e}^{-\mathrm{i}\mathbf{k}\cdot\mathbf{r}_a} . \tag{6.95}$$

The linear polarization is assumed. This allows us to consider $\mathbf{e}^* = \mathbf{e}$. The sum is carried out over N target electrons, $\mathbf{r}_a$ is the position vector of the ath electron.

The quantity A^{ν} stands for the four-potential of the field created by the moving projectile:

$$A^{\nu} = Z_0\, e \sum_{a=1}^{N} \int \mathrm{d}\mathbf{r}\, \Psi^{(-)\dagger}_{\mathbf{p}_2\mu_2}(\mathbf{r})\, \gamma^{\mu} D_{\mu\nu}(\omega, \mathbf{r} - \mathbf{r}_a)\, \Psi^{(+)}_{\mathbf{p}_1\mu_1}(\mathbf{r}) \tag{6.96}$$

[5] The normalization factor $\sqrt{2\pi/\omega}$ is not included into the definition of $\mathbf{A}_{(\gamma)}$.

Here, ω and $\mathbf{q}$ are the energy and momentum of the virtual photon, the quantity $D_{\mu\nu}(\omega, \mathbf{r}-\mathbf{r}_a)$ stands for the photon propagator.[6]

The bi-spinor wavefunctions $\Psi^{(\pm)}_{\mathbf{p}\mu}(\mathbf{r})$ of the asymptotic momentum $\mathbf{p}$ and polarization μ correspond to the out- (the superscript '+') and in- ('−') scattering states. In a more general case they can be considered as the solutions of the Dirac equation with Hamiltonian $\hat{h} = \varepsilon\gamma^0 - \boldsymbol{\gamma}\cdot\hat{\mathbf{p}} - Z_0 e\gamma^0 V(r) - m_0$ where $V(r)$ is the potential created by a (spherically symmetric) target. The partial-wave series for these functions are given by the following formula [2]:

$$\Psi^{(\pm)}_{\mathbf{p}\mu}(\mathbf{r}) = \frac{4\pi}{p}\sum_{jlm}\left(\Omega^{\dagger}_{jlm}(\mathbf{n_p})\,\chi_\mu(\mathbf{n_p})\right)\mathrm{e}^{\pm\mathrm{i}\delta_{jl}(\varepsilon)}\Psi_{\varepsilon jlm}(\mathbf{r})\,. \tag{6.97}$$

Here the sum is carried over the total angular momentum j, the orbital momentum $l = j \pm 1/2$, and projection of total angular momentum, m. The notation $\Omega_{jlm}(\mathbf{n})$ stands for a spherical spinor defined as in [420]. The superscript '†' marks the Hermitian conjugation. The two-component spinors χ_μ satisfy the orthogonality condition $\chi^{\dagger}_\mu\chi_\nu = \delta_{\mu\nu}$. The quantities $\delta_{jl}(\varepsilon)$ are the scattering phaseshifts. In (6.97) a general notation $\mathbf{n_a}$ is introduced for the unit vector along the direction $\mathbf{a}$.

The details of the evaluation of the PBrS amplitude (6.94) (and of the PBrS cross section) within the framework of relativistic DPWA one finds in [258].[7] In the supplementary Sect. 6.5.5 we present the key steps of this derivation, the final formulae for the amplitude and cross section as well as the analysis of the limiting cases.

Aiming to point out the qualitative differences, which are model-independent, between the relativistic case and the non-relativistic one, below we discuss the PBrS cross section within the framework of the *relativistic plane-wave Born approximation*. In this limit the initial and final states of projectile are described by wavefunctions of the following type:

$$\Psi^{(\pm)}_{\mathbf{p}\mu}(\mathbf{r}) = u_\mu(\varepsilon, \mathbf{p})\,\mathrm{e}^{\mathrm{i}\mathbf{p}\cdot\mathbf{r}}\,. \tag{6.98}$$

The bi-spinor amplitude $u_\mu(\varepsilon, \mathbf{p})$ is defined in accordance with [81]:

$$u_\mu(\varepsilon, \mathbf{p}) = \begin{pmatrix} \sqrt{\varepsilon + m_0}\,\chi_\mu(\mathbf{n_p}) \\ \sqrt{\varepsilon - m_0}\,(\boldsymbol{\sigma}\cdot\mathbf{n_p})\,\chi_\mu(\mathbf{n_p}) \end{pmatrix}. \tag{6.99}$$

Here $\boldsymbol{\sigma}$ are the Pauli matrices.

Calculated with the help of wavefunctions (6.98) the four-potential (6.96) reads[8]

[6] We adopt the Coulomb gauge for the propagator.

[7] The corresponding formalism for the OBrS process was developed much earlier [409, 410] (see also [233, 407]).

[8] The subscript 'B' in the formulae presented in this section as well as in Sects. 6.5.3 and 6.5.4 in indicates that the quantity marked was obtained in the Born limit.

$$A_{\mathrm{B}}^{\nu} = Z_0\, e \sum_{a=1}^{N} \int \frac{\mathrm{d}\mathbf{q}}{(2\pi)^3}\, \mathrm{e}^{\mathrm{i}\mathbf{q}\cdot\mathbf{r}_a}\, D_{\mu\nu}(\omega,\mathbf{q})\, B_{21}^{\mu}(\mathbf{q})\,. \tag{6.100}$$

The four-vector $B_{21}^{\mu}(\mathbf{q})$ stands for the Fourier image of the current of projectile:

$$B_{21}^{\mu}(\mathbf{q}) = \int \mathrm{d}\mathbf{r}\, \Psi_{\mathbf{p}_2\mu_2}^{(-)\dagger}(\mathbf{r})\, \gamma^0\gamma^{\mu}\, \mathrm{e}^{-\mathrm{i}\mathbf{q}\cdot\mathbf{r}}\, \Psi_{\mathbf{p}_1\mu_1}^{(+)}(\mathbf{r}) = (2\pi)^3\, b^{\mu}\, \delta(\mathbf{p}_1 - \mathbf{p}_2 - \mathbf{q}) \tag{6.101}$$

where

$$b^{\mu} = u_{\mu_2}^{\dagger}(\varepsilon_2,\mathbf{p}_2)\,\gamma^0\gamma^{\mu}\, u_{\mu_1}(\varepsilon_1,\mathbf{p}_1)\,. \tag{6.102}$$

The delta-function $\delta(\mathbf{p}_1 - \mathbf{p}_2 - \mathbf{q})$ fixes the momentum of the virtual photon reducing it to the momentum transfer of the projectile:

$$\mathbf{q} = \mathbf{p}_1 - \mathbf{p}_2\,. \tag{6.103}$$

For a fixed photon energy ω of the virtual photon the minimum and maximum values of q are equal to $q_{\min} = p_1 - p_2$, $q_{\max} = p_1 + p_2$.

By analogy with the approach used in the Hartree-Fock-Dirac approximation (see, for example, [287]), we assume that the target states $|0\rangle$, $|n\rangle$ can be described by one-electron wavefunctions corresponding to states with definite relativistic energy ε, total angular momentum j, orbital angular momentum $l = j \pm 1/2$, and total angular momentum projection m. Introducing (6.95)–(6.100) into (6.94), one can replace the sum over atomic electrons with the sums over the quantum numbers $\varepsilon_{\mathrm{i}}, j_{\mathrm{i}}, l_{\mathrm{i}}, m_{\mathrm{i}}$ of the atomic shells in the ground (i $= 0$) and excited (i $= n$). One-electron wavefunctions $\Psi_{\varepsilon_{\mathrm{i}} j_{\mathrm{i}} l_{\mathrm{i}} m_{\mathrm{i}}}(\mathbf{r})$ have the following bi-spinor structure:

$$\Psi_{\varepsilon_{\mathrm{i}} j_{\mathrm{i}} l_{\mathrm{i}} m_{\mathrm{i}}}(\mathbf{r}) = \frac{1}{r}\begin{pmatrix} g(r)\,\Omega_{j_{\mathrm{i}} l_{\mathrm{i}} m_{\mathrm{i}}}(\mathbf{n}) \\ -\mathrm{i} f(r)\,(\boldsymbol{\sigma}\cdot\mathbf{n})\,\Omega_{j_{\mathrm{i}} l_{\mathrm{i}} m_{\mathrm{i}}}(\mathbf{n}) \end{pmatrix}. \tag{6.104}$$

Here $g(r) \equiv g_{\varepsilon_{\mathrm{i}} j_{\mathrm{i}} l_{\mathrm{i}}}(r)$ and $f(r) \equiv f_{\varepsilon_{\mathrm{i}} j_{\mathrm{i}} l_{\mathrm{i}}}(r)$ are, respectively, the large and the small components of the relativistic wavefunction. They can be obtained by numerically solving the system of Hartree-Fock-Dirac equation (see, for example, [108]).

After some transformations (see [255, 258] for the details) the PBrS amplitude is written as follows:

$$f_{\mathrm{pol,B}} = Z_0 e\, b^{\mu}\, D_{\mu\nu}(\omega,\mathbf{q})\, M^{\nu}(\omega,\mathbf{k},\mathbf{e};\mathbf{q})\,. \tag{6.105}$$

The last factor describes the dynamic response of the target

$$M^{\nu}(\omega,\mathbf{k},\mathbf{e};\mathbf{q}) = e^2 \sum_{\{0,n\}} \left[\frac{\langle 0|\mathbf{e}\cdot\boldsymbol{\gamma}\,\mathrm{e}^{-\mathrm{i}\mathbf{k}\cdot\mathbf{r}}|n\rangle\langle n|\gamma^{\nu}\mathrm{e}^{\mathrm{i}\mathbf{q}\cdot\mathbf{r}}|0\rangle}{\omega_{n0} - \omega} \right.$$

$$+\frac{\langle 0|\gamma^{\nu} \mathrm{e}^{\mathrm{i}\mathbf{q}\cdot\mathbf{r}}|n\rangle\langle n|\mathbf{e}\cdot\boldsymbol{\gamma}\,\mathrm{e}^{-\mathrm{i}\mathbf{k}\cdot\mathbf{r}}|0\rangle}{\omega_{n0}+\omega}\Bigg]. \tag{6.106}$$

Here $\omega_{n0} = \varepsilon_n(1 - \mathrm{i}0) - \varepsilon_0$ is the transition energy, the sum $\sum_{\{0\}}$ is carried out over the quantum numbers of the core subshells, each characterized with the energy, ε_0, the total angular momentum, j_0, the orbital momentum, l_0, and the projection of the total momentum, m_0. We assume that all subshells are the filled ones, hence $m_0 = -l_0 \ldots l_0$. The sum $\sum_{\{n\}}$ is evaluated over the quantum numbers ε_n, j_n, l_n, m_n of the intermediate states.

Separating the contributions of the scalar and the vector parts of the four-potential A^{ν} (see (6.100)), one represents the amplitude in the form

$$f_{\mathrm{pol,B}} = (4\pi)^2\, Z_0 e\,(\mathbf{e}\cdot\mathbf{F}) = (4\pi)^2\, Z_0 e\,\mathbf{e}\cdot\left(\mathbf{F}^{(\mathrm{s})} + \mathbf{F}^{(\mathrm{e})} + \mathbf{F}^{(\mathrm{m})}\right). \tag{6.107}$$

The terms $\mathbf{F}^{(\mathrm{s,e,m})}$ (the meaning of the superscripts is given below in the section) are as follows:

$$\mathbf{e}\cdot\mathbf{F}^{(\mathrm{s})} = -\frac{1}{4\pi}\frac{e^2}{q^2}\,b^0\sum_{\{0,n\}}\left[\frac{\mathscr{F}_{0n}(\mathbf{e},-\mathbf{k})\,F_{n0}(\mathbf{q})}{\omega_{n0}-\omega} + \frac{F_{0n}(\mathbf{q})\,\mathscr{F}_{n0}(\mathbf{e},-\mathbf{k})}{\omega_{n0}+\omega}\right] \tag{6.108}$$

$$\mathbf{e}\cdot\left(\mathbf{F}^{(\mathrm{e})} + \mathbf{F}^{(\mathrm{m})}\right) = -\frac{1}{4\pi}\frac{e^2}{\omega^2 - q^2}\sum_{\{0,n\}}\left[\frac{\mathscr{F}_{0n}(\mathbf{e},-\mathbf{k})\,\mathscr{F}_{n0}(\mathbf{R},\mathbf{q})}{\omega_{n0}-\omega}\right.$$
$$\left.+\frac{\mathscr{F}_{0n}(\mathbf{R},\mathbf{q})\,\mathscr{F}_{n0}(\mathbf{e},-\mathbf{k})}{\omega_{n0}+\omega}\right]. \tag{6.109}$$

The vector $\mathbf{R}$, which is orthogonal to $\mathbf{q}$, is defined through

$$\mathbf{R} = \mathbf{B} - \frac{(\mathbf{B}\cdot\mathbf{q})\,\mathbf{q}}{q^2} \tag{6.110}$$

The matrix element $\mathscr{F}_{n0}(\mathbf{e},-\mathbf{k})$ corresponds to the atomic transition $0 \to n$ under the action of the operator $\mathbf{e}\cdot\boldsymbol{\gamma}\exp(-\mathrm{i}\mathbf{k}\cdot\mathbf{r})$

$$\mathscr{F}_{n0}(\mathbf{e},-\mathbf{k}) = \int \mathrm{d}\mathbf{r}\,\Psi^{\dagger}_{\varepsilon_n j_n l_n m_n}(\mathbf{r})\,\gamma^0\,\mathbf{e}\cdot\boldsymbol{\gamma}\,\mathrm{e}^{-\mathrm{i}\mathbf{k}\cdot\mathbf{r}}\,\Psi_{\varepsilon_0 j_0 l_0 m_0}(\mathbf{r}). \tag{6.111}$$

The matrix elements $F_{n0}(\mathbf{q})$ and $\mathscr{F}_{n0}(\mathbf{R},\mathbf{q})$ describe the $0 \to n$ transition due to the atomic electron coupling to the scalar part (proportional to $\gamma^0\exp(\mathrm{i}\mathbf{q}\cdot\mathbf{r})$) and the vector part ($\propto \boldsymbol{\gamma}\exp(\mathrm{i}\mathbf{q}\cdot\mathbf{r})$) of the virtual-photon field created by the projectile:

$$F_{n0}(\mathbf{q}) = \int \mathrm{d}\mathbf{r}\,\Psi^{\dagger}_{\varepsilon_n j_n l_n m_n}(\mathbf{r})\,\mathrm{e}^{\mathrm{i}\mathbf{q}\cdot\mathbf{r}}\,\Psi_{\varepsilon_0 j_0 l_0 m_0}(\mathbf{r}) \tag{6.112}$$

$$\mathscr{F}_{n0}(\mathbf{R},\mathbf{q}) = \int \mathrm{d}\mathbf{r}\, \Psi^{\dagger}_{\varepsilon_n j_n l_n m_n}(\mathbf{r})\, \gamma^0\, \mathbf{R}\cdot\boldsymbol{\gamma}\, \mathrm{e}^{\mathrm{i}\mathbf{q}\cdot\mathbf{r}}\, \Psi_{\varepsilon_0 j_0 l_0 m_0}(\mathbf{r}) \tag{6.113}$$

The matrix elements $\mathscr{F}_{0n}(\mathbf{e},-\mathbf{k})$, $F_{0n}(\mathbf{q})$, and $\mathscr{F}_{0n}(\mathbf{R},\mathbf{q})$, corresponding to the virtual transition $n \to 0$ one obtains from (6.111) to (6.113) by interchanging n and 0.

Introducing the multipole expansions of the factors $\mathbf{e}\cdot\boldsymbol{\gamma}\,\mathrm{e}^{-\mathrm{i}\mathbf{k}\cdot\mathbf{r}}$ and $\mathrm{e}^{-\mathrm{i}\mathbf{q}\cdot\mathbf{r}}$ in terms of vector spherical harmonics $\mathbf{Y}^{(\lambda)}_{lm}(\mathbf{n})$ ($\lambda = 0, 1$) and spherical harmonics $Y_{lm}(\mathbf{n}) = \mathbf{n}\cdot\mathbf{Y}^{(-1)}_{lm}(\mathbf{n})$, respectively and carrying out the angular algebra (this can be done using the relations from Sects. 5 and 7 of the book [420]) one obtains the multipole series of the vectors $\mathbf{F}^{(\mathrm{s,e,m})}$. Due to different selection rules for the magnetic multipole transitions ($\lambda = 0$), and for the electric ($\lambda = 1$) and longitudinal ($\lambda = -1$) multipole transitions,

$$l_n + l_0 + l = \begin{cases} \text{odd} & \text{for } \lambda = 0 \\ \text{even} & \text{for } \lambda = \pm 1, \end{cases} \tag{6.114}$$

the term $\mathbf{F}^{(\mathrm{s})}$ contains only the components proportional to the electric multipoles of the emitted photon field, while the terms $\mathbf{F}^{(\mathrm{e})}$ and $\mathbf{F}^{(\mathrm{m})}$ are diagonal in the indices λ of the virtual and the real photons.

Therefore, in the case of spherically symmetric atomic target the PBrS amplitude can be expressed in terms of partial generalized polarizabilities of three different types corresponding to the following allowed combinations of the virtual–real photons: longitudinal–electric, electric–electric and magnetic–magnetic. Each of these polarizabilities depends on the photon energy, ω, the orbital momentum, l, (these are the same for both photons), and on the magnitudes of the momenta q and k of the virtual and real photons.

Such a form for representing the amplitude has advantages of both general and computational nature. The possibility of expressing f in terms of the aforementioned polarizabilities, which contain all necessary information on the dynamic structure of the many-electron target, simplifies the problem and allows computational time for calculating the characteristics of PBrS to be substantially reduced.

The analytic expressions for $\mathbf{F}^{(\mathrm{s,e,m})}$ are:

$$\mathbf{F}^{(\mathrm{s})} = -\frac{\mathbf{b}\cdot\mathbf{q}}{q} \sum_{lm} \frac{\sqrt{l(l+1)}}{2l+1}\, Y^*_{lm}(\mathbf{n_q})\, \mathbf{Y}^{(1)}_{lm}(\mathbf{n_k})\, \alpha_l(\omega, q, k) \tag{6.115}$$

$$\mathbf{F}^{(\mathrm{e})} = \frac{1}{\omega^2 - q^2} \sum_{lm} \frac{l(l+1)}{2l+1} \left(\mathbf{b}\cdot\mathbf{Y}^{(1)*}_{lm}(\mathbf{n_q})\right) \mathbf{Y}^{(1)}_{lm}(\mathbf{n_k})\, \beta^{(1)}_l(\omega, q, k) \tag{6.116}$$

$$\mathbf{F}^{(\mathrm{m})} = \frac{1}{\omega^2 - q^2} \sum_{lm} \frac{l(l+1)}{2l+1} \left(\mathbf{b}\cdot\mathbf{Y}^{(0)}_{lm}(\mathbf{n_q})\right) \mathbf{Y}^{(0)}_{lm}(\mathbf{n_k})\, \beta^{(0)}_l(\omega, q, k) \tag{6.117}$$

In (6.115) it is accounted for the relation $b^0\omega = \mathbf{b}\cdot\mathbf{q}$ which follows from the law of conservation of the four-vector current of the projectile. The sums over l are carried out from $l = 1$ up to infinity.

The partial generalized polarizabilities appearing in the right-hand sides are defined as follows

$$\alpha_l(\omega, q, k) = \frac{2l+1}{q} e^2 \sum_{\varepsilon_0 j_0 l_0} \sum_{\varepsilon_n j_n l_n} \frac{2C_{0n}^{(-1)}(l) f_{0n}^{(1)}(k; l) f_{n0}^{(-1)}(q; l)}{\omega_{n0}^2 - \omega^2} \tag{6.118}$$

$$\beta_l^{(\lambda)}(\omega, q, k) = (2l+1) e^2 \sum_{\varepsilon_0 j_0 l_0} \sum_{\varepsilon_n j_n l_n} \frac{2\omega_{n0} C_{0n}^{(\lambda)}(l) f_{0n}^{(\lambda)}(k; l) f_{n0}^{(\lambda)}(q; l)}{\omega_{n0}^2 - \omega^2}\Bigg|_{\lambda=0,1} . \tag{6.119}$$

Here the atomic radial integrals $f_{ba}^{(\lambda)}(x; l)$ (with $\lambda = -1, 0, 1$, $(b, a) = (0, n)$, $x = (k, q)$) are given by

$$f_{ba}^{(-1)}(x; l) = \int_0^\infty \mathrm{d}r \left(g_b^*(r) g_a(r) + f_b^*(r) f_a(r) \right) j_l(xr) \tag{6.120}$$

$$f_{ba}^{(0)}(x; l) = \int_0^\infty \mathrm{d}r \left(g_b^*(r) f_a(r) + f_b^*(r) g_a(r) \right) j_l(xr) \tag{6.121}$$

$$\begin{aligned} f_{ba}^{(1)}(x; l) = \int_0^\infty \mathrm{d}r \Bigg[&\left(g_b^*(r) f_a(r) - f_b^*(r) g_a(r) \right) \frac{j_l(xr)}{xr} \\ &- \frac{\kappa_b - \kappa_a}{l(l+1)} \left(g_b^*(r) f_a(r) + f_b^*(r) g_a(r) \right) \left(\frac{1}{x} \frac{\mathrm{d} j_l(xr)}{\mathrm{d}r} + \frac{j_l(xr)}{xr} \right) \Bigg] , \end{aligned} \tag{6.122}$$

where $j_n(z)$ is the spherical Bessel function and $\kappa = l(l+1) - j(j+1) - \frac{1}{4}$.

The coefficients $C_{0n}^{(\lambda)}(l)$ (with $\lambda = -1, 0, 1$) are defined as follows:

$$C_{0n}^{(\lambda)}(l) = \xi(l_n l_0 l \lambda 1) \Pi_{j_0 j_n}^2 \begin{pmatrix} j_0 & j_n & l \\ \frac{1}{2} & -\frac{1}{2} & 0 \end{pmatrix}^2 \times \begin{cases} 1, & \lambda = \pm 1 \\ -\dfrac{(\kappa_0 + \kappa_n)^2}{l^2(l+1)^2}, & \lambda = 0 \end{cases} \tag{6.123}$$

where the $3j$-symbol is defined in accordance to [420], the shorthand notation $\Pi_{l_a l_b \ldots} = (2l_a + 1)^{1/2} (2l_b + 1)^{1/2} \ldots$ is used. The coefficient $\xi(l_a l_b l_c l_d l_e)$ is equal to 1 if $l_a + l_b + l_c + l_d + l_e$ is even and to 0 if otherwise. The notation κ_a is defined as follows:

$$\kappa_a = \begin{cases} -(l_a + 1) \text{ for } j_a = l_a + 1/2 \\ \quad\; l_a \text{ for } j_a = l_a - 1/2 \end{cases} \tag{6.124}$$

The important feature of the formulae (6.117), (6.115)–(6.117) is that they clearly demonstrate that in the relativistic case the PBrS amplitude is expressed in terms of the multipole generalized dynamic polarizabilities of three different types corresponding to the allowed combinations of the types of the virtual/real photons: (a) $\alpha_l(\omega, q, k)$ corresponds to the longitudinal/electric combination, (b) $\beta_l^{(1)}(\omega, q, k)$ to the electric/electric one, and (c) $\beta_l^{(0)}(\omega, q, k)$ to the magnetic/magnetic. Each of these polarizabilities depends on the photon energy ω, its orbital momentum l (which defines the multipolarity of the polarizability), and on the magnitudes of the momenta q and k of the virtual and real photons.

The right-hand sides of (6.118)–(6.119) can be to a form more convenient for computations by introducing the relativistic one-particle Green's function to describe excited electronic states of the target. The corresponding equations are listed in Sect. 6.5.6.

6.5.2 Limiting Cases of the Relativistic PBrS Amplitude

Let us demonstrate, that general relativistic formula (6.94), accompanied with its DPWA and multipole series representation (6.145), allow one to derive expressions for $f_{\rm pol}$ which follow from simpler theories. In more detail this consideration is carried out in [258].

6.5.2.1 Non-relativistic Dipole-Photon Limit

The non-relativistic ($v_1/c \to 0$) dipole-photon ($k \to 0$) limit of (6.94), where the wavefunctions (6.97) are used, one can obtain using the following transformations. Firstly, carrying out the non-relativistic limit with respect to the projectile, one notices that the vector potential **A** (which is due to the interaction retardation) can be neglected because of the relation $|\mathbf{A}|/|\Phi| \sim v_{1,2}/c \ll 1$. Therefore, only the Coulomb component the 4-potential A_ν survives, where one uses $D_{00}(\omega, \mathbf{r} - \mathbf{r}_a) = 1/|\mathbf{r} - \mathbf{r}_a|$. Secondly, carrying out the limit $k \to 0$ in (6.95) one makes a substitution $\gamma \cdot \mathbf{A}_\gamma \to \mathbf{e} \cdot \sum_{a=1}^{N} \hat{\mathbf{p}}_a \equiv \mathbf{e} \cdot \hat{\mathbf{P}}$ where $\hat{\mathbf{P}}$ is the momentum operator of all atomic electrons. The third step is to substitute $\gamma^\nu A_\nu$ with its non-relativistic analogue equal to $Z_0 \sum_a \langle \mathbf{p}_2^{(-)} |1/|\mathbf{r} - \mathbf{r}_a|| \, \mathbf{p}_1^{(+)} \rangle$. As a result of these transformations the sum over the negative-energy continuum in (6.94) becomes identically equal to zero, whereas the positive-energy sum reduces to the non-relativistic limit of $f_{\rm pol}$ calculated within the DPWA (see (2.18)–(2.21)).

6.5.2.2 Relativistic Born-Approximation and Non-relativistic Dipole Limit for Atom

The PBrS amplitude within the framework of relativistic plane-wave first Born approximation is given by (6.107) supplemented with (6.115)–(6.123).

If, in addition, one carries out (a) the non-relativistic limit with respect to the target's states, which implies the omission of the small component in the bi-spinor function (6.104), and, (b) the dipole-photon limit $k \to 0$, then the resulting formula reduces to expression (6.48).

6.5.2.3 High-Energy Photon Limit

Finally, let us briefly discuss the high-energy photon limit of the PBrS amplitude, which corresponds to the photon energies beyond the K-shell ionization potential $\omega \gg I_{1s}$. In the non-relativistic dipole-photon theory this limit is called the 'striping' approximation, see Sect. 4.6.

In [258] this limit was considered within the framework of the following approach: relativistic DPWA for the projectile and non-relativistic treatment of the target. The effects of retardation and the emission into higher multipoles were taken into account.

Similar to the non-relativistic dipole-photon case (see Sect. 4.6.2 and [236]) the PBrS amplitude can be represented as a matrix element of the effective operator V_{eff} calculated between the scattering states (6.97) of the projectile. Omitting the details of evaluation we present the final result for f_{pol} in this limit:

$$f_{\mathrm{pol}} = \langle \mathbf{p}_2 \mu_2 \,| V_{\mathrm{eff}}(\mathbf{r}) |\, \mathbf{p}_1 \mu_1 \rangle \tag{6.125}$$

where the operator of $V_{\mathrm{eff}}(\mathbf{r})$ is given by

$$\begin{aligned} V_{\mathrm{eff}}(\mathbf{r}) = &\frac{\mathrm{i}\, e^2}{m\omega^2} \int \mathrm{d}\mathbf{r}' \rho(r') \mathrm{e}^{-\mathrm{i}\mathbf{k}\cdot\mathbf{r}'} \frac{\mathrm{e}^{\mathrm{i}k|\mathbf{r}-\mathbf{r}'|}}{|\mathbf{r}-\mathbf{r}'|} \\ &\times \left[\gamma_0 \frac{\mathbf{e}\cdot(\mathbf{r}-\mathbf{r}')}{|\mathbf{r}-\mathbf{r}'|} \left(k + \frac{\mathrm{i}}{|\mathbf{r}-\mathbf{r}'|} \right) + k\, \mathbf{e}\cdot\boldsymbol{\gamma} \right] \end{aligned} \tag{6.126}$$

Here $\rho(r')$ is the electron density in the target atom.

Formulae (6.125) and (6.126) generalize the result of the non-relativistic dipole photon treatment of the PBrS process within the frame of the 'stripping' approximation. Indeed, carrying out the limit $k = 0$ in (6.126) and substituting in (6.125) the relativistic wavefunctions with the non-relativistic ones one obtains $f_{\mathrm{pol}} = -\omega^{-2} \langle \mathbf{p}_2 \,| \mathbf{e}\cdot\mathbf{a}_{\mathrm{el}} |\, \mathbf{p}_1 \rangle$ where $\mathbf{a}_{\mathrm{el}}$ is the acceleration of the projectile due to the static field created by the atomic electrons (see 4.6.2).

6.5.3 *Cross Section of PBrS*

To calculate spectral-angular distribution of PBrS one substitutes the amplitude (6.117) into (6.93). Then, carrying out the summation over the photon polarizations (this is done accounting for the relation $\mathbf{k}\cdot\mathbf{F}^{(\mathrm{s,e,m})} \propto \mathbf{k}\cdot\mathbf{Y}_{lm}^{(0,1)}(\mathbf{n_k}) = 0$) and representing the element $\mathrm{d}\Omega_{\mathbf{p}_2}$ as $\mathrm{d}\phi_{\mathbf{q}}\, q\, \mathrm{d}q/(p_1 p_2)$, one writes

$$\frac{\mathrm{d}^2\sigma_{\mathrm{pol,B}}}{\mathrm{d}\omega \mathrm{d}\Omega_{\mathbf{k}}} = Z_0^2 \alpha \frac{2\omega}{p_1^2} \sum_{\mu_1,\mu_2} \int_0^{2\pi} \mathrm{d}\varphi_{\mathbf{q}} \int_{q_{\min}}^{q_{\max}} q\, \mathrm{d}q\, \mathbf{F}\cdot\mathbf{F}^{\dagger} \tag{6.127}$$

Here $\alpha = e^2/\hbar c \approx 1/137$ is the fine structure constant.

Further transformation of expression (6.127) includes (1) the summation over the polarizations of the projectile, (2) the multipole expansion of the quadratic forms $\mathbf{Y}^{(\lambda)}_{lm}(\mathbf{n_k}) \cdot \mathbf{Y}^{(\lambda')*}_{l'm'}(\mathbf{n_k})$, and (3) the integration over $\mathrm{d}\phi_{\mathbf{q}}$.

The final result for the spectral-angular distribution within the framework of the relativistic Born approximation reads:

$$\frac{\mathrm{d}^2\sigma_{\mathrm{pol,B}}}{\mathrm{d}\omega \mathrm{d}\Omega_{\mathbf{k}}} = \sum_{l_k=0}^{\infty} (2l_k+1)\, P_{l_k}(\cos\theta_{\mathbf{k}})\, \frac{\mathrm{d}\sigma_{l_k}}{\mathrm{d}\omega}, \tag{6.128}$$

$$\frac{\mathrm{d}\sigma_{l_k}}{\mathrm{d}\omega} = Z_0^2\alpha\, \frac{\omega}{8\pi p_1^2} \int_{q_{\min}}^{q_{\max}} \mathrm{d}q^2 \left[A_{l_k} + B_{l_k} + C_{l_k} + D_{l_k}\right]. \tag{6.129}$$

Here

$$A_{l_k} = \left(\frac{q^2-\omega^2}{4} + (\mathbf{p}_1\cdot\mathbf{n_q})(\mathbf{p}_2\cdot\mathbf{n_q})\right) P_{l_k}(y) \sum_{l,l'=1}^{\infty} \begin{pmatrix} l' & l & l_k \\ 0 & 0 & 0 \end{pmatrix}^2$$
$$\times \Big(l(l+1) + l'(l'+1) - l_k(l_k+1)\Big) \mathrm{Re}\left[\alpha_l(\omega,q,k)\, \alpha^*_{l'}(\omega,q,k)\right] \tag{6.130}$$

$$B_{l_k} = \left(\frac{q^2-\omega^2}{2} + p_1^2 \sin^2\theta_{\mathbf{q}}\right) \frac{P_{l_k}(y)}{(q^2-\omega^2)^2} \sum_{l,l'=1}^{\infty} \sum_{\lambda,\lambda'=0}^{1} \xi(l_k l l' \lambda\lambda')$$
$$\times l l'\,(l+1)\,(l'+1) \begin{pmatrix} l' & l & l_k \\ 1 & -1 & 0 \end{pmatrix}^2 \mathrm{Re}\left[\beta_l^{(\lambda)}(\omega,q,k)\, \beta_{l'}^{(\lambda')*}(\omega,q,k)\right] \tag{6.131}$$

$$C_{l_k} = \frac{p_1^2 \sin^2\theta_{\mathbf{q}}}{(q^2-\omega^2)^2}\, \frac{2y\, P'_{l_k}(y) - l_k(l_k+1)\, P_{l_k}(y)}{\sqrt{(l_k-1)l_k(l_k+1)(l_k+2)}}$$
$$\times \sum_{l,l'=1}^{\infty} \sum_{\lambda,\lambda'=0}^{1} (-1)^{\lambda\lambda'}\, \xi(l_k l l' \lambda\lambda') l l'\,(l+1)(l'+1) \begin{pmatrix} l' & l & l_k \\ 1 & -1 & 0 \end{pmatrix} \begin{pmatrix} l & l_k & l' \\ 1 & -2 & 1 \end{pmatrix}$$
$$\times \mathrm{Re}\left[\beta_l^{(\lambda)}(\omega,q,k)\, \beta_{l'}^{(\lambda')*}(\omega,q,k)\right] \tag{6.132}$$

$$D_{l_k} = \frac{4p_1\,(\mathbf{p}_2\cdot\mathbf{n_q})\, \sin^2\theta_{\mathbf{q}}}{q^2-\omega^2}\, \frac{P'_{l_k}(y)}{\sqrt{l_k(l_k+1)}} \sum_{l,l'=1}^{\infty} \sum_{\lambda=0}^{1} \xi(l_k l l' \lambda 1) l(l+1)$$
$$\times \sqrt{l'(l'+1)} \begin{pmatrix} l' & l & l_k \\ 1 & -1 & 0 \end{pmatrix} \begin{pmatrix} l' & l_k & l \\ 0 & -1 & 1 \end{pmatrix} \mathrm{Re}\left[\beta_l^{(\lambda)}(\omega,q,k)\, \alpha^*_{l'}(\omega,q,k)\right] \tag{6.133}$$

In these formulae the notation $P'_{l_k}(y)$ stands for the derivative of the Legendre polynomial with respect to its argument, $y = \mathbf{n_q}\cdot\mathbf{n_{p_1}}$. Using the energy conservation law and (6.103), one expresses the kinematic factors from (6.130) to (6.131) in terms of ω, q, $q_{\min} = p_1 - p_2$ and $q_{\max} = p_1 + p_2$:

$$
\begin{aligned}
&y = \cos\theta_{\mathbf{q}} = \left(q_{\min}q_{\max} + q^2\right)/2p_1q \\
&\sin^2\theta_{\mathbf{q}} = \left(q^2 - q_{\min}^2\right)\left(q_{\max}^2 - q^2\right)/4p_1^2q^2 \\
&2\mathbf{p}_1\cdot\mathbf{q} = q_{\min}q_{\max} + q^2 \\
&2\mathbf{p}_2\cdot\mathbf{q} = q_{\min}q_{\max} - q^2 \\
&q^2 - \omega^2 + 4(\mathbf{p}_1\cdot\mathbf{n_q})(\mathbf{p}_2\cdot\mathbf{n_q}) = \left(q_{\min}^2q_{\max}^2 - q^2\omega^2\right)/q^2 .
\end{aligned}
\tag{6.134}
$$

Let us point out that the terms $C_{0,1}$ and D_0 in (6.132) and (6.133) are equal to zero due to vanishing of the combinations of the Legendre polynomials and/or its derivatives. Hence, the summation over l_k in (6.128) starts with $l_k = 0$ for the A_{l_k} and B_{l_k}, with $l_k = 2$ for C_{l_k}, and with $l_k = 1$ for D_{l_k}.

The terms C_{l_k} and D_{l_k} do not contribute to the spectral distribution of the radiation. The latter is obtained by integrating (6.134) over the solid angle of the photon emission. This leaves only the terms with $l_k = 0$. The final result can be written as follows [255]:

$$
\begin{aligned}
\frac{d\sigma_{\mathrm{pol,B}}}{d\omega} = Z_0^2\alpha\,\frac{2\omega}{p_1^2}\sum_{l=1}^{\infty}\frac{l(l+1)}{2l+1}\int_{q_{\min}}^{q_{\max}}\frac{dq}{q}\Bigg[&\left(q_{\min}^2\,q_{\max}^2 - q^2k^2\right)\left|\alpha_l(\omega,q,k)\right|^2 \\
&+ l(l+1)\,\frac{2q^2(q^2-k^2) + (q^2-q_{\min}^2)(q_{\max}^2-q^2)}{2(q^2-k^2)^2}\sum_{\lambda=0,1}\left|\beta_l^{(\lambda)}(\omega,q,k)\right|^2\Bigg] .
\end{aligned}
\tag{6.135}
$$

The first term in the integrand, proportional to $|\alpha_l(\omega,q,k)|^2$, is due to the dynamic polarization of the target by the Coulomb (scalar) part of the field of the projectile, whereas the second term, containing the polarizabilities $\beta_l^{(\lambda)}(\omega,q,k)$ appears as the result of the retardation in the vector part of the interaction. In the non-relativistic limit (with respect to the projectile and the target) and in the dipole-photon regime all polarizabilities but $\alpha_1(\omega,q,k)$ are equal to zero. In this limit $\alpha_1(\omega,q,k) \to \alpha(\omega,q)$, which is the non-relativistic dipole generalized polarizability, and the right-hand side of (6.135) reduces to the non-relativistic PBrS cross section (see Chap. 2). Similar to the non-relativistic result the cross section (6.135) weakly depends on the mass of a projectile.

Expressions (6.128)–(6.135) correspond to the relativistic plane-wave Born approximation. More accurate description of the scattering process can be achieved within the relativistic DPWA. In [258] closed analytical expressions were obtained for the spectral and spectral-angular distribution of the PBrS radiation using the relativistic partial-wave expansion of the initial and final state wavefunctions of projectile. It was demonstrated that the formulae obtained reduce to (6.128)–(6.135) in the Born limit.

An important distinguishing feature of the relativistic PBrS cross section is its logarithmic growth with incident energy ε_1 [35, 57]. Qualitatively, the reason for this is as follows. As was already mentioned, unlike a non-relativistic particle, a relativistic one interacts with the target not only by its Coulomb field but also (in the ultra-relativistic case, predominantly) by the field of transverse virtual photons. The effective radius of this field, $R_{\rm eff}$, increases infinitely with the energy of the projectile. The magnitude of $R_{\rm eff}$ can be estimated as $R_{\rm eff} \sim 1/q^{\perp}_{\rm min}$, where $q^{\perp}_{\rm min} \approx \gamma^{-1} q_{\rm min}$, with $\gamma = \varepsilon_1/m$ and $q_{\rm min} = p_1 - p_2 \approx \omega/v_1$ [35]. As a result, the range of distances at which a projectile effectively polarizes a target increases. This leads to the growth of $d\sigma_{\rm pol,B}(\omega)$ with ε_1.

Quantitative description of this effect can be carried out directly from (6.135) by analyzing the contribution of the region $q_0 \geq q \geq q_{\rm min} \approx \omega/v_1$ to the integral (here $q_0 = R_{\rm at}^{-1}$ and $R_{\rm at}$ is the radius of the target). The result reads [255] (see also [262]):

$$\left.\frac{d\sigma_{\rm pol,B}}{d\omega}\right|_{q\sim q_{\rm min}} \approx \alpha \frac{4Z_0^2\,\omega^4}{v_1^2}\left(A \ln\frac{q_{\rm at}}{q_{\rm min}} + \frac{B}{\omega^4}\ln\frac{\varepsilon_1}{m}\right) \tag{6.136}$$

where

$$A = \sum_{l=1}^{\infty} \frac{2l(l+1)}{2l+1}\left|\alpha_l(\omega, q_{\rm min}, k)\right|^2, \quad B = \sum_{l=1}^{\infty} \frac{l^2(l+1)^2}{2l+1}\sum\left|\beta_l^{(\lambda)}(\omega, q_{\rm min}, k)\right|^2. \tag{6.137}$$

The right-hand side of (6.136) is written in a so-called 'logarithmic' approximation which implies that the magnitude of the resulting logarithmic factors is large enough to neglect other, non-logarithmic, terms. The first term in term in the brackets corresponds to the contribution of the Coulomb part of the interaction, and it remains constant as the energy ε_1 grows. The second term, which is due to the interaction retardation, logarithmically increases with ε_1. This feature was first noted in [35, 57], where the PBrS of a relativistic projectile was studied within the framework of the non-relativistic dipole-photon description of the target (see also Sect. 6.3). The result of these studies follows from more general expressions (6.136) and (6.137). Carrying out the non-relativistic dipole photon limit, and accounting for the relations $\lim_{k\to 0}\alpha_1(\omega, q_{\rm min}, k) = \lim_{k\to 0}\omega^{-2}\beta_1^{(1)}(\omega, q_{\rm min}, k) \approx \alpha_{\rm d}(\omega)$ if $q_{\rm min}R_{\rm at} \ll 1$ one derives formula (6.50) as the limiting case of (6.136).

The retardation, as well as the relativistic and higher multipoles effects, strongly modify not only the PBrS spectral distribution but the spectral-angular distribution as well [255]. The spectral-angular distribution (6.128) can be written in the following form:

$$\frac{d^2\sigma_{\rm pol}}{d\omega d\Omega} = \frac{1}{4\pi}\frac{d\sigma_{\rm pol}}{d\omega}\left(1 + \sum_{l_k=1}^{\infty} a_{l_k}(\omega)\, P_{l_k}(\cos\theta)\right). \tag{6.138}$$

Let us point out the differences between formula (6.138) and its non-relativistic limit (2.28).

The first one, already discussed above, concerns the different dependence of the cross section $d\sigma_{\rm pol}(\omega)/d\omega$ in the relativistic and non-relativistic domain on ε_1. In the former the cross section increases proportionally to $\ln(\varepsilon_1)$ and this is due to the retardation.

The second, rather obvious, difference, is reflected by an infinite number of terms on the right-hand side of (6.138), which are due to the infinite number of the photon multipoles taken into account, in contrast to the two terms in (2.28).

Finally, there is a 'hidden' difference which is related to the relativistic description of the internal dynamics of the target rather than to the multipole character of the radiation. The easiest way to trace the origin of this difference is to consider the contribution of only dipole photons to the series in (6.138). The momentum of a dipole photon is $l = 1$, therefore only these term on the right-hand sides of (6.115)–(6.117) contribute to the PBrS amplitude (6.107). Then, instead of (6.138) one obtains [255]:

$$\left.\frac{d^2\sigma_{pol}}{d\omega d\Omega}\right|_{l=1} = \frac{1}{4\pi}\left.\frac{d\sigma_{pol}}{d\omega}\right|_{l=1}\Big(1 + a_1(\omega)\, P_1(\cos\theta) + a_2(\omega)\, P_2(\cos\theta)\Big). \tag{6.139}$$

It is seen that in contrast to non-relativistic formula (2.28) the angular distribution of the relativistic *dipole* radiation contains the term proportional to $P_1(\cos\theta)$. As a result, the distribution becomes asymmetric with respect to the transformation $\theta \longrightarrow \pi - \theta$.

The reason for this effect is as follows. If the target is treated within the non-relativistic framework then the dipole photons emitted via the polarizational mechanism belong to the 'electric' type [81]. This statement, valid for arbitrary non-relativistic velocities, implies that the PBrS amplitude is proportional to $\alpha(\omega, q) \equiv \alpha_1(\omega, q, k)$ (see (6.118)) which couples the Coulomb field of a projectile and the field of 'electric' dipole photon. For relativistic velocities, $v_1 \leq c$, the amplitude f_{pol} contains two polarizabilities, $\alpha(\omega, q)$ and $\beta(\omega, q) \equiv \beta_1^{(1)}(\omega, q, k)$ (see (6.119)). The former couples the 'electric' dipole photon with the 'electric' dipole virtual photon. Due to the selection rules, the angular dependent part of the intensity of the 'electric' dipole radiation contains the term proportional to $P_2(\cos\theta)$ only [81]. In the most explicit form this can be illustrated if one evaluates the angular distribution of PBrS using the 'logarithmic approximation'. Then, either for a non-relativistic projectile [454] or a relativistic one [35] the PBrS angular distribution is proportional to $1 + P_2(\cos\theta)/2 \propto 1 + \cos^2\theta$, which coincides with the angular distribution of radiation emitted by a rotating electric dipole [278].

In the case when the internal dynamics of the target is treated within the relativistic theory, there appears a possibility to emit a photon of the 'magnetic' type. If the 'magnetic' dipole radiation is treated separately, its angular dependent part also contains the term proportional to $P_2(\cos\theta)$ only [81]. However, if both types of the dipole photons can be emitted in a process, then their interference results in the term proportional to $P_1(\cos\theta)$. This is exactly what happens in the relativistic PBrS. Using the general formulae presented in [255, 258] one can establish, that the coefficient $a_1(\omega)$ in (6.139) is proportional to the cross-terms containing the products of the relativistic polarizability $\beta_1^{(0)}(\omega, q, k)$ of the 'magnetic'-type with the polarizabilities

$\beta_1^{(1)}(\omega, q, k)$ and $\alpha_1(\omega, q, k)$ which belong to the 'electric'-type. The coefficient $a_1(\omega)$ contains the quadratic forms of the polarizabilities of the same type, i.e. the terms proportional to $\left|\beta_1^{(\lambda)}(\omega, q, k)\right|^2$ (with $\lambda = 0, 1$) and to $\alpha_1(\omega, q, k)\,\beta_1^{(1)}(\omega, q, k)$. The magnitude of $a_1(\omega)$ relative to $a_2(\omega)$ is defined by the ratios $\left|\beta_1^{(0)}(\omega, q, k)/\beta_1^{(1)}(\omega, q, k)\right|$ and $\left|\beta_1^{(0)}(\omega, q, k)/\alpha_1(\omega, q, k)\right|$ which increase with the photon energy ω or/and the charge of the nucleus Z, i.e. the factors responsible for the magnitude of the relativistic effects in atomic radiative processes.

6.5.4 Numerical Results

The results presented in this section (except those shown in Fig. 6.5) refer to proton collisions with Al^{+12}, Ag^{+46} and Au^{+78} [255, 259]. Because of the large mass of the proton, the OBrS mechanism is suppressed by the factor $(m_e/m_p)^2 \sim 10^{-6}$ and, therefore, can be excluded from consideration in a wide range of emitted photon energies. In addition, in the ω-range considered here (photon energies close to the ionization potentials of atomic inner shells), the PBrS channel dominates over other radiation mechanisms such as secondary electron BrS [386], inelastic BrS [36, 37, 339], and molecular orbital radiation [388].

The calculation of the BrS cross sections was carried out within the framework of relativistic Born approximation using (6.128) and (6.135) for the spectra-angular and spectral distributions. The contributions of first five multipoles were taken into account. To calculate the partial generalized dynamic polarizabilities (6.118) and (6.119) the approach based on the relativistic Coulomb Green's function was used (see Sect. 6.5.6). The non-relativistic dependencies were obtained using the approach proposed in [253] (see Sect. 4.5).

In the photon energy range $\omega \gtrsim I_{1s}$ the main contribution to PBrS comes from the K-shell electrons. Therefore, the results presented in this section can be easily generalized to neutral Al, Ag and Au atoms by multiplying the cross sections obtained for the hydrogen-like ions by a factor of 4, which takes into account a (approximately) twofold increase in the polarizabilities of filled K-shells.

The spectral dependencies $\omega \mathrm{d}\sigma_{\mathrm{pol,B}}/\mathrm{d}\omega$ calculated for two proton energies (as indicated) are presented in Fig. 6.3. Let us note two specific features clearly seen on these plots.

First, the contribution of the term proportional to the sum $\sum_l (\dots)|\beta_l^{(1)}(\omega, q, k)|^2$ on the right-hand side of (6.135) becomes more noticeable as the energy of the incident particle increases. This is because of a relative increase of the role of the exchange with the transverse virtual photon between the incident particle and target electrons, whereas the contribution of the Coulomb interaction part is only determined by the particle velocity and is virtually independent on ε_1 at $v_1 \approx c$.

The second feature to be mentioned is the ratio between the relativistic and non-relativistic curves. At intermediate relativistic energies, $\varepsilon_1 \approx m_p c^2$, the cross section

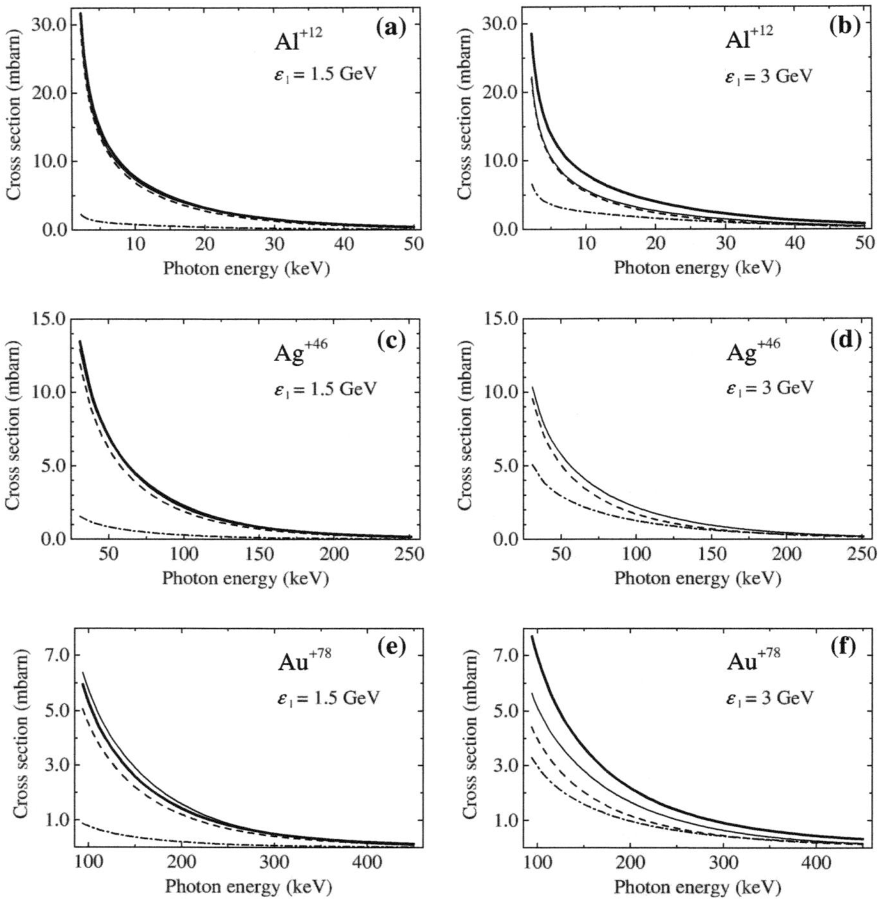

Fig. 6.3 Spectral dependencies $\omega \mathrm{d}\sigma_{\mathrm{pol,B}}/\mathrm{d}\omega$ for PBrS in collisions of 1.5 GeV (graphs a,c and e) and 3 GeV (graphs b,d and f) proton with ions Al^{+12}, Ag^{+46} and Au^{+78}. The *thick solid curve* stands for the relativistic cross section (6.135), the think *solid line* represents the non-relativistic dipole-photon limit (2.28). The contributions of the terms in (6.135) containing $\sum_l(\ldots)|\alpha_l(\omega, q, k)|^2$ and $\sum_l(\ldots)|\beta_l^{(1)}(\omega, q, k)|^2$ are shown by the *dashed* and *chained lines*, respectively. The contribution of the term $\propto \sum_l(\ldots)|\beta_l^{(0)}(\omega, q, k)|^2$ is small and not shown

calculated in the non-relativistic dipole approximation exceeds the relativistic one, since at a given photon frequency the non-relativistic dipole polarizability $\alpha(\omega, q)$ exceeds the corresponding relativistic component $\alpha_1(\omega, q, k)$, which makes the largest contribution to $\omega \,\mathrm{d}\sigma_{\mathrm{pol,B}}/\mathrm{d}\omega$ in this energy range (see the discussion below). The contribution of the terms with $|\beta_l^{(1)}(\omega, q, k)|^2$ becomes more significant as the incident particle energy increases. As a result, the relativistic cross section increases and exceeds the non-relativistic one.

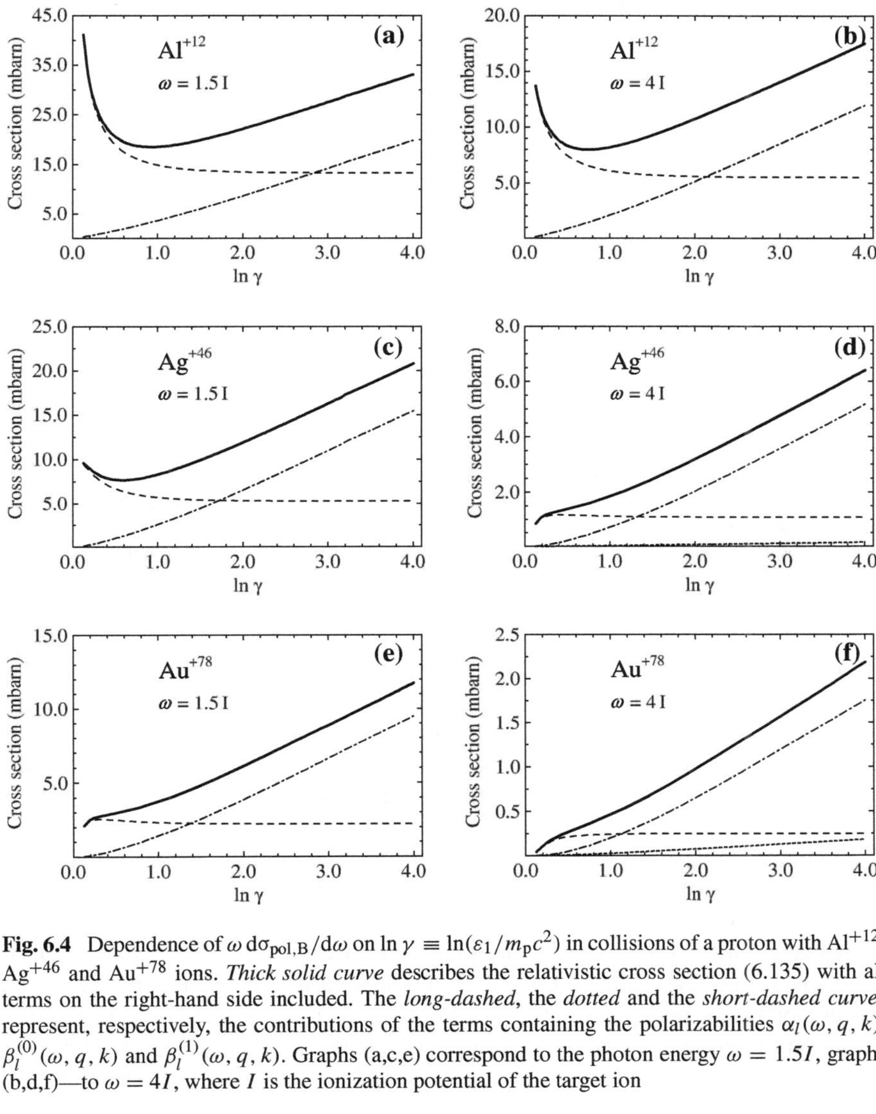

Fig. 6.4 Dependence of $\omega\, \mathrm{d}\sigma_{\mathrm{pol,B}}/\mathrm{d}\omega$ on $\ln \gamma \equiv \ln(\varepsilon_1/m_{\mathrm{p}}c^2)$ in collisions of a proton with Al^{+12}, Ag^{+46} and Au^{+78} ions. *Thick solid curve* describes the relativistic cross section (6.135) with all terms on the right-hand side included. The *long-dashed*, the *dotted* and the *short-dashed curves* represent, respectively, the contributions of the terms containing the polarizabilities $\alpha_l(\omega, q, k)$, $\beta_l^{(0)}(\omega, q, k)$ and $\beta_l^{(1)}(\omega, q, k)$. Graphs (a,c,e) correspond to the photon energy $\omega = 1.5I$, graphs (b,d,f)—to $\omega = 4I$, where I is the ionization potential of the target ion

The cross section $\omega\, \mathrm{d}\sigma_{\mathrm{pol,B}}/\mathrm{d}\omega$ as well as the contributions of the terms containing the moduli squares of the corresponding polarizabilities are shown in Fig. 6.4 as function of the relativistic Lorentz factor $\gamma = \varepsilon_1/m_{\mathrm{p}}c^2$ of the incident proton. For each target the dependencies are presented for two values of the photon energy, $\omega = 1.5I$ and $4I$ where I is the K-shell ionization potential of the target: $I \approx 2.3$ keV for Al^{+12}, $I \approx 31$ keV for Ag^{+46} and $I \approx 93.5$ keV for Au^{+78}. The growth of the cross

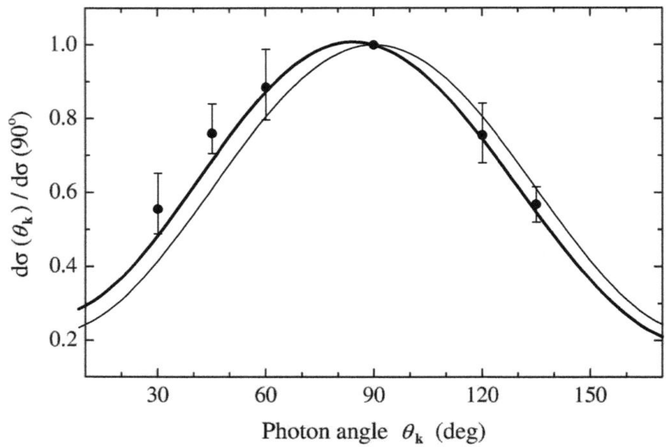

Fig. 6.5 Angular distribution of the PBrS radiation (normalized to that value of $d^2\sigma/d\omega d\Omega_{\mathbf{k}}$ at $\theta_{\mathbf{k}} = 90°$) for a proton with kinetic energy 1.5 MeV on an aluminum atom and emitted photon energy in the range 5.18–5.67 keV. The *thick solid curve* corresponds to relativistic cross section (6.136), the *thin solid curve* represents the non-relativistic dipole-photon approximation (2.28). Experimental data are from [253]

section, which is nearly linear for high values of $\ln(\gamma)$ in accordance with (6.136), is due to the contribution of the terms containing $\beta_l^{(\lambda)}(\omega, q, k)$ ($\lambda = 0, 1$) terms.

As it was discussed in Sect. 6.5.3, the important feature of the relativistic PBrS process is that it incorporates the effects of retardation and radiation in high multipoles. The multipole character of radiation becomes pronounced even at intermediate relativistic energies, and its manifests itself in the asymmetry of the angular distribution of the emitted photons. Figure 6.5 presents the profiles of the angular distribution (characterized by the ratio $d^2\sigma(\theta_{\mathbf{k}})/d^2\sigma(\theta_{\mathbf{k}} = 90°)$) formed in the collision of a proton with a medieval kinetic energy with an Al atom. Thick solid curve stands for the calculation with account for the relativistic effects, (6.136), whereas the thin solid curve represents the non-relativistic dipole-photon approximation (2.28). The calculations of the polarizabilities were performed with the hydrogen-like wavefunctions corresponding to the effective charge of nucleus $Z_{\text{eff}} = 12.7$ (see Sect. 5.4 for details). Filled circles with error bars are the available experimental data [204] which correspond to the photon energies in the range 5.18–5.67 keV, i.e. well-above the K-shell ionization threshold in the Al atom ($I \approx 2.3$ keV). We note that the non-relativistic dipole-photon approximation exhibits the well-known symmetric shape $\propto (1 + \beta(\omega) P_2(\cos\theta_{\mathbf{k}}))$ whereas the relativistic calculation reproduces experimentally observed forward-shifted distribution.

The role of relativistic effects in modifying the PBrS spectral-angular distribution is illustrated further by Fig. 6.6, where the profiles of $\omega d^2\sigma_{\text{pol,B}}/d\omega d\Omega_{\mathbf{k}} \equiv d^2\sigma_{\text{pol,B}}(\omega, \Omega_{\mathbf{k}})$, calculated for a 3 GeV proton collision with low-Z (Al^{+12}), intermediate-Z (Au^{+46}) and high-Z (Au^{+78}) hydrogen-like ions, are presented. For

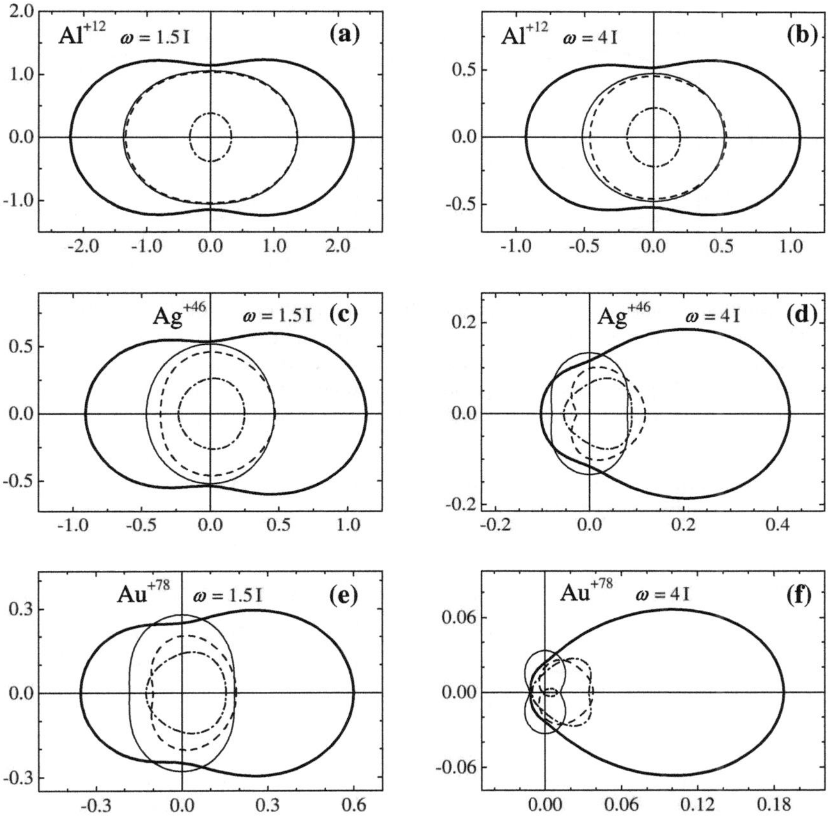

Fig. 6.6 Profiles of the angular distribution of PBrS $\omega d^2\sigma_{pol,B}/d\omega d\Omega_{\mathbf{k}}$ for collisions of 3 GeV protons with Al^{+12}, Ag^{+46} and Au^{+78} ions calculated for two photon energies as indicated. *Thick solid curves* stand for the full relativistic calculations ((6.136)), The *dashed, dotted,* and *chained curves* correspond to the contributions of the right-hand side terms containing $|\alpha_l(\omega,q,k)|^2$, $|\beta_l^{(0)}(\omega,q,k)$ and $|\beta_l^{(1)}(\omega,q,k)|^2$, respectively. *Thin solid curves* correspond to the non-relativistic dipole-photon approximation (2.28)

each ion the calculations were performed for two photon energies equal to $1.5I$ and $4I$. In these plots (as well in Figs. 6.7 and 6.8) the length of the segment connecting the origin and a curve point equals to the value of $d^2\sigma_{pol,B}(\omega,\Omega)$ (in millibarn/srad) in the corresponding direction. The horizontal axis ($\theta_{\mathbf{k}}=0$) is directed along the initial momentum $\mathbf{p}_1$. The thick solid curve stand for a full relativistic calculation of the cross section (6.128)). The contributions of the terms containing $\left|\beta_l^{(0)}(\omega,q,k)\right|^2$, $\left|\beta_l^{(1)}(\omega,q,k)\right|^2$, and $|\alpha_l(\omega,q,k)|^2$ are also shown in the figure. Note, that the sum of these contribution is not equal to $d^2\sigma_{pol,B}(\omega,\Omega)$ which, in addition, contains the cross-terms. The thin solid curve in each graphs represents the non-relativistic

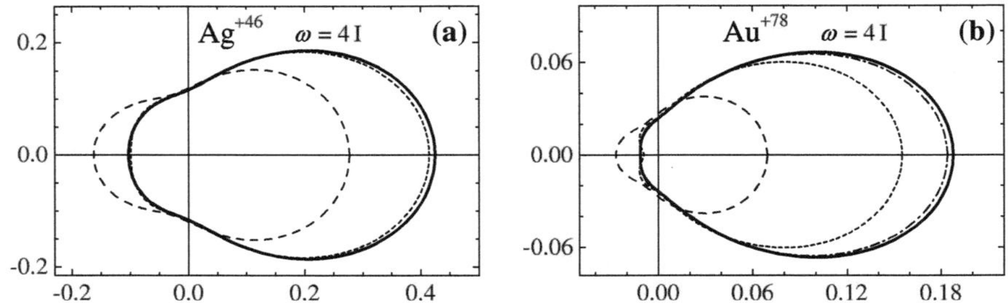

Fig. 6.7 Multipole contributions to the angular distribution $\omega d^2\sigma_{\mathrm{pol,B}}/d\omega d\Omega_{\mathbf{k}}$ of PBrS, (6.128), in a 3 GeV proton collision with Ag^{+46} and Au^{+78} for the indicated photon energies. The *long-dashed*, *dashed* and *chained curves* show the profiles of the contributions which account for the dipole, dipole+quadrupole and dipole+quadrupole+octupole terms, respectively. *Thick solid line* corresponds to the exact result which accounts for the sum of the terms up to $l = 5$

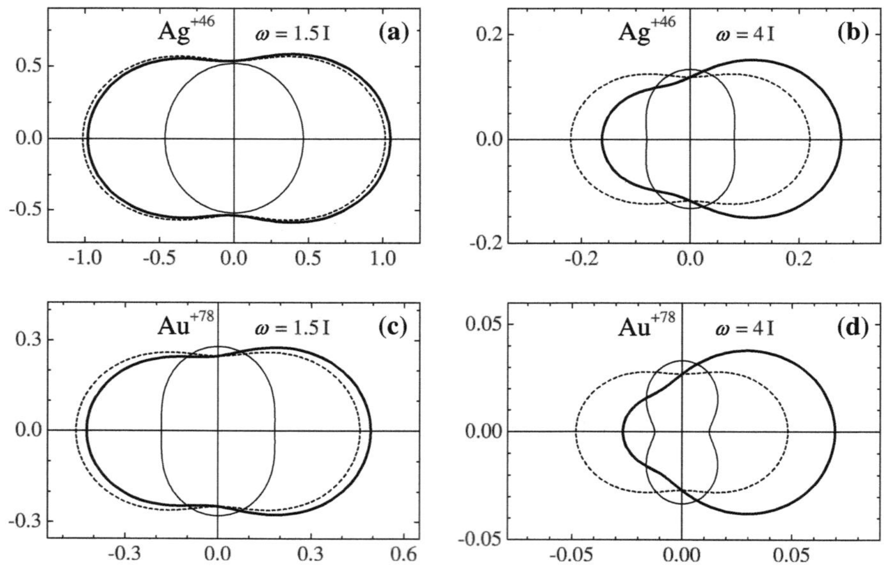

Fig. 6.8 Angular distribution $\omega d^2\sigma_{\mathrm{pol,B}}/d\omega d\Omega_{\mathbf{k}}$ of PBrS in a 3 GeV proton collision with Ag^{+46} and Au^{+78} for two indicated energies of the emitted photon. *Thick solid curves* corresponds to the relativistic dipole approximation, (6.139); *thin solid curves*—to the non-relativistic dipole approximation (2.28). The *dashed line* presents the contributions of the terms with $l_k = 0$ and $l_k = 2$ to the right-hand side of (6.139).

dipole-photon cross section (2.28). In contrast to the symmetric shape (with respect to the substitution $\theta_{\mathbf{k}} \leftrightarrow \pi - \theta_{\mathbf{k}}$) of the non-relativistic cross section, the relativistic angular distribution is asymmetric, being enhanced in the forward direction. The asymmetry increases with ω and Z .

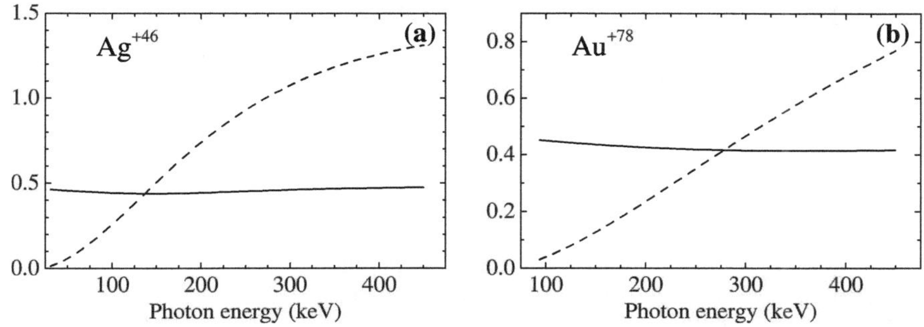

Fig. 6.9 Angular asymmetry parameters $a_1(\omega)$ and $a_2(\omega)$ from relativistic dipole formula (6.139) versus ω for a 3 GeV proton collision with Ag^{+46} and Au^{+78} ions. *Dashed curves* correspond to the $a_1(\omega)$ dependence, *solid curves*—to $a_2(\omega)$

The contribution of the dipole (terms with $l, l' = 1$ in the sums on the right-hand sides of (6.130)–(6.133)), dipole-and-quadrupole (the terms with $l, l' = 1, 2$) and dipole-and-quadrupole-octupole (the terms with $l, l' = 1, 2, 3$) polarizabilities to the angular distribution of radiation formed in p+Ag^{+46} and p+Au^{+78} (for $\varepsilon_1 = 3\,\text{GeV}$ and $\omega = 4I$) are presented in Fig. 6.7.

Figure 6.7 illustrates that in contrast to the non-relativistic dipole-photon approximation, the relativistic curves corresponding to the dipole contribution are strongly asymmetric (see (6.139) and the subsequent discussion).

The angular asymmetry parameters $a_1(\omega)$ and $a_2(\omega)$, entering the relativistic dipole angular distribution (6.139), as functions of the photon energy are presented in Fig. 6.9 for $\varepsilon_1 = 3\,\text{GeV}$ proton collision with Ag^{+46} and Au^{+78} ions. The presented dependencies elucidate the asymmetry in the profile of $\left.d^2\sigma_{\text{pol}}/d\omega d\Omega_{\mathbf{k}}\right|_{l=1}$. The asymmetry is related to the coefficient $a_1(\omega)$ which is very small for low photon energies but increases rapidly with ω and exceeds the value of $a_2(\omega)$. Thus, the term $a_1(\omega)P_1(\theta_{\mathbf{k}})$ dominates in the spectrum in the region of large ω.

6.5.5 Appendix A: Relativistic DPWA Formalism for PBrS

The fully relativistic formalism describing the BrS process in a charged particle–many-electron atom collision was presented in [258]. Using the relativistic DPWA, (6.97), to describe the motion of a projectile in the field of a target, and treating the target states within the framework of the one-electron relativistic self-consistent field approximation, the DPWA and the multipole series for the PBrS amplitude and the cross sections were evaluated. Thus, the effects of both the radiation retardation and the emission into higher multipoles were taken into account.

In this Supplementary section we present the key steps of the formalism, write down the explicit expressions for the amplitude and cross section series, and evaluate the limiting cases of the general formulae to demonstrate that they reproduce correctly the results of simpler theories.

6.5.5.1 DPWA and Multipole Series for the PBrS Amplitude

Introducing the incoming and outgoing series (6.97) for the projectile's wavefunctions into (6.94), one represents the PBrS amplitude as follows:

$$f_{\rm pol} = Z_0 e \int \frac{d\mathbf{q}}{(2\pi)^3} \left(I_{\rm s}(\omega, \mathbf{k}, \mathbf{e}; \mathbf{q}) + I_{\rm v}(\omega, \mathbf{k}, \mathbf{e}; \mathbf{q}) \right) . \tag{6.140}$$

The subscripts "s" and "v" indicate that the corresponding term originates due to the scalar and the 3-vector parts of the 4-potential A^{ν}, (6.96).

Using the Coulomb gauge for the photon propagator, one can write the terms $I_{\rm s} \equiv I_{\rm s}(\omega, \mathbf{k}, \mathbf{e}; \mathbf{q})$ and $I_{\rm v} = I_{\rm v}(\omega, \mathbf{k}, \mathbf{e}; \mathbf{q})$ as follows

$$I_{\rm s} = -\frac{4\pi e^2}{q^2} B_0 \sum_{\{0,n\}} \left[\frac{\mathcal{F}_{0n}(\mathbf{e}, -\mathbf{k})\, F_{n0}(\mathbf{q})}{\omega_{n0} - \omega} + \frac{F_{0n}(\mathbf{q})\, \mathcal{F}_{n0}(\mathbf{e}, -\mathbf{k})}{\omega_{n0} + \omega} \right], \tag{6.141}$$

$$I_{\rm v} = -\frac{4\pi e^2}{\omega^2 - q^2} \sum_{\{0,n\}} \left[\frac{\mathcal{F}_{0n}(\mathbf{e}, -\mathbf{k})\, \mathcal{F}_{n0}(\mathbf{R}, \mathbf{q})}{\omega_{n0} - \omega} + \frac{\mathcal{F}_{0n}(\mathbf{R}, \mathbf{q})\, \mathcal{F}_{n0}(\mathbf{e}, -\mathbf{k})}{\omega_{n0} + \omega} \right]. \tag{6.142}$$

See the text below (6.106) for the explanation of the summations conventions.

The notations used in these formulae are identical to those in (6.108)–(6.113) with the following two exceptions. First, the four-vector $B^{\nu}_{21}(\mathbf{q})$, which enters the short-hand notation $B_{\mu} \equiv g_{\mu\nu} B^{\nu}_{21}(\mathbf{q})$, is defined by the first equality in (6.101) only since the second equality does not hold beyond the BA limit. Second, the vector $\mathbf{q}$ is not fixed by the momentum conservation law (6.103), but enters the right-hand side of (6.140) as the variable of integration.

Evaluation of the multipole series for $I_{\rm s}$ and $I_{\rm v}$ is carried out similar to the evaluation of (6.115)–(6.117). The result reads

$$I_{\rm s} = -(4\pi)^2 \frac{B_0\, \omega}{q} \sum_{lm} \frac{\sqrt{l(l+1)}}{2l+1} Y^{*}_{lm}(\mathbf{n}_{\mathbf{q}})\, \mathbf{e} \cdot \mathbf{Y}^{(1)}_{lm}(\mathbf{n}_{\mathbf{k}})\, \alpha_l(\omega, q, k) \tag{6.143}$$

$$I^{(\lambda)}_{\rm v} = \frac{(4\pi)^2}{\omega^2 - q^2 + {\rm i}0} \sum_{lm} \frac{l(l+1)}{2l+1} \mathbf{B} \cdot \mathbf{Y}^{(\lambda)\,*}_{lm}(\mathbf{n}_{\mathbf{q}})\, \mathbf{e} \cdot \mathbf{Y}^{(\lambda)}_{lm}(\mathbf{n}_{\mathbf{k}})\, \beta^{(\lambda)}_l(\omega, q, k) \Big|_{\lambda=0,1} . \tag{6.144}$$

The partial generalized polarizabilities $\alpha_l(\omega, q, k)$ and $\beta^{(\lambda)}_l(\omega, q, k)$ are defined by (6.118) and (6.119).

To obtain the DPWA series of the amplitude $f_{\rm pol}$ one first evaluates the partial-wave representation of the 4-vector $B^{\mu}_{21}(\mathbf{q})$ by introducing the expansions (6.97) into the integrand in (6.101). Then, the obtained result is used in (6.143), (6.144) and further in (6.140). In the end one obtains the following expression for the PBrS amplitude in terms of the multipole series over $\mathbf{Y}^{(\lambda)}_{lm}(\mathbf{n}_{\mathbf{k}})$ and the DPWA series with respect to the partial-wave expansions of the initial and the final states of the projectile [258]:

$$f_{\rm pol} = {\rm i}\,\frac{(4\pi)^{5/2}\,Z_0}{p_1 p_2} \sum_{\substack{j_1 l_1 m_1\\ j_2 l_2 m_2}} \sum_{l=1}^{\infty} \sum_{m=-l}^{l} \sum_{\lambda=0,1} (-1)^{m_1+\frac{1}{2}}\,{\rm i}^{-l-\lambda}\,\phi_{21}\,\Pi_{j_1 j_2 l}\,\xi(l_2 l_1 l \lambda 1)$$
$$\times \begin{pmatrix} j_2 & j_1 & l \\ -m_2 & m_1 & -m \end{pmatrix} \begin{pmatrix} j_2 & j_1 & l \\ \frac{1}{2} & -\frac{1}{2} & 0 \end{pmatrix} \mathbf{e}\cdot\mathbf{Y}^{(\lambda)}_{lm}(\mathbf{n_k})\,\mathscr{P}^{(\lambda)}_{21}(\omega,k,l)\,. \tag{6.145}$$

For the explanation of the notations $\Pi_{j_1 j_2 l}$ and $\xi(l_2 l_1 l \lambda 1)$ see (6.123). The short-hand notation ϕ_{21} stands for

$$\phi_{21} = {\rm e}^{{\rm i}\delta_{j_1 l_1}(\varepsilon_1)+{\rm i}\delta_{j_2 l_2}(\varepsilon_2)} \left(\chi^{\dagger}_{\mu_2}(\mathbf{n_{p_2}})\,\Omega_{j_2 l_2 m_2}(\mathbf{n_{p_2}})\right) \left(\Omega^{\dagger}_{j_1 l_1 m_1}(\mathbf{n_{p_1}})\,\chi_{\mu_1}(\mathbf{n_{p_1}})\right) \tag{6.146}$$

The quantity $\mathscr{P}^{(1)}_{21}(\omega,k,l)$ is the partial amplitude of the PBrS corresponding to the emission of the electric-type photon. It consists of two terms, $\mathscr{P}^{(l)}_{21}(\omega,k,l) + \mathscr{P}^{(e)}_{21}(\omega,k,l)$, which represent the contributions of the longitudinal, superscript "(l)", and the electric, superscript "(e)", multipoles of the virtual photon. The partial amplitude $\mathscr{P}^{(0)}_{21}(\omega,k,l)$ corresponds to the emission of the magnetic-type photon. These quantities are defined as follows

$$\mathscr{P}^{(l)}_{21}(\omega,k,l) = -\frac{2}{\pi}\,\frac{\sqrt{l(l+1)}}{2l+1} \int_0^{\infty} q\,{\rm d}q\,\alpha_l(\omega,q,k)\,f^{(-1)}_{21}(q;l) \tag{6.147}$$

$$\mathscr{P}^{(e)}_{21}(\omega,k,l) = \frac{2}{\pi}\,\frac{[l(l+1)]^{3/2}}{2l+1} \int_0^{\infty} \frac{q^2\,{\rm d}q}{\omega^2-q^2+{\rm i}0}\,\beta^{(1)}_l(\omega,q,k)\,f^{(1)}_{21}(q;l) \tag{6.148}$$

$$\mathscr{P}^{(0)}_{21}(\omega,k,l) = -\frac{2}{\pi}\,(\kappa_2+\kappa_1)\,\frac{\sqrt{l(l+1)}}{2l+1}$$
$$\times \int_0^{\infty} \frac{q^2\,{\rm d}q}{\omega^2-q^2+{\rm i}0}\,\beta^{(0)}_l(\omega,q,k)\,f^{(0)}_{21}(q;l)\,. \tag{6.149}$$

The radial integrals $f^{(-1,0,1)}_{21}(q;l)$ are defined in accordance with (6.120), (6.121) and (6.122). The notation κ_j $(j=1,2)$ is defined in (6.124).

6.5.5.2 Total BrS Amplitude and Cross Section

The OBrS amplitude is given by the matrix element of the operator $\mathbf{e}\cdot\boldsymbol{\gamma}\,\exp(-{\rm i}\mathbf{k}\cdot\mathbf{r})$ between the scattering states (6.97):

$$f_{\rm ord} = Z_0 \int {\rm d}\mathbf{r}\,\Psi^{(-)\dagger}_{\mathbf{p}_2\mu_2}(\mathbf{r})\,\gamma^0\,\mathbf{e}\cdot\boldsymbol{\gamma}\,{\rm e}^{-{\rm i}\mathbf{k}\cdot\mathbf{r}}\,\Psi^{(+)}_{\mathbf{p}_1\mu_1}(\mathbf{r})\,. \tag{6.150}$$

The DPWA and multipole series structure of the polarizational and ordinary terms in the total amplitude is similar, hence

$$f_{\text{tot}} = f_{\text{ord}} + f_{\text{pol}} = \mathrm{i}\,\frac{(4\pi)^{5/2}\,Z_0}{p_1 p_2} \sum_{\substack{j_1 l_1 m_1\\ j_2 l_2 m_2}} \sum_{l=1}^{\infty} \sum_{m=-l}^{l} \sum_{\lambda=0,1} (-1)^{m_1+\frac{1}{2}}\, \mathrm{i}^{-l-\lambda}\, \phi_{21}\, \Pi_{j_2 j_1 l}$$
$$\times \xi(l_1 l_2 l \lambda 1) \begin{pmatrix} j_2 & j_1 & l \\ -m_2 & m_1 & -m \end{pmatrix} \begin{pmatrix} j_2 & j_1 & l \\ \frac{1}{2} & -\frac{1}{2} & 0 \end{pmatrix} \mathbf{e}\cdot \mathbf{Y}_{lm}^{(\lambda)}(\mathbf{n_k})\, \mathscr{T}_{21}^{(\lambda)}(\omega, k, l) \tag{6.151}$$

where the partial amplitudes of the total BrS are introduced

$$\mathscr{T}_{21}^{(\lambda)}(\omega, k, l) = \mathscr{O}_{21}^{(\lambda)}(\omega, k, l) + \mathscr{P}_{21}^{(\lambda)}(\omega, k, l)\,. \tag{6.152}$$

The OBrS partial amplitudes $\mathscr{O}_{21}^{(\lambda)}(\omega, k, l)$ (with $\lambda = 0, 1$) are related to the radial integrals (6.121) and (6.122) as follows:

$$\mathscr{O}_{21}^{(\lambda)}(\omega, k, l) = \begin{cases} \dfrac{\kappa_2+\kappa_1}{\sqrt{l(l+1)}}\, f_{21}^{(0)}(k; l) \text{ for } \lambda = 0 \\ \sqrt{l(l+1)}\, f_{21}^{(1)}(k; l) \text{ for } \lambda = 1 \end{cases}. \tag{6.153}$$

Substituting these formulae into (6.93) one derives the cross section differential with respect to the scattering angle, the emission angle and calculated for the fixed polarizations of both the projectile (μ_1, μ_2) and the photon (λ_p), allowing one thus to obtain the most detailed information on the BrS process. If one is interested only in the spectral-angular distribution of the radiation, then the modulus squared of f_{tot} must be integrated over the scattering solid angle and summed over the polarizations of the particles. Carrying the angular algebra (this can be done, for example, by making use of the collection of the formulae presented in [420]) one obtains [258]

$$\frac{d^2\sigma}{d\omega\, d\Omega_{\mathbf{k}}} = \frac{Z_0^2\,\alpha}{2\pi}\,\frac{\omega}{p_1^3 p_2} \sum_{\substack{j_1 l_1\\ j_1' l_1'}} \sum_{j_2 l_2 L} \sum_{l,l'=1}^{\infty} \sum_{\lambda,\lambda'=0,1} (-1)^{\varphi}\, \mathrm{i}^{l'+\lambda'-l-\lambda}\, \mathrm{e}^{\mathrm{i}(\delta_{j_1 l_1}(\varepsilon_1) - \delta_{j_1' l_1'}(\varepsilon_1))}$$
$$\times \Pi^2_{j_2 j_1 j_1' l l' L}\, \xi(L l l' \lambda \lambda')\, \xi(L l_1 l_1' 0 0)\, \xi(l_2 l_1 l \lambda 1)\, \xi(l_2 l_1' l' \lambda' 1)$$
$$\times \begin{pmatrix} l' & l & L \\ 1 & -1 & 0 \end{pmatrix} \begin{pmatrix} j_1' & j_1 & l \\ \frac{1}{2} & -\frac{1}{2} & 0 \end{pmatrix} \begin{pmatrix} j_2 & j_1 & l \\ \frac{1}{2} & -\frac{1}{2} & 0 \end{pmatrix} \begin{pmatrix} j_2 & j_1' & l' \\ \frac{1}{2} & -\frac{1}{2} & 0 \end{pmatrix} \begin{Bmatrix} L & l' & l \\ j_2 & j_1 & j_1' \end{Bmatrix}$$
$$\times \mathscr{T}_{21}^{(\lambda)}(\omega, k, l)\, \mathscr{T}_{21'}^{(\lambda')*}(\omega, k, l')\, P_L(\cos\theta_{\mathbf{k}}) \tag{6.154}$$

Here $\varphi = l_1 + l_1' - j_1 - j_1' - j_2 - l - l' - \frac{1}{2}$, $P_L(x)$ is the Legendre polynomial and $\theta_{\mathbf{k}}$ is the emission angle with respect to $\mathbf{p}_1$.

Integrating (6.154) over the emission angles one obtains the spectral distribution of radiation:

$$\frac{d\sigma}{d\omega} = Z_0^2\alpha\frac{2\omega}{p_1^3 p_2}\sum_{\substack{j_1 l_1\\ j_2 l_2}}\sum_{l=1}^{\infty}\sum_{\lambda}\Pi^2_{j_2 j_1 l}\xi(l_2 l_1 l \lambda 1)\begin{pmatrix} j_2 & j_1 & l\\ \frac{1}{2} & -\frac{1}{2} & 0\end{pmatrix}^2 \left|\mathscr{T}_{21}^{(\lambda)}(\omega,k,l)\right|^2. \tag{6.155}$$

6.5.6 Appendix B: Generalized Polarizabilities Expressed in Terms of Relativistic Green's Function

To avoid the summation (integration) over the intermediate-state energy ε_n in the right-hand sides of (6.118) and (6.119) (this procedure implies the evaluation of $\sum_{\varepsilon_n}$ for both the positive-energy, $\varepsilon_n > 0$, and the negative-energy, $\varepsilon_n < 0$, states) one can re-write these formulae by introducing the relativistic radial Green's function to describe an electron excited from the core's subshell [258]. Assuming that the electron propagates in a central field created by the nucleus and other electrons, one can apply the standard formalism of the single-particle relativistic Green's function (see, for example, [444]). The definition and the multipole bi-spinor structure of the relativistic Green's function $G_E(\mathbf{r}_2, \mathbf{r}_1)$ is given by

$$\begin{aligned} G_E(\mathbf{r}_2,\mathbf{r}_1) &= \sum_{\varepsilon jlm}\frac{\Psi_{\varepsilon jlm}(\mathbf{r}_2)\,\Psi^{\dagger}_{\varepsilon jlm}(\mathbf{r}_1)\,\gamma^0}{E-\varepsilon(1-\mathrm{i}0)} \\ &= \frac{1}{r_2 r_1}\sum_{jl}\begin{pmatrix} G^{(1)}_{Ejl}(r_2,r_1)\,Y_{ll}(\mathbf{n}_2,\mathbf{n}_1) & \mathrm{i}G^{(2)}_{Ejl}(r_2,r_1)\,Y_{ll'}(\mathbf{n}_2,\mathbf{n}_1)\\ \mathrm{i}G^{(3)}_{Ejl}(r_2,r_1)\,Y_{l'l}(\mathbf{n}_2,\mathbf{n}_1) & -G^{(4)}_{Ejl}(r_2,r_1)\,Y_{l'l'}(\mathbf{n}_2,\mathbf{n}_1)\end{pmatrix}. \end{aligned} \tag{6.156}$$

Here $Y_{l_1 l_2}(\mathbf{n}_2,\mathbf{n}_1) = \sum_{m=-j}^{j}\Omega_{jl_1 m}(\mathbf{n}_2)\,\Omega^{\dagger}_{jl_2 m}(\mathbf{n}_1)$ and the radial parts of the Green's function, $G^{(1-4)}_{Ejl}(r_2,r_1)$, can be expressed in terms of large $g_{\varepsilon jl}(r)$ and small $f_{\varepsilon jl}(r)$ components of the (radial) single-particle wavefunction of the excited electron:

$$\begin{aligned} &\begin{pmatrix} G^{(1)}_{Ejl}(r_2,r_1) G^{(2)}_{Ejl}(r_2,r_1)\\ G^{(3)}_{Ejl}(r_2,r_1) G^{(4)}_{Ejl}(r_2,r_1)\end{pmatrix} \\ &\quad = \sum_{\varepsilon}\frac{1}{E-\varepsilon(1-\mathrm{i}0)}\begin{pmatrix} g_{\varepsilon jl}(r_2)g^{*}_{\varepsilon jl}(r_1) g_{\varepsilon jl}(r_2)f^{*}_{\varepsilon jl}(r_1)\\ f_{\varepsilon jl}(r_2)g^{*}_{\varepsilon jl}(r_1) f_{\varepsilon jl}(r_2)f^{*}_{\varepsilon jl}(r_1)\end{pmatrix} \end{aligned} \tag{6.157}$$

Using (6.157) in (6.118)–(6.119) one obtains the following expressions for the $\alpha_l(\omega,q,k)$, $\beta_l^{(0,1)}(\omega,q,k)$ in terms of the quantities $G^{(1-4)}_{Ejl}(r_2,r_1)$:

$$\alpha_l(\omega,q,k) = \frac{e^2}{q\omega}(2l+1)\sum_{\substack{\varepsilon_0 j_0 l_0\\ j_n l_n}} C^{(-1)}_{0n}(l)\sum_{m=1,2}(-1)^m\int_0^{\infty}\!\!\int_0^{\infty}\mathrm{d}r_1\mathrm{d}r_2\; j_l(qr_1)\Big(f_0^{*}(r_2),\, g_0^{*}(r_2)\Big)$$

$$\times \begin{pmatrix} I_k^-(r_2)\, G^{(1)}_{\varepsilon_m j_n l_n}(r_2,r_1)\; I_k^-(r_2)\, G^{(2)}_{\varepsilon_m j_n l_n}(r_2,r_1) \\ I_k^+(r_2)\, G^{(3)}_{\varepsilon_m j_n l_n}(r_2,r_1)\; I_k^+(r_2)\, G^{(4)}_{\varepsilon_m j_n l_n}(r_2,r_1) \end{pmatrix} \begin{pmatrix} g_0(r_1) \\ f_0(r_1) \end{pmatrix} \tag{6.158}$$

$$\beta_l^{(0)}(\omega,q,k) = e^2(2l+1) \sum_{\substack{\varepsilon_0 j_0 l_0 \\ j_n l_n}} C^{(0)}_{0n}(l) \sum_{m=1,2} \int_0^\infty \int_0^\infty \mathrm{d}r_1 \mathrm{d}r_2\, j_l(qr_1) j_l(kr_2)$$
$$\times \left(f_0^*(r_2),\, g_0^*(r_2)\right) \begin{pmatrix} G^{(1)}_{\varepsilon_m j_n l_n}(r_2,r_1) G^{(2)}_{\varepsilon_m j_n l_n}(r_2,r_1) \\ G^{(3)}_{\varepsilon_m j_n l_n}(r_2,r_1) G^{(4)}_{\varepsilon_m j_n l_n}(r_2,r_1) \end{pmatrix} \begin{pmatrix} f_0(r_1) \\ g_0(r_1) \end{pmatrix} \tag{6.159}$$

$$\beta_l^{(1)}(\omega,q,k) = e^2(2l+1) \sum_{\substack{\varepsilon_0 j_0 l_0 \\ j_n l_n}} C^{(1)}_{0n}(l) \sum_{m=1,2} \int_0^\infty \int_0^\infty \mathrm{d}r_1 \mathrm{d}r_2 \left(f_0^*(r_2),\, g_0^*(r_2)\right)$$
$$\times \begin{pmatrix} I_q^+(r_1)\, I_k^-(r_2)\, G^{(1)}_{\varepsilon_i j_n l_n}(r_2,r_1)\; I_q^-(r_1)\, I_k^-(r_2)\, G^{(2)}_{\varepsilon_i j_n l_n}(r_2,r_1) \\ I_q^+(r_1)\, I_k^+(r_2)\, G^{(3)}_{\varepsilon_i j_n l_n}(r_2,r_1)\; I_q^-(r_1)\, I_k^+(r_2)\, G^{(4)}_{\varepsilon_i j_n l_n}(r_2,r_1) \end{pmatrix} \begin{pmatrix} f_0(r_1) \\ g_0(r_1) \end{pmatrix}. \tag{6.160}$$

Here $\varepsilon_1 = \varepsilon_0 - \omega$, $\varepsilon_2 = \varepsilon_0 + \omega$ and the functions $I_s^{\pm}(r)$ $(s = q, k)$ are defined as follows

$$I_s^{\pm}(r) = \frac{\kappa_0 - \kappa_n}{l(l+1)} \left(\frac{1}{s} \frac{\mathrm{d} j_l(sr)}{\mathrm{d}r} + \frac{j_l(sr)}{sr} \right) \pm \frac{j_l(sr)}{sr}, \tag{6.161}$$

with $j_l(sr)$ being the spherical Bessel function and $\kappa_m = l_m(l_m+1) - j_m(j_m+1) - \frac{1}{4}$ $(m = 0, n)$.

Chapter 7
PBrS from Atomic Clusters and Fullerenes

7.1 Introduction

This chapter is devoted to the description of the PBrS process in collisions involving atomic clusters.

A group of atoms bound together by interatomic forces is called an atomic cluster (AC). There is no qualitative distinction between small clusters and molecules, except perhaps that the binding forces must be such as to permit the molecule to grow much larger by stacking more atoms or molecules of the same type if the system is to be called a cluster. As the number of atoms in the system increases, ACs acquire more and more specific properties, making them unique physical objects different from both single molecules and from the solid state.

In Nature, there are many different types of AC: van der Waals clusters, metallic clusters, fullerenes, molecular, semiconductor, mixed clusters, and their shapes can depart considerably from the common spherical form: arborescent, linear, spirals, etc. Usually, one can distinguish between different types of clusters by the nature of the forces between the atoms, or by the principles of spatial organization within the clusters. Clusters can exist in all forms of matter: solid state, liquid, gases and plasmas.

ACs have been recognized as new physical objects with their own properties relatively recently. This became clear after such experimental successes as the discovery of electron shell structure in metal clusters [128, 232], observation of plasmon resonances in metal clusters [93, 383, 384] and fullerenes [83, 195], formation of singly and doubly charged negative cluster ions [194] and many others. Comprehensive survey of the field, with emphasis on the topics related closely to the phenomena discussed below in this chapter, can be found in review papers and books; see, e.g. [92, 94, 117, 119, 130, 138, 179, 191, 218, 314, 325, 370].

In this chapter we concentrate on the role of the dynamic response of a target in formation of the photon emission spectrum in the region of the giant dipole resonances. This range of frequencies is of particular interest since the polarizability of a target is very large. This is borne out by the extremely large photoionization

A. V. Korol and A. V. Solov'yov, *Polarization Bremsstrahlung*,
Springer Series on Atomic, Optical, and Plasma Physics 80,
DOI: 10.1007/978-3-642-45224-6_7, © Springer-Verlag Berlin Heidelberg 2014

cross sections of the giant resonances in AC as compared with atoms (see, e.g., [94, 394] for metal clusters, and [83, 195, 212, 213, 361], for fullerenes). Electrons of a suitable energy can thus excite a giant dipole oscillation, which will radiate, and this radiation is superimposed on ordinary BrS background. The significance of the dynamic polarization in the radiative processes with metal clusters has been stressed in [114, 115] (see also [148]) where the processes of radiative electron capture and BrS was considered. The photon emission spectrum generated in electron–fullerene collisions was described in the main logarithmic approach in [31] and, more accurately, in [154]. Recently, the detection of the contribution of a cooperative effect was reported in experimentally measured PBrS spectra of 0.7 keV electrons scattered from xenon clusters [162, 163].

7.2 Plasmon Resonance Approximation

Contemporary software for the quantum-chemical calculations provides an accurate quantitative description of the ground state of many-particle systems (clusters and fullerenes, in particular), and allows one to obtain information on geometrical and chemical properties of the system. Meanwhile, the description of dynamic properties by means of such programs faces significant difficulties. In many cases the properties are governed by a collective excitation of electrons and formation of the plasmon resonances in the excitation spectra [391]. In various systems, plasmon resonances lie either below the ionization threshold (in metal clusters) or above it (e.g., in fullerenes). Calculation of the optical response of metal clusters with the help of quantum-chemical programs was made possible nearly two decades ago (see, e.g., [368, 394]). However, it is still a challenging task to apply these codes to calculate collective electron excitations in fullerenes, which lie in the continuous spectrum [109, 110, 423].

However, this can be achieved within simplified model approximations. The advantage of such approaches is that they allow one to overcome significant computational difficulties but at the same time take into account the essential features of the processes providing clear physical insight into the phenomena. One of the well-known and widely used approaches is based on the jellium model [128]. The main feature of the model is that the detailed ionic structure of the system is smeared out and substituted by the uniform spherically symmetric distribution of the positive charge. Delocalized electrons of the system, moving in the field created by the uniform positive charge distribution, are considered within either the self-consistent HF or the local density approximations. The jellium model was applied frequently to the description of ground state properties of metal clusters and fullerenes as well as to the investigation of photoexcitation and electron scattering processes in these systems (see, e.g., [79, 80, 216, 217, 226, 306, 421, 422, 424] and [391] for earlier references). Many-electron correlations can be taken into account, where it is necessary, using the perturbation theory, the RPAE, or the time-dependent local density approximation. Model approaches provide quite accurate quantitative description of a variety of the collisional processes in which collective electron excitations (the plasmons) and/or polarization effects play a dominant role. From this viewpoint, the frameworks based on simplified but physically solid models have clear advantage in comparison with the standard quantum-chemical methods.

Delocalized electrons of atomic clusters may form two different types of collective excitations, namely the surface and volume plasmons [149, 191, 268]. The dipole surface plasmon is responsible for the formation of the giant resonance in photoabsorption spectra of metal clusters [130, 182] and fullerenes [195], and plays also an important role in the process of inelastic scattering of electrons [391]. The electron impact ionization of metal clusters revealed the existence of the volume plasmon which has a higher resonance frequency and is essential for the formation of the ionization cross section [149].

While the photoionization experiments allow one to investigate only optically allowed dipole excitations, the electron energy loss (EEL) experiments, where an electron scattered at a certain angle is detected, probe plasmon excitations of higher angular momenta. The first experimental observation of plasmon excitations in the EEL spectra of gaseous C_{60} [224] revealed a maximum at about 22 eV, which was consistent with previous theoretical predictions [83] and with the later experimental observations [195, 378] of the surface plasmon in the photoionization spectra.

Recent experiments on photoionization of neutral [361] and charged [378] C_{60} molecules revealed the presence of the second collective resonance at about 40 eV which initially was interpreted as a volume plasmon [378]. Later on it was argued that the second resonance was associated not with the volume plasmon but with the antisymmetric mode of the surface plasmon [263] (see also [379]). While only the surface plasmons can be excited by the uniform field of an electromagnetic wave in the process of photoionization, the volume plasmon may be formed in the system under the exposure to a non-uniform field of the charged projectile.

Theoretical investigations of fast electrons scattering on metal clusters and fullerenes, carried out the RPAE as well as in the Plasmon Resonance Approximation (PRA), revealed the existence of the diffraction phenomena (see [391]). The first experimental observation of the electron diffraction on C_{60} was reported in [147]. It was shown that plasmon modes of different angular momenta provide dominating contributions to the differential cross section at different electron scattering angles. In [116] the PRA was applied to describe the dynamic response of atomic clusters in the processes of radiative electron capture, polarizational BrS and multiphoton excitation.

7.2.1 Polarizabilities of Metal Clusters and Fullerenes

General formulae (2.8–2.21) can be applied to determine the amplitude of the BrS process in a charged particle scattering (for the sake of clarity, below we refer to electron scattering) from a metallic cluster or/and a fullerene.

The only difference is in the definition of the ground and excited states wavefunctions, $|0\rangle$ and $|n\rangle$, which in the case of a cluster target can be obtained within the jellium model [128, 180, 210, 211, 297, 311]. The wavefunctions of the projectile electron, $|\mathbf{p}^{\pm}_{1,2}\rangle$, can be considered within the BA or, similar to (2.11), sought in the form of the DPWA series, with the radial waves calculated in the frozen field of the cluster.

As well as in the case of atomic target (see Sects. 4.1, 4.3), an accurate treatment of the generalized polarizability, $\alpha(\omega, q)$ is important for the correct description of the PBrS amplitude, (2.16).

In [154] a method was suggested to describe the dynamic response of a system possessing a collective giant resonance. The method was applied to evaluate simple but sufficiently accurate expression for the PBrS cross section for electron collision with metal clusters or/and fullerenes in the vicinity of plasmon resonances where the surface plasmon excitations provide the main contribution to $\alpha(\omega, q)$. In what follows, we outline the main ideas of the method which relies on the small parameter, $a/R \ll 1$, where $a \sim R_{\rm at}$ is the thickness of the fullerene shell or the mono-atomic surface layer in the case of a metal cluster, and R is the cluster radius. The width a characterizes the layer within which the surface plasmon is excited. This assumption is rather clear for a fullerene due to its hull-like structure. However, this statement is correct for metal clusters too [268, 293, 294], where the electron density inside the cluster is well compensated by the oppositely charged density of the ionic background.

Let us first consider the dipole polarizability $\alpha(\omega)$, which defines the dynamic response of the system to a uniform field $\mathbf{E}$ of the electromagnetic wave of a frequency ω. According to Kubo linear response theory [269], alteration of the electron density of a cluster, $\delta\rho(\mathbf{r})$, under the action of an external electric field is given by

$$\delta\rho(\mathbf{r}) = \int \Pi(\mathbf{r}, \mathbf{r}')\, \mathbf{E}\cdot\mathbf{r}'\, \mathrm{d}\mathbf{r}'. \tag{7.1}$$

Here $\Pi(\mathbf{r}, \mathbf{r}')$ is the polarization operator

$$\Pi(\mathbf{r}, \mathbf{r}') = \sum_n \left(\frac{\langle 0|\hat{\rho}(\mathbf{r})|n\rangle\langle n|\hat{\rho}(\mathbf{r}')|0\rangle}{\omega - \omega_{n0} + i\delta} - \frac{\langle 0|\hat{\rho}(\mathbf{r}')|n\rangle\langle n|\hat{\rho}(\mathbf{r})|0\rangle}{\omega + \omega_{n0} + i\delta} \right), \tag{7.2}$$

where $\hat{\rho}(\mathbf{r}) = \sum_a \delta(\mathbf{r} - \mathbf{r}_a)$ denotes the operator of the electron density. The summation is carried out over the whole spectrum of the system, including the term with $|n\rangle = |0\rangle$. However, it does not contribute to the sum in the case of a spherically-symmetric target since $\langle 0|\sum_a \mathbf{E}\cdot\mathbf{r}_a|0\rangle = 0$.

To proceed further one accounts for the relation $\lambda_{\rm e} \ll R$, where $\lambda_{\rm e}$ is the de Broglie wave length of a cluster electron. It can be estimated to be equal, by the order of magnitude, to the scale $a \sim R_{\rm at}$. Since the electron wave functions oscillate frequently over the size R, the main contribution to the sum $\sum_n(\dots)$ from (7.2) comes from the dipole transitions $0 \to n$ between the states with the energies close to the Fermi energy $\varepsilon_{\rm F}$, and close principal quantum numbers. Thus, the characteristic momenta of such states differ approximately by $1/R$. Therefore, the transition energy

ω_{n0} can be estimated as $\omega_{n0} \sim \varepsilon_{\rm F}(a/R)$. The energy of the plasmon excitation $\omega_{\rm p}$ and the Fermi energy $\varepsilon_{\rm F}$ are equal by the order of magnitude. Therefore, to calculate the dynamic response in the photon energy range $\omega \sim \omega_{\rm p}$ one can expand the right-hand side of (7.2) in powers of $\omega_{n0}/\omega \sim (\varepsilon_{\rm F}/\omega)(a/R) \sim (\omega_{\rm p}/\omega)(a/R) \sim a/R \ll 1$ and keep the first non-vanishing term, which is:

$$\delta\rho(\mathbf{r}) = \frac{1}{\omega^2}\sum_n \omega_{n0}\left(\langle 0|\hat{\rho}(\mathbf{r})|n\rangle\langle n|\sum_a \mathbf{E}\cdot\mathbf{r}_a|0\rangle + \langle 0|\sum_a \mathbf{E}\cdot\mathbf{r}_a|n\rangle\langle n|\hat{\rho}(\mathbf{r})|0\rangle \right). \tag{7.3}$$

Using the standard rules (see, e.g., [385]), one substitutes the dipole matrix elements in the form of 'length' with the equivalent expressions written in the form of 'velocity': $\mathbf{r}_{n0} = -\mathrm{i}\omega_{n0}^{-1}\hat{\mathbf{p}}_{n0}$, where $\hat{\mathbf{p}}$ is the momentum operator of an electron. Then, making use of the completeness condition $\sum_n |n\rangle\langle n| = \hat{1}$, one arrives at

$$\delta\rho(\mathbf{r}) = -\frac{\mathbf{E}\cdot\nabla n(\mathbf{r})}{\omega^2}\,. \tag{7.4}$$

Here $n(\mathbf{r}) = \langle 0\,|\hat{\rho}(\mathbf{r})|\,0\rangle$ is the ground state electron density of the cluster. Quantum mechanical calculations, performed within the framework of the jellium model, show that $n(\mathbf{r})$ is approximately constant inside the cluster but rapidly changes at the surface within the a-thick layer (see, e.g., [268]). Hence, (7.4) indicates that the plasmon-type excitation is localized at the cluster surface.

Equations (7.1–7.4) describe the cluster response in the first-order of the perturbation theory with respect to the interaction potential $\phi = -\sum_a \mathbf{E}\cdot\mathbf{r}_a$ of the cluster electrons with a uniform external field $\mathbf{E}$. Thus, they do not account for large depolarization due to strong electron correlations in a metal cluster. More accurate result can be obtained within the RPAE scheme [83, 366, 367]:

$$\delta\rho(\mathbf{r}) = \int \mathrm{d}\mathbf{r}'\,\Pi(\mathbf{r},\mathbf{r}')\,\mathbf{E}\cdot\mathbf{r}' + \int \mathrm{d}\mathbf{r}'\int \mathrm{d}\mathbf{r}''\,\Pi(\mathbf{r},\mathbf{r}')\frac{\Pi(\mathbf{r}',\mathbf{r}'')}{|\mathbf{r}'-\mathbf{r}''|}\,\mathbf{E}\cdot\mathbf{r}'' + \cdots\,. \tag{7.5}$$

As long as electric charge is localized within the layer $a \ll R$, each term in this series is factorized so that the right-hand side can be calculated. The result reads [293]:

$$\delta\rho(\mathbf{r}) = -\frac{\mathbf{E}\cdot\nabla n(\mathbf{r})}{\omega^2 - \omega_{\rm p}^2 + \mathrm{i}\omega\Gamma}, \tag{7.6}$$

where $\omega_{\rm p}$ is the plasmon frequency, and Γ stands for the width of the plasmon resonance.

Similar considerations can be applied to a fullerene. The electron motion can be treated as the superposition of rather rapid motion normal to the fullerene surface and relatively slow angular motion along the fullerene surface [366, 367]. The normal motion is localized within the fullerene shell of the thickness a by the attractive carbon potential. The surface motion can be treated as that of a rotator with the electron mass and the fullerene radius. As well as in the case of a metal cluster, the main contribution to the perturbation of the electron density is due to the dipole transitions between electronic states in the vicinity of the Fermi level. The final result reads as given by (7.6).

Within the PRA, the dipole dynamic polarizability of either a metal cluster or a fullerene one obtains from (7.6) as a ratio of the induced dipole moment to the strength of the external electric field. This leads to the following result:

$$\alpha(\omega) = R^3 \frac{\omega_p^2}{\omega_p^2 - \omega^2 - i\omega\Gamma}. \tag{7.7}$$

The plasmon frequency ω_p, calculated within the framework of the classical Mie theory (see, e.g., [268]), can be expressed via the cluster/fullerene radius R and the number N of the delocalized electrons:

$$\omega_p = \begin{cases} \sqrt{\dfrac{N}{R^3}} & \text{for a metal cluster} \\ \sqrt{\dfrac{2N}{3R^3}} & \text{for a fullerene} \end{cases}. \tag{7.8}$$

The resonance width Γ is connected with the plasmon damping. Damping of the plasmon oscillations is connected with the decay of the collective electron excitations to the single-particle ones similar to the mechanism of Landau damping in infinite electron gas (see, for example, [280]). Various theoretical approaches to calculate the widths of surface and volume plasmons in clusters and fullerenes are discussed in [391].

Within the PRA the width enter the right-hand side of (7.7) as the parameter which can be deduced from the experimental data on the photoabsorption cross section $\sigma_\gamma(\omega)$ which is proportional to the imaginary part of $\alpha(\omega)$, (4.10). For example, the experimental data for $\sigma_\gamma(\omega)$ for the C_{60} fullerene [195] can be reproduced if one inserts in (7.7) the frequency $\omega_p = 19\,\text{eV}$ and the width $\Gamma = 11\,\text{eV}$. Equation (7.8), applied to the fullerene C_{60} ($R = 3.53$ Å, $N = 240$), produces close result for the resonance frequency: $\omega_p = 18.5\,\text{eV}$.

We note that the PRA was successfully applied to describe also multipole dynamic polarizabilities of clusters and fullerenes [116, 118, 146, 147, 151, 391]. This approach has been also widely used to analyze various processes (photoabsorption, electron scattering, Auger decay) involving pristine and endohedral fullerenes within the model, which treats a fullerene as as a spherical layer of a finite thickness [47, 48, 277, 289, 290, 338, 421–423].

To evaluate the expression for the generalized dynamic polarizability $\alpha(\omega, q)$ of a cluster or a fullerene one notices, that the arguments used to transform the polarization operator $\Pi(\mathbf{r}, \mathbf{r}')$, (7.2), are applicable to the matrix elements on the right-hand side of (2.14). Namely, in the vicinity of the plasmon resonance, the main contribution to the matrix elements comes from a thin layer of width a near the surface, where the oscillation of the charge density takes place. Therefore, one can assume $r_a \approx R$ in

(2.14). Using this proximity, one expresses the generalized dynamic polarizability via the dipole polarizability [154]:

$$\alpha(\omega, q) = \frac{3j_1(qR)}{qR}\,\alpha(\omega). \qquad (7.9)$$

Here $j_1(x) = \sin x/x^2 - \cos x/x$ is the spherical Bessel function of the first order.

Equation (7.9) is valid if the transferred momentum q is smaller than the characteristic atomic momentum a^{-1}, i.e., if $q \ll 1/a$.

7.2.2 BrS Cross Section

The projectile electron effectively excites a plasmon, when the time of the collision, R/v_1, is close to the period of the plasmon oscillation. This occurs at the collision velocity $v_1 \sim \omega_p R$, which is larger than the characteristic velocities of the valence electrons in a cluster. Therefore, in describing photon emission in the vicinity of the plasmon resonance, one can use expression (2.17) for the PBrS amplitude within the framework of the BA.

To construct the total BrS amplitude $f_{\rm tot} = f_{\rm ord} + f_{\rm pol}$ one substitutes (7.9) into (2.17) and, then, adds the OBrS amplitude calculated within the BA (see (2.1)). Using the result in (2.25) and carrying out all necessary summations and integration, one arrives at the following formula for the double differential cross section which characterizes spectral-angular distribution of the total BrS in the vicinity of the giant resonance (cf. (2.31–2.33)):

$$\frac{{\rm d}^2\sigma_{\rm tot}}{{\rm d}\omega{\rm d}\Omega_{\bf k}} = \frac{4}{3\pi c^3 v_1^2 \omega}\int\limits_{q_{\min}}^{q_{\max}} \frac{{\rm d}q}{q}\left|F(q) - \omega^2\alpha(\omega, q)\right|^2 \Big(1 - P_2(\cos\theta_{\bf q})\,P_2(\cos\theta_{\bf k})\Big)$$

$$\approx \frac{{\rm d}^2\sigma_{\rm ord}}{{\rm d}\omega{\rm d}\Omega_{\bf k}} + \frac{{\rm d}^2\sigma_{\rm tot}}{{\rm d}\omega{\rm d}\Omega_{\bf k}}\,. \qquad (7.10)$$

Here $P_2(\cos\theta_{\bf k})$ and $P_2(\cos\theta_{\bf q})$ are the Legendre polynomials, the quantity $\theta_{\bf q} = \widehat{{\bf p}_1, {\bf q}}$ is the angle between ${\bf p}_1$ and the vector ${\bf q} = {\bf p}_1 - {\bf p}_2$ of the transferred momentum, $q_{\min} = p_1(1 - \sqrt{1 - \omega/\varepsilon_1})$ and $q_{\max} = p_1(1 + \sqrt{1 - \omega/\varepsilon_1})$ are the minimum and maximum values of q. The explicit expression for $\cos\theta_{\bf q}$ is presented in (6.134). The quantity $F(q)$ denote the total form-factor of the charge distribution in the cluster. It includes the contribution from the electrons, $F_-(q)$, as well as that of the ionic background $F_+(q)$:

$$F(q) = \int {\rm e}^{{\rm i}{\bf q}\cdot{\bf r}}\Big(\rho_+({\bf r}) - \rho_-({\bf r})\Big){\rm d}{\bf r} = F_+(q) - F_-(q)\,, \qquad (7.11)$$

where $\pm\rho_{\pm}(\mathbf{r})$ are the charge densities of the positive and negative charges.

The approximate equality in (7.10) stresses that in the vicinity of the plasmon resonance, $\omega \approx \omega_{\rm p}$, the interference part of the spectrum can be omitted. A similar feature was discussed in detail with respect to the BrS spectrum formed in electron–atom collisions in the vicinity of the giant dipole resonance, see Sect. 4.1.

In the case of a cluster target, the polarizability of which is enhanced by a factor $\sim (R/R_{\rm at})^3$ as compared to the atomic target, the relative yield of the OBrS background radiation for $\omega \approx \omega_{\rm p}$ is suppressed even more. The OBrS is mainly generated in the region of large transferred momenta, $q > R_{\rm at}^{-1} \sim a^{-1}$, which corresponds to the range of distances smaller than cluster or even atomic size, where the electric field accelerating the projectile electron is relatively strong. Therefore, to estimate the OBrS cross section for an electron–cluster/fullerene collision one can consider it as an incoherent sum of the OBrS contributions generated in collisions of the projectile with the cluster atoms [31, 114, 115]. Thus, the cross section $\mathrm{d}\sigma_{\rm ord}/\mathrm{d}\omega$ can be estimated as follows:

$$\omega\frac{\mathrm{d}\sigma_{\rm ord}}{\mathrm{d}\omega} \approx N\omega\frac{\mathrm{d}\sigma_{\rm ord}^{\rm (at)}}{\mathrm{d}\omega} \sim \frac{16e^4 N_e}{3c^3 v_1^2}. \tag{7.12}$$

Here $\mathrm{d}\sigma_{\rm ord}^{\rm (at)}/\mathrm{d}\omega$ is the atomic OBrS, N is the number of atoms and N_e is the number of electrons in a cluster with ionization potential less then the energy $\varepsilon_1 = p_1^2/2$ of the incoming electron.

To be noted is that the OBrS cross sections depends on the model chosen for the description of a cluster target. For example, in the optimized jellium model [211], the OBrS contribution is totally absent due to the condition $W_+(q) = W_-(q)$, which is fulfilled for all q in this model. In this case the role of the PBrS becomes even more prominent.[1]

In general, the relative role of the PBrS mechanism in the cross section (7.10) is increased in the vicinity of the roots of the equation $W_-(q) = W_+(q)$. The solutions of this equation exist in all models of cluster. For example, the point $q = 0$ is the most trivial solution in the case of a neutral cluster. We further demonstrate that this equation has non-trivial roots as well.

With the help of (7.9), the polarizational part of the cross Section (7.10) can be written as follows

$$\frac{\mathrm{d}^2\sigma_{\rm pol}}{\mathrm{d}\omega \mathrm{d}\Omega_{\mathbf{k}}} = \frac{4\omega^3}{3\pi c^3 v_1^2}\,|\alpha(\omega)|^2 \Big(S(x_{\rm min}) - S(x_{\rm max})\Big) \times \Big(1 + \beta(\omega, v_1, R) P_2(\cos\theta_{\mathbf{k}})\Big). \tag{7.13}$$

[1] In atomic physics, the absence of the OBrS due to identical distribution of the positive and negative charges in the target occurs in electron (or positron) scattering from a positronium [32]. Other objects which potentially reveal this property are electron-positron clusters [308, 392, 441].

Here the short-hand notations $x_{\min} = q_{\min}R$ and $x_{\max} = q_{\max}R$ are introduced. The function S and the angular asymmetry parameter $\beta(\omega, v_1, R)$ are defined as follows:

$$S(x) = \frac{9x^2 + 6 + 2x(2x^4 - x^2 - 6)\sin 2x - (2x^4 - 3x^2 + 6)\cos 2x}{8x^6} - \mathrm{Ci}(2x), \tag{7.14}$$

$$\beta(\omega, v_1, R) = [S(x_{\min}) - S(x_{\max})]^{-1}\,\frac{27}{2}\int_{x_{\min}}^{x_{\max}} \mathrm{d}x\, \frac{j_1^2(x)}{x^5}\left(\frac{\omega R}{v_1} + \frac{x^2}{2v_1R}\right)^2 - \frac{1}{2}. \tag{7.15}$$

In (7.15) the notation Ci(.) stands for the cosine integral function.

Integrating (7.13) over the emission angles, one derives the spectral distribution [154]:

$$\omega\,\frac{\mathrm{d}\sigma_{\mathrm{pol}}}{\mathrm{d}\omega} = \frac{16\omega^4}{3c^3v_1^2}\,|\alpha(\omega)|^2\left(S(x_{\min}) - S(x_{\max})\right). \tag{7.16}$$

For large v_1, when $q_{\min} \approx \omega/v_1 \ll 1/R$, the PBrS process occurs at distant collisions. In this limit, $q_{\max}R \approx 2p_1R \gg 1$ so that the term $S(x_{\max})$ can be omitted on the right-hand sides of (7.13, 7.15–7.16). Evaluating the limit of $S(q_{\min}R)$ for $q_{\min}R \ll 1$, one finds: $S(q_{\min}R) \approx \ln(\kappa/q_{\min}R) \approx \ln(v_1/\omega R)$, where $\kappa \approx 7/4 - 1.61549\,\gamma$, $\gamma = 0.577216$ is the Euler constant. In this limit (7.16) reduces to the PBrS cross section in electron–fullerene collision obtained in [31] in the so-called logarithmic approximation (see [45]):

$$\omega\,\frac{\mathrm{d}\sigma_{\mathrm{pol}}}{\mathrm{d}\omega}\bigg|_{v_1 \gg \omega R} \approx \frac{16\omega^4}{3c^3v_1^2}\,|\alpha(\omega)|^2 \ln\frac{v_1}{\omega R}\,. \tag{7.17}$$

In the limit, $x \gg 1$, the asymptotic formula for $S(x)$ includes the oscillatory terms:

$$S(x) \approx \frac{9}{8x^4}\left(1 + \frac{2}{3x^2} - \frac{2\sin 2x}{x} + \frac{\cos 2x}{x^2}\right). \tag{7.18}$$

The origin of these terms in the cross section is physically clear [114, 115]. Large $x \gg 1$ correspond to the limit when the wave length $\pi/q_{\min}^{-1}$ becomes less than the size R of the cluster. The diffraction of the projectile electron at the edge of the cluster results in the terms in (7.13) and (7.16) which have oscillatory character as functions of the parameter $x = \omega R/v_1$. The oscillations take place in a wide range of frequencies, $v_1/R \ll \omega \ll v_1/a$.

Let us compare the OBrS and PBrS contributions to the spectrum of radiation. Using (7.12) and (7.16), one derives the following expressions for the ratio $A = \mathrm{d}\sigma_{\mathrm{pol}}/\mathrm{d}\sigma_{\mathrm{ord}}$ at $\omega = \omega_{\mathrm{p}}$ in electron collisions with metallic clusters and fullerenes:

$$A = \frac{\mathrm{d}\sigma_{\mathrm{pol}}}{\mathrm{d}\sigma_{\mathrm{ord}}} \sim N\left(\frac{\omega_{\mathrm{p}}}{\Gamma}\right)^2 S\left(\frac{\omega_p R}{v_1}\right) \times \begin{cases} 1 & \text{for a metal cluster} \\ \frac{4}{9} & \text{for a fullerene}\,. \end{cases} \tag{7.19}$$

For metal clusters the ratio $\omega_\mathrm{p}/\Gamma \sim 10$ is rather large and does not, virtually, depend on the cluster size. For a fullerene C_{60}, $\omega_p/\Gamma \approx 2$. Thus, the PBrS dominates by orders of magnitude over the OBrS background in the vicinity of the plasmon resonance either for metal clusters or for fullerenes [31, 114, 115].

The origin of the factors on the right-hand side of (7.19) is physically clear. The factor N characterizes the coherence of radiation by the delocalized electrons in the PBrS process. Factor $(\omega_\mathrm{p}/\Gamma)^2$ is due to the resonance nature of PBrS in the vicinity of the plasmon excitation. The function $S(\omega_p R/v_1)$ describes the kinematics of the collision and depends on the scaling parameter $\omega_p R/v_1$. The coefficient 4/9 in the fullerene case is a result of the topological difference between metal clusters and hollow fullerenes.

In Sects. 7.2.3 and 7.2.4 we present results of numerical calculations of the PBrS spectra carried out within the PRA for electron–fullerene and electron–metal-cluster collisions.

Prior to doing this we conclude the theoretical section by considering the cross section in two limiting cases, $\omega \gg \omega_\mathrm{p}$ and $\omega \ll \omega_\mathrm{p}$, as well as by commenting on the BrS spectrum formed in cluster–cluster collisions.

7.2.2.1 High-Frequency Limit, $\omega \gg \omega_\mathrm{p}$

The total BrS cross section of an electron on a cluster (a fullerene) exhibits peculiar features in the photon energy range far above the plasmon resonance.[2]

In the limit $\omega \gg \omega_p$ the generalized polarizability $\alpha(\omega, q)$ can be replaced by its asymptotic value (cf. (2.4) for an atomic target):

$$\alpha(\omega, q) \approx -\frac{F_-(q)}{\omega^2}\,, \tag{7.20}$$

where $F_-(q)$ is the form-factor of the delocalized electrons (see (7.11)).

Substituting (7.20) into (7.10), and carrying out integration over the emission angles, one derives the following expression for the total BrS cross section differential in the photon energy and the transferred momentum:

$$\frac{\mathrm{d}^2\sigma_\mathrm{tot}}{\mathrm{d}\omega \mathrm{d}q} = \frac{16}{3c^3 p_1^2 \omega}\,\frac{|F_+(q)|^2}{q}\,. \tag{7.21}$$

The right-hand side contains only the form-factor of the ionic core positive charge distribution. Thus, the total photon emission is formed due to the projectile electron acceleration in the non-screened field of the core. The origin of this de-screening effect is the same in electron–atom collisions (see Sect. 4.6): the system 'projectile

[2] It is assumed below that being large compared to ω_p, the photon energy ω is much less than the ionization potential of the atomic core electrons, which are not collectivized in the cluster/fullerene target.

electron + quasi-free electrons of the target' does not possess a dipole moment and thus does not radiate [12]. Hence, the target is effectively 'stripped off' those electrons the ionization potential of which is much smaller than ω (compare (7.21) with (4.21)).

The form-factor of the cluster ionic background can be calculated within the framework of the most widely used jellium model [128], which implies the step-function distribution of the positive charge: $\rho_+ = const$ for $r \le R$ and $\rho_+ = 0$ if otherwise. This model, being used in (7.11), produces

$$W_+(q) = 3N \left(\frac{\sin qR}{(qR)^3} - \frac{\cos qR}{(qR)^2} \right) = 3N \, \frac{j_1(qR)}{qR} \, . \tag{7.22}$$

The cross section (7.21), being proportional to the oscillating factor $j_1^2(qR)$, has a series of zeros which are the zeroes of the spherical Bessel function of the first order. These values are known (see, for example, [1]). The first four zeroes are found as $q_j = x_j/R$, where $x_{1,2,3} = 4.493;\, 7.725;\, 10.904$. For the large $j \gg 1$, one has $x_j \approx (2j+1)\pi/2$. The distance between two neighboring zeroes in this case is equal to $\Delta q = q_{i+1} - q_i = (x_{i+1} - x_i)/R \approx \pi/R$, i.e., it is determined by the radius of the cluster.

Similar consideration applies to other cluster models as well. In all cases, the diffraction behavior of the cross section is determined by the radius of the cluster, providing thus a method for experimental determination of cluster sizes and verification of various cluster models.

To illustrate this feature of the double differential cross section we present Fig. 7.1 where the dimensionless quantity $\omega q^3 \mathrm{d}^2\sigma/\mathrm{d}\omega \mathrm{d}q$ for Na_{20}^+ is plotted as a function of the transferred momentum q [114]. The full curve is the result of calculation based on (7.10). The broken curves were obtained using the asymptotic equation (7.21) for the two ω values indicated in the caption. The oscillatory character of the double cross section is clearly seen. Also worth to note, that the deviation of the asymptotic result from (7.10) decreases with increasing ω.

Substituting (7.22) into (7.21) and carrying out the integration over q, one arrives at the spectral distribution of radiation in the asymptotic region:

$$\frac{\mathrm{d}\sigma_{\mathrm{tot}}}{\mathrm{d}\omega} = \frac{16}{3c^3 v_1^2} \frac{N^2}{\omega} \Big(S(x_{\mathrm{min}}) - S(x_{\mathrm{max}}) \Big) . \tag{7.23}$$

7.2.2.2 Low-Frequency Limit

In the limit $\omega \ll \omega_\mathrm{p}$, the polarizability (7.7) reduces to the static polarizability of a metal sphere, $\alpha(0) = R^3$. At the same time, this is a correct value of the static polarizability of a metal cluster [191]. Therefore, it is interesting to compare the PBrS cross section (7.16) with the result obtained within the model which treat the cluster as a metal sphere of the radius R. In this model, the PBrS appears due to alteration

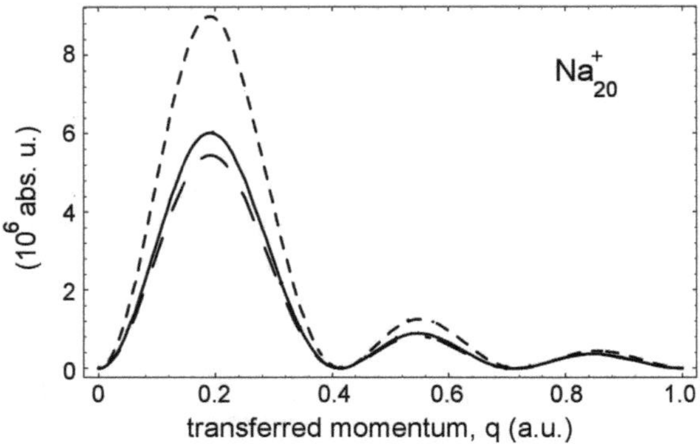

Fig. 7.1 Dimensionless cross section $\omega q^3 \mathrm{d}^2\sigma/\mathrm{d}\omega \mathrm{d}q$ versus transferred momentum q calculated for Na_{20}^{+} at $\omega = 5$ and 10 eV. The *full curve* is the result of a calculation based on (7.10). The *short-broken* and the *long-dashed curves* are derived from the asymptotic equation (7.21) $\omega = 5$ and 10 eV, respectively, [114]

of the dipole moment induced on the sphere by a moving charge. The intensity of radiation will be proportional to the squared Fourier image of the dipole moment of the projectile electron and its image due to the sphere. The result for the cross section of PBrS within the model reads [300]:

$$\frac{\mathrm{d}^2\sigma_{\mathrm{pol}}}{\mathrm{d}\rho \mathrm{d}\omega} = \frac{16\rho\omega^5 R^6}{3c^3 v_1^4} \left[K_0^2(x) + K_1^2(x) \right]_{x=\omega\rho/v_1} . \tag{7.24}$$

Here $K_0(x)$ and $K_0(x)$ are the modified Bessel functions of the second type (the Macdonald function) [1], ρ is the impact parameter.

Comparing results obtained within the metal sphere approximation and within PRA, (7.16) and (7.24), one notices that both formulae produce close results only in the region of $\omega \ll \omega_{\mathrm{p}}$ when $x \ll 1$. Using asymptotic formulae for $K_1(x)$, $K_0(x)$ [1], and integrating (7.24) over ρ from R to v_1/ω, one derives the main logarithmic term which appears on the right-hand side of (7.16).

In the vicinity of the resonance $\omega = \omega_{\mathrm{p}}$ (7.24) underestimates the cross section by the resonance enhancement factor $(\omega_{\mathrm{p}}/\Gamma)^2 \gg 1$.

In the region of $\omega \gg v_1/R$, the metal sphere approximation gives an exponentially small result, while PRA leads to the much slower decrease, $\propto 1/\omega^5$.

7.2.2.3 BrS in Cluster–Cluster Collisions

If the projectile particle has its own internal structure, as occurs, for example, in cluster-cluster collisions, then the dynamic polarization of the projectile cluster must be taken into account along with that of the target. Analytically, the evaluation of the BrS amplitude and the cross section within the framework of the non-relativistic BA is similar to the case of atom–atom collisions, see Sect. 5.2 and [33, 34]. The result for the spectrum of radiation reads (cf. (5.18)):

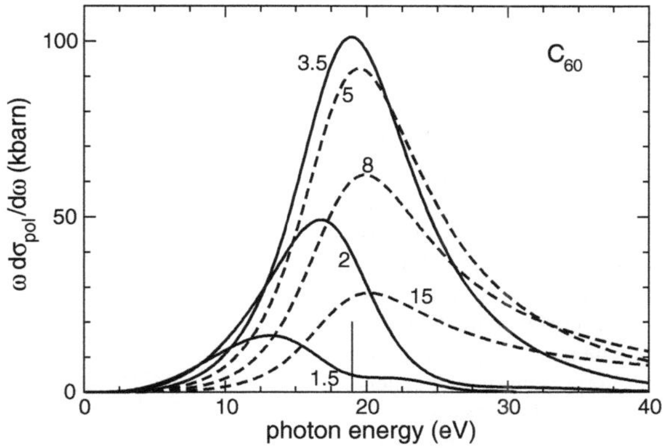

Fig. 7.2 Spectral distribution of PBrS in electron–C_{60} collision calculated for indicated values (in a.u.) of initial velocities v_1. *Vertical dash* marks the plasmon resonance frequency $\omega_p = 19\,\text{eV}$

$$\frac{d\sigma_{cl-cl}}{d\omega} = \frac{16\omega^3}{3c^3 v_1^2} \int_{q_{min}}^{q_{max}} \frac{dq}{q} \left| F^{(1)}(q)\,\alpha^{(2)}(\omega, q) - F^{(2)}(q)\,\alpha^{(1)}(\omega, q) \right|^2 . \quad (7.25)$$

Here the subscripts $j = 1, 2$ denote the form-factors and polarizabilities of the projectile cluster and the target cluster, respectively.

It is rather interesting to investigate the behaviour of the cross section (7.25) particularly in the case when both colliding clusters have close-lying giant resonances.

Similar to the atom–atom case, Sect. 5.2, no dipole-photon radiation appears in the collision of two identical clusters, since the total induced dipole moment is identically equal to zero. In this case quadrupole radiation contribution to the cross section becomes important.

7.2.3 BrS in Electron–Fullerene Collisions

In this section we present results of numerical calculations of the PBrS spectral and angular dependencies for electron collisions with a C_{60} fullerene. The fullerene radius, plasmon resonance energy and the width of the plasmon peak used in the calculations are equal to 6.67 a.u., 19 and 11 eV, respectively.

The PBrS spectra (7.16), calculated for different values of the initial velocity v_1 (as indicated), are presented in Fig. 7.2. The shape of the spectral distribution is determined by two factors. First, it is the modulus squared of the fullerene dynamic polarizability, $|\alpha(\omega)|^2$, which has the resonance at $\omega \approx \omega_p$. For the chosen values of v_1, the second factor, $S(q_{min}R) - S(q_{max}R) \approx S(q_{min}R)$ (see (7.14)) depends mainly on the parameter $q_{min}R \approx \omega R/v_1$, which is the ratio of the fullerene radius to the distance $\rho = v_1/\omega$ at which the dynamic polarization of fullerene occurs. The factor $S(q_{min}R)$ logarithmically diverges if $\rho \gg R$ and rapidly decreases in the domain $\rho \lesssim R$. This factor sets the cutoffs of the spectrum at $\omega \sim v_1/R$. It is seen from the figure, that for the collision velocities close to $\omega_p R \approx 4.7$ a.u. the

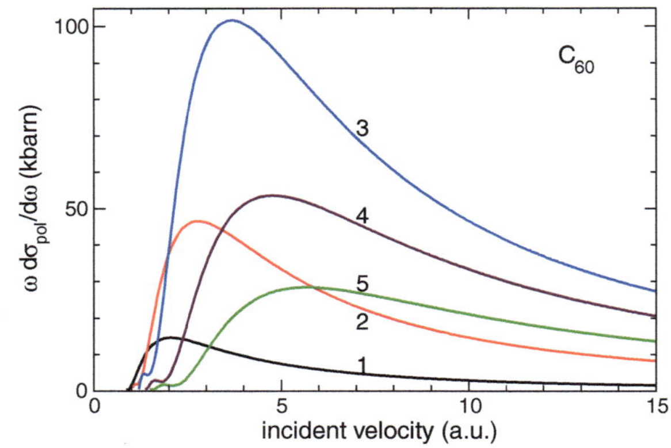

Fig. 7.3 PBrS cross section $\omega \mathrm{d}\sigma_{\mathrm{pol}}/\mathrm{d}\omega$, (7.16), for electron–$C_{60}$ collision calculated as a function of the incident electron velocity v_1 for several values of the photon energy ω. *Curves 1* and *2* stand, respectively for $\omega = 10$ and 14 eV which are below the resonance energy $\omega_{\mathrm{p}} = 19$ eV, *curve 3* corresponds to the resonance $\omega = \omega_{\mathrm{p}}$, *curves 4* and *5* represent the dependencies for ω values 25 and 30 eV, respectively, which are above the resonance

intensity has a clearly resonance character.This is illustrated by the curves standing for $v_1 = 3.5$ and 5 a.u. At smaller collision velocities, the position of the maximum shifts towards lower values of photon energies (the curves for $v_1 = 1.5$ and 2 a.u.). It is worth noting that for small collision velocities the BrS cross section acquires the oscillatory character: the curve for $v_1 = 1.5$ a.u. has the second, less pronounced, maximum at $\omega \approx 22.5$ eV. At higher velocities $v_1 > \omega_p R$ the cutoff values lies at higher frequencies than the polarizability maximum, which raises the right shoulder of the resonance maximum (the curves $v_1 = 8$ and 15 a.u.).

Note that the dependence of the position of the maximum of the giant resonance in the photon emission spectrum on collision velocity in electron–atom collisions been reported experimentally for the Xe atomic target [428] and studied theoretically [28, 428].

Dependence of the PBrS cross section on the initial velocity of the electron is presented in Fig. 7.3 where each curve corresponds to the particular photon energy ω which is either below (curves 1 and 2) or above (curves 4 and 5) the resonance value $\omega = \omega_{\mathrm{p}}$ (curve 3). Curve 3 reaches its maximum value (which exceeds the maximum value of $\omega \mathrm{d}\sigma_{\mathrm{pol}}/\mathrm{d}\omega$ for any other velocity v_1) in the point where the characteristic time for the plasmon excitation, R/v_1, is close to ω_{p}^{-1}. For lower velocities, the characteristic collision distance is smaller then the fullerene radius and the cross section decreases since the factor $S(q_{\min} R) - S(q_{\max} R) \approx S(q_{\min} R) \ll 1$ on the right-hand side of (7.16)). For larger velocities, $\omega \mathrm{d}\sigma_{\mathrm{pol}}/\mathrm{d}\omega$ decreases due to the factor v_1^{-2} which reflects the decreasing of the interaction time with the growth of the collision velocity. For the off-resonance photon energies, the BrS curves lie mostly below the resonance one with their maxima shifted towards the value $v_{\max} \approx \omega R$. For $\omega > \omega_{\mathrm{p}}$ this is true for the whole range of velocities, whereas for $\omega < \omega_{\mathrm{p}}$ there is the interval $\sqrt{2\omega} \le v_1 < v_{\max}$ where the off-resonance spectra exceed the values of the resonance one. Let us not that within this interval the curves corresponding to $\omega < \omega_{\mathrm{p}}$ exhibit the oscillatory behaviour.

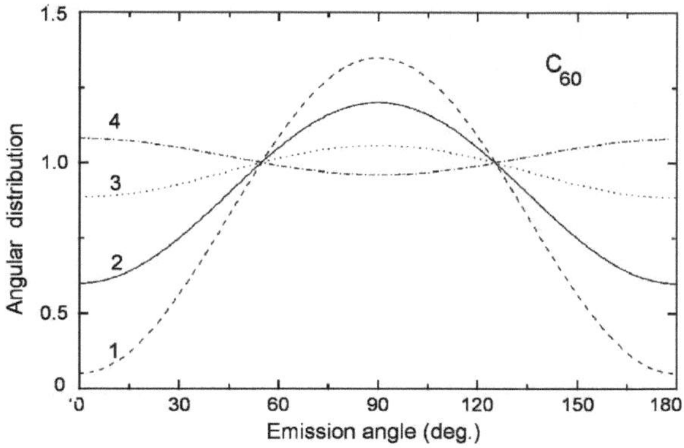

Fig. 7.4 Angular distribution, $1 + \beta(\omega, v_1, R)P_2(\cos\theta_{\mathbf{k}})$, of PBrS for electron–$C_{60}$ collision calculated as a function of emission angle $\theta_{\mathbf{k}}$ at the resonance frequency $\omega = \omega_p$ for several values of the incident velocities. *Curves 1, 2, 3* and *4* for $v_1 = 2$, 3.5, 6 and 15 a.u., respectively [154]

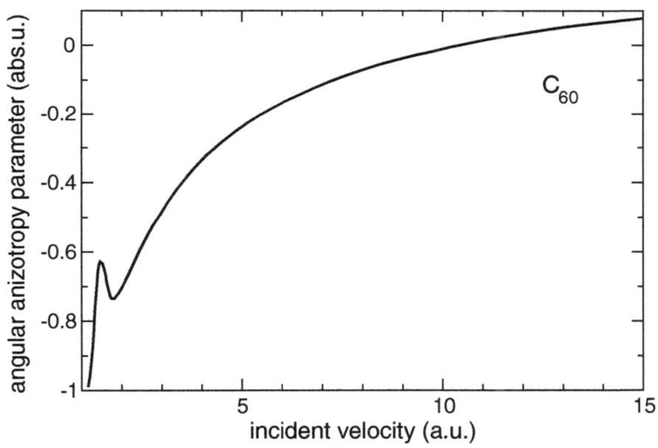

Fig. 7.5 Angular anisotropy parameter, $\beta(\omega, v_1, R)$ of PBrS for electron–C_{60} collision calculated as a function of the incident velocity at the plasmon resonance energy $\omega = \omega_p = 19\,\text{eV}$ [154]

The angular distribution of photons, proportional to $1 + \beta(\omega, v_1, R)P_2(\cos\theta_{\mathbf{k}})$ (see (7.13)) is shown in Fig. 7.4. The angular distribution, calculated at the resonance frequency $\omega = \omega_p$, is plotted for several values of the incident electron velocity, as indicated in the caption. It is seen that the shape of the angular distribution depends on the collision velocity. At large velocities, large impact parameters are kinematically allowed, $\rho \sim v_1/\omega \gg R$. A strong polarization effect in this case is caused by the component $\mathbf{q}_\perp$ of transferred momenta perpendicular $\mathbf{v}_1$. The induced dipole moment, changing along the perpendicular direction, increases the yield of the radiation in the forward and backward directions. At small collision velocities, small transferred momenta parallel to $\mathbf{v}_1$ contribute mainly to the cross section. Hence, the plasmon oscillations occur virtually in the direction of the projectile electron, resulting in the photon emission in the perpendicular direction.

The shape of the angular distribution is determined by the angular anisotropy parameter $\beta(\omega, v_1, R)$, (7.15). In Fig. 7.5 its dependence on the collision velocity is presented calculated for the resonance frequency $\omega = \omega_p$. At large velocities $\beta(\omega, v_1, R)$ approaches logarithmically the asymptotic value 1/2, transforming thus

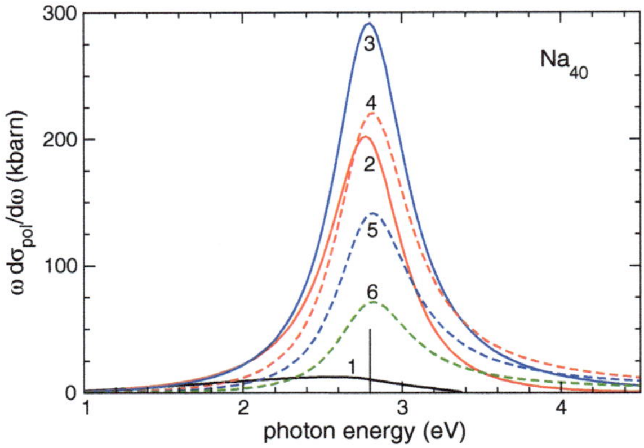

Fig. 7.6 Spectral distribution of PBrS in electron–Na_{40} collision calculated for several values of initial velocity v_1. *Curves 1, 2, 3, 4, 5* and *6* stand for $v_1 = 0.5, 0.75, 1.0, 2.0, 3.0$ and 5.0 a.u., respectively. *Vertical dash* marks the plasmon resonance frequency $\omega_p = 2.8$ eV

the shape of the angular distribution to $(1 + \cos^2\theta_{\mathbf{k}})$, which coincides with that of a rotating dipole [81]. For smaller velocities, the momentum transfer along $\mathbf{v}_1$ becomes more important. As a result, $\beta(\omega, v_1, R)$ decreases reaching its lowest allowed value of -1 at the threshold, $v_1 = \sqrt{2\omega}$, where the shape is proportional to $\sin^2\theta_{\mathbf{k}}$.

7.2.4 Electron–Metal-Cluster Collisions

Results of the calculations carried out for electron collision with metallic Na_{40} cluster are presented in Figs. 7.6 and 7.7. The parameters used in the calculations are as follows. The cluster radius $R = 13.54$ a.u. was defined from (see, for example, [268])

$$R = r_s N^{1/3}, \tag{7.26}$$

where N is the number of atoms in the cluster and r_s is the Wigner-Seits radius, which is equal to 3.96 a.u. for the bulk sodium [231]. The plasmon resonance frequency and the width of the peak were set to $\omega_p = 2.8$ eV and $\Gamma = 0.2\omega_p = 0.56$ eV [191, 384].

The spectral dependencies $\omega d\sigma/d\omega$, shown in Fig. 7.6, correspond to several velocities of the incident electron, as indicated in the caption. The PBrS spectra have resonances at $\omega_{res} \approx \omega_p$ which are narrower than in the case of the fullerene target (cf. Fig. 7.2). This feature is due the smaller value of the ratio $\Gamma/\omega_p = 0.13$ for Na_{40} than that equal to 0.6 for C_{60}. The plasmon resonance frequency in clusters lies below the ionization potential I, and the resonance width is determined only by the damping processes. In contrast, $\omega_p > I$ for the fullerene resulting in the large value of Γ due to the photoionization process. As a consequence, the shape of the emission spectrum in electron–cluster collisions is governed mainly by the dependence of the dipole polarizability on ω. In this case, the variation of the maximum value of the

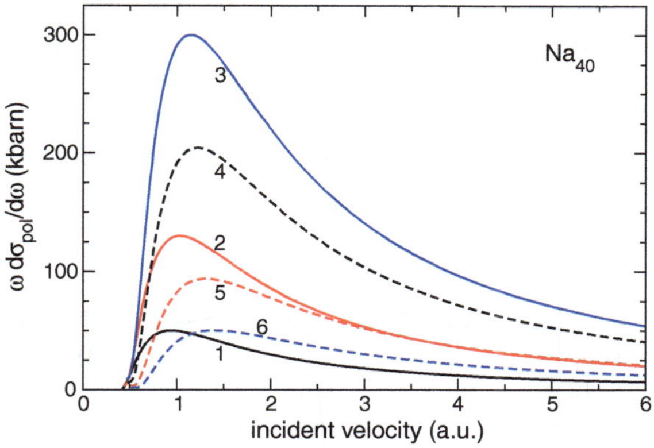

Fig. 7.7 PBrS cross section for electron–Na_{40} collision calculated as a function of the incident electron velocity v_1 for several values of the photon energy ω. *Full curves 1* and *2* stand, respectively, for $\omega = 2.25$ and 2.5 eV below the resonance energy $\omega_p = 2.8$ eV (*curve 3*). *Broken curves 4, 5* and *6* represent the dependencies for ω values 3.0, 3.25 and 3.5 eV, respectively, which are above the resonance

PBrS cross section with the collision velocity is much more pronounced than the shift in the resonance position.

The shape of the angular distribution, $1+\beta(\omega, v_1, R)P_2(\cos\theta_{\mathbf{k}})$, for metal clusters is much alike to the fullerene case (see Fig. 7.4). Therefore, we do not present any additional figures for its illustration.

The velocity dependence of the PBrS spectrum calculated at several values of ω in the vicinity of the plasmon resonance is shown in Fig. 7.7. The full curves correspond to $\omega \leq \omega_p = 2.8$ eV, the broken curves represent the dependencies for higher proton energies. In contrast to the C_{60} case, the oscillatory behaviour of the cross section just above the threshold $v_1 = \sqrt{2\omega}$ is hardly visible. This is due to a much smaller value of the plasmon frequency for the metal cluster than for the fullerene. The oscillations appear, however, with the increase of the cluster size.

The size of a metal cluster can be varied from several up to several hundreds Ångstrom. Let us illustrate the oscillatory behaviour of the PBrS cross section for the sodium clusters much larger than Na_{40}. When doing so, we do not account for the variation of the width of the plasmon resonance peak with the number N of atoms in the cluster. Even so, the figures presented below allow us to stress important features which will persist in the results of more accurate calculations.

Figure 7.8 shows photon emission spectra in the region $\omega \approx \omega_p = 2.85$ eV in collision of a $\varepsilon_1 = 13.6$ eV ($v_1 = 1$ a.u.) electron with sodium clusters of $N =$ 100, 1000, 10000, 100000 atoms [154]. Figure 7.8 exposes the features in the high-frequency part of the spectrum for these collisions. It is seen that the number of oscillations increases in both low- and high-frequency parts of the spectrum with the cluster size (Fig. 7.9).

Figure 7.10 illustrates the velocity dependence of 5 section at $\omega = \omega_p$ calculated for the same set of the sodium clusters. It is shown that oscillations appear in the velocity dependence of the cross section. The location and the number of oscillations depend on the cluster size R. Their number grows with R and the position of the main maximum shifts towards higher velocities.

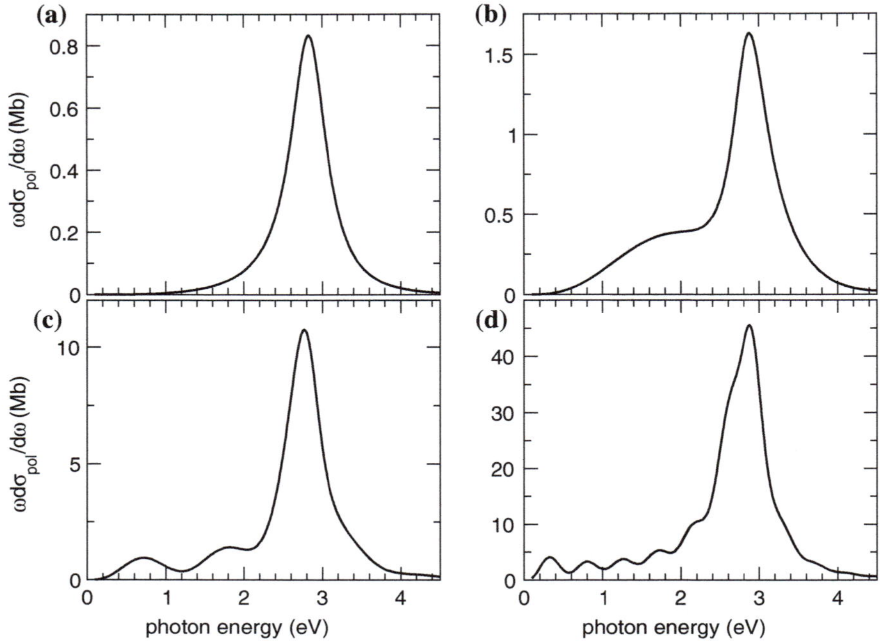

Fig. 7.8 PBrS cross section versus photon energy in the domain $\omega \sim \omega_p$ calculated for $\varepsilon_1 = 13.6\,\text{eV}$ ($v_1 = 1\,\text{a.u.}$) electron collision with sodium cluster targets of various number N of atoms. The graphs, **a–d** correspond to $N = 10^2$, 10^3, 10^4 and 10^5, respectively

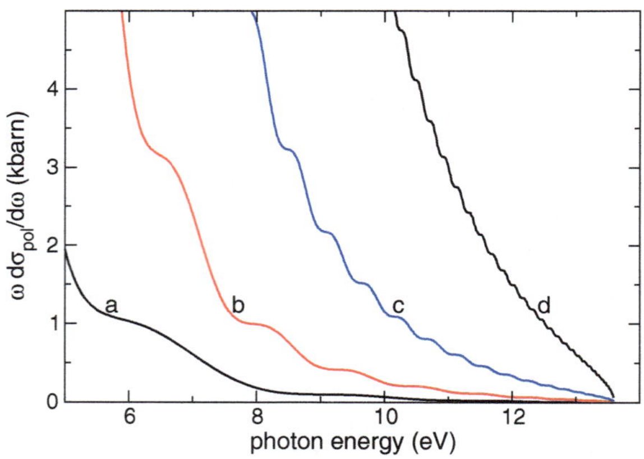

Fig. 7.9 Same as in Fig. 7.8 but for the photon energies much above the plasmon resonance energy $\omega = 2.85\,\text{eV}$

The oscillatory behaviour of the cross section is a feature typical of cluster targets, and it is due to the existence of a (sharp) cut-off radius in the electron and ion density distributions. It is natural that this feature is absent for atomic targets.

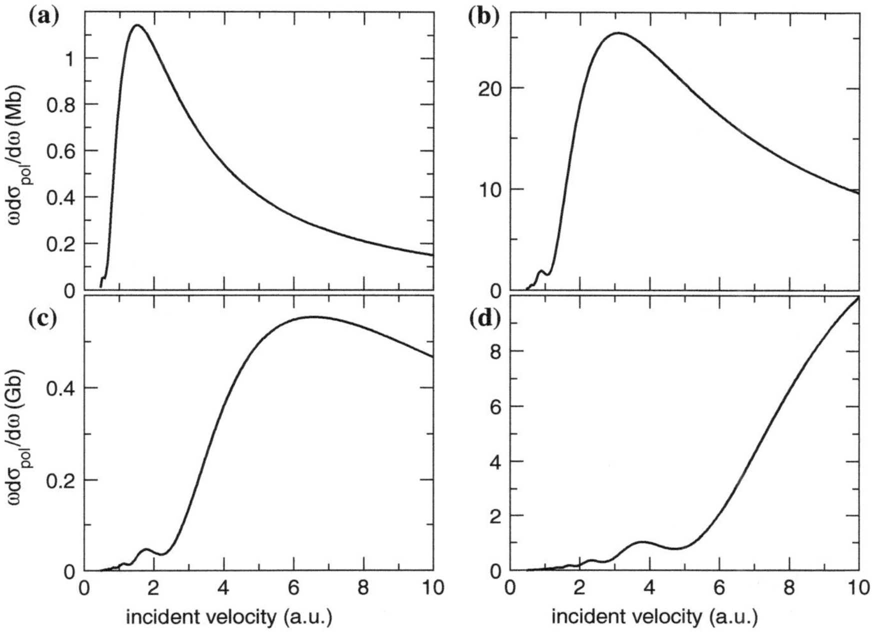

Fig. 7.10 PBrS cross section for electron–Na_N collision versus the incident electron velocity v_1 at the photon energy equal to $\omega_p = 2.25$. The graphs, **a**–**d** correspond to $N = 10^2$, 10^3, 10^4 and 10^5, respectively

Finally, let us note that the oscillating behaviour of the cross section arises in the region, where it becomes relatively small compared to its resonance value. Therefore, the experimental study of this phenomenon could be difficult due to the presence of the radiative background formed by other processes not considered here. Investigation of this issue can be a subject for further work.

7.3 Calculation of PBrS Cross Section by Means of Many-Body Theory

The formalism of the many-body treatment of the BrS radiation in collisions of fast electrons with cluster targets and the results of numerical calculations of the spectral and angular distributions for the Na_{92} cluster were reported in [148]. The scattering process was considered within the plane-wave BA. The generalized dynamic polarizability $\alpha(\omega)$ of the cluster was calculated within RPAE using the HF wavefunctions obtained within the jellium model. When calculating the contribution of the discrete electron excitations $0 \to n$ to $\alpha(\omega)$ the widths $\Gamma_{n0} = 0.09\omega_{n0}$ [440] was added to

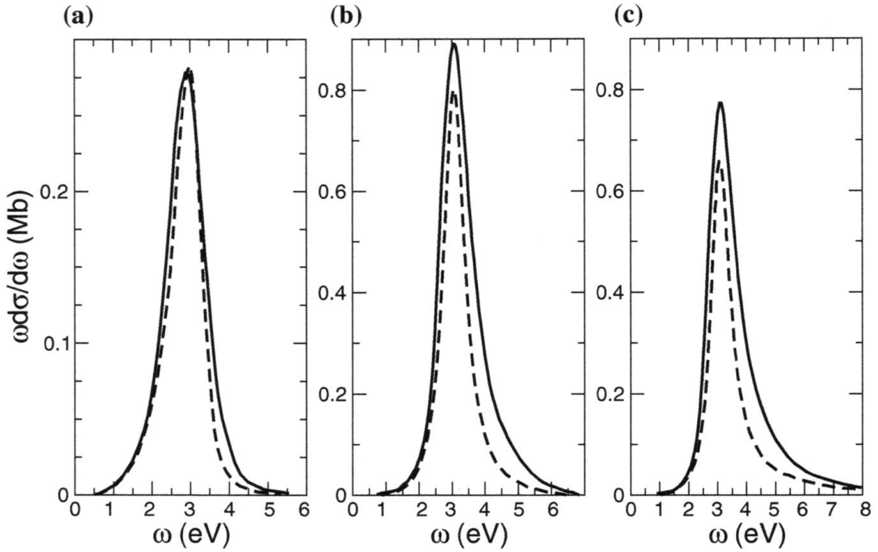

Fig. 7.11 BrS cross section $\omega d\sigma/d\omega$ versus emitted photon energy ω in the e^{-}–Na_{92}-cluster collision calculated for different incident electron energies: **a** $\varepsilon_1 = 10\,eV$, **b** $\varepsilon_1 = 30\,eV$, **c** $\varepsilon_1 = 90\,eV$. *Full curves* stand for the RPAE results, *broken curves* correspond to PRA (7.16) [148]

the denominator on the right-hand side of (2.14) to account the finite life-time of the excited state.

The BrS spectral dependencies in the vicinity of the plasmon resonance $\omega_p \approx 3\,eV$ are presented in Fig. 7.11a–c which correspond to different energies ε_1 of the incoming electron (see the caption). The full curves represent the RPAE results, the broken curves correspond to the plasmon resonance approximation. In accordance with the general qualitative consideration presented in Sect. 7.2.1 the highest yield of the BrS radiation at the resonance $\omega \approx \omega_p$ can be expected at the collision velocities $v_1 \approx \omega_p R \approx 1.9\,a.u.$ This value corresponds, approximately, to Fig. 7.11b where the data for $\varepsilon_1 = 30\,eV$ ($v_1 \approx 1.5\,a.u.$) are presented. For such velocities the collision time R/v_1 is close to the period of the plasmon oscillation. At smaller (Fig. 7.11a) and higher (Fig. 7.11c) velocities the peak intensity decreases. For small collision velocities the position of the BrS resonance is shifted towards smaller frequencies $\omega \approx v_1/R$.

The BrS cross sections within the PRA (the broken curves) were calculated in accordance with (7.16) where the above cited value of ω_p and $\Gamma = 0.75\,eV$ were used. The comparison with the RPAE results shows that the collective electron excitations provide the main contribution to the emission spectrum. The small discrepancy can be attributed to the contribution of single-electron excitations which are not accounted for within the PRA framework.

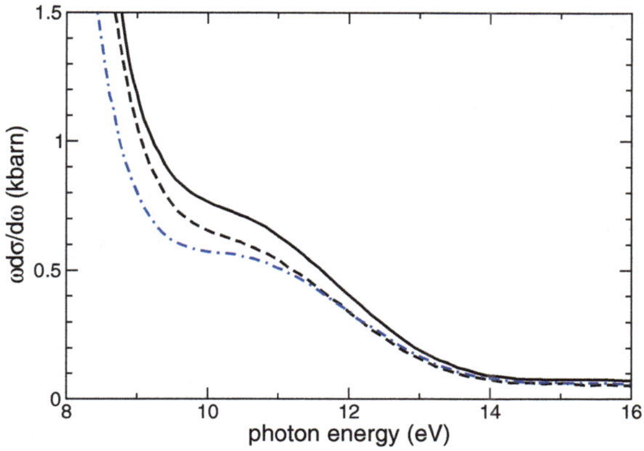

Fig. 7.12 BrS spectrum in the domain $\omega \gg \omega_p$ for $\varepsilon_1 = 30$ eV electron scattering on Na_{92} calculated in the RPAE with (*full curve*) and without (*broken curve*) the OBrS contribution to the amplitude. The *chained curve* stands for the asymptotic result (7.23) [148]

Figure 7.12 presents the BrS spectral dependence in the high-energy photon range, $\omega \gg \omega_p$. The data refer to the incident electron energy $\varepsilon_1 = 30$ eV. The RPAE results were obtained from (7.10) with (full curve) and without (broken curve) the OBrS contribution to the total amplitude. The two curves are close to each other thus indicating, that the OBrS provides relatively small contribution even in the region of large ω. This feature is due to the compensation of the electron and ion densities inside the cluster, $r \leq R$. The chained curve stands for the asymptotic behaviour of the spectrum which is defined by (7.23). It is seen, that as ω increases, the approximate result merges with the exact one.

As a function of the collision velocity v_1 the BrS spectra calculated for Na_{92} within the RPAE for fixed ω [148] exhibit the features similar to those discussed above in connection with the PRA results, see Fig. 7.7. The cross section has an absolute maximum when characteristic plasmon excitation time R/v_1 is close to ω_p^{-1}. At smaller velocities, the characteristic collision distance v_1/ω_p is smaller than the cluster radius and the cross section decreases. At higher velocities, the decrease is due to the factor v_1^{-2} in (7.10), which reflects the decrease in the interaction time with the velocity. For the non-resonance frequencies, the position of the maximum shifts towards the value ωR. The oscillations in the cross sections in the domain of small v_1 are also visible. The comparison of the RPAE and PRA results, carried out in the cited paper, reveals the best agreement at the plasmon resonance frequency $\omega = \omega_p$.

Figure 7.13 shows the angular anisotropy parameter versus incident electron velocity calculated for $e^- + Na_{92}$ collision at the plasmon resonance energy [148]. The comparison of the RPAE (full curve) and the PRA (broken curve) data shows that the latter approximation provides a rather accurate result for the angular distribution of photons. The figure demonstrates also that the diffraction effect appears in the v_1-dependence of β (note the oscillation in the domain of low v_1 values). The calculation performed showed that parameter β is negative for collision energies in the range $\varepsilon_1 < 300$ eV ($v_1 \lesssim 4.7$ a.u) resulting in a more intensive radiation in the

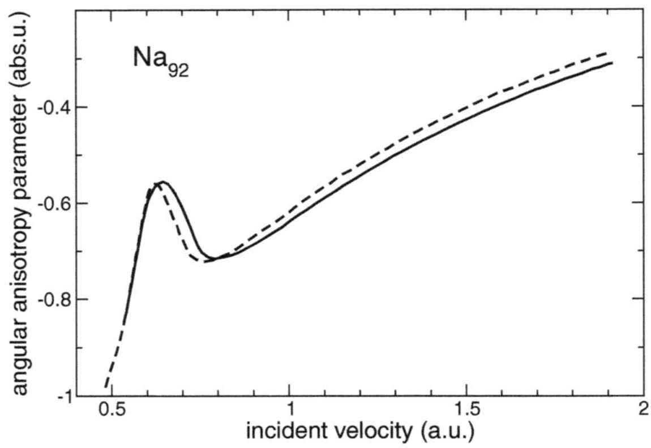

Fig. 7.13 Angular anisotropy parameter, $\beta(\omega, v_1, R)$ of PBrS for electron–Na_{92} collision calculated as a function of the incident velocity at $\omega = 3\,\text{eV}$. The *full curve* represents the calculations within the RPAE, the *dashed curve* stands for PRA. [148]

direction perpendicular to $\mathbf{v}_1$. This occurs because the transferred momenta $\mathbf{q}$ parallel to $\mathbf{v}_1$ provide the main contribution to the cross section. At the threshold, $\varepsilon_1 \approx \omega$, the parameter is close to -1 leading to the profile of the angular distribution proportional to $\sin^2 \theta_{\mathbf{k}}$. At large collision energies, $\varepsilon_1 > 300\,\text{eV}$, the anisotropy parameter becomes positive. In this case, the transferred momenta perpendicular to the direction of the incoming electron projectile become important, increasing the radiation in the forward and backward directions. As v_1 increases, the parameter β logarithmically approaches 1, leading to the angular distribution profile $\propto (1 + \cos^2 \theta_{bfk})$.

Let us now demonstrate that semi-empirical "G-function" approximation (see Sect. 4.3) is a powerful tool for simulations of the PBrS from atomic clusters. The approximation suggests to calculate the generalized polarizability as $\alpha(\omega, q) = \alpha(\omega)G(\omega, q)$, (4.8), where the dipole dynamic polarizability $\alpha(\omega)$ is very sensitive to the many-electron correlations, while the factor $G(\omega, q)$, being not that sensitive to the correlations, shows weak dependence on ω. As indicated in Sect. 4.3, the imaginary and real parts of $\alpha(\omega)$ one can restore from the data on photoabsorption cross section and the Kramers-Kronig dispersion relation. Photoabsorption cross section can be obtained either from experiment or from the direct *ab initio* calculations. The latter approach was adopted in [300] where the time-dependent density functional theory and molecular dynamics simulations were used to calculate the optical response and the PBrS from the small Mg clusters. The B3PW91/6-311+G(d) functional was employed following the methodology of the earlier works on the structure and optical response of Mg clusters [303, 394] and utilize the 6-311+G(d) basis set of primitive Gaussian functions to expand the cluster orbitals [136, 141]. Our calculations have been performed with the use of the Gaussian 98 software package [141].

The total BrS spectra calculated according to (7.12) and (7.16) for a $\varepsilon_1 = 27.2\,\text{eV}$ (1 a.u.) electron on small Mg clusters are presented in Fig. 7.14. The cluster geometries, corresponding to the lowest cluster energy, were calculated in [303] and are shown in the insets.

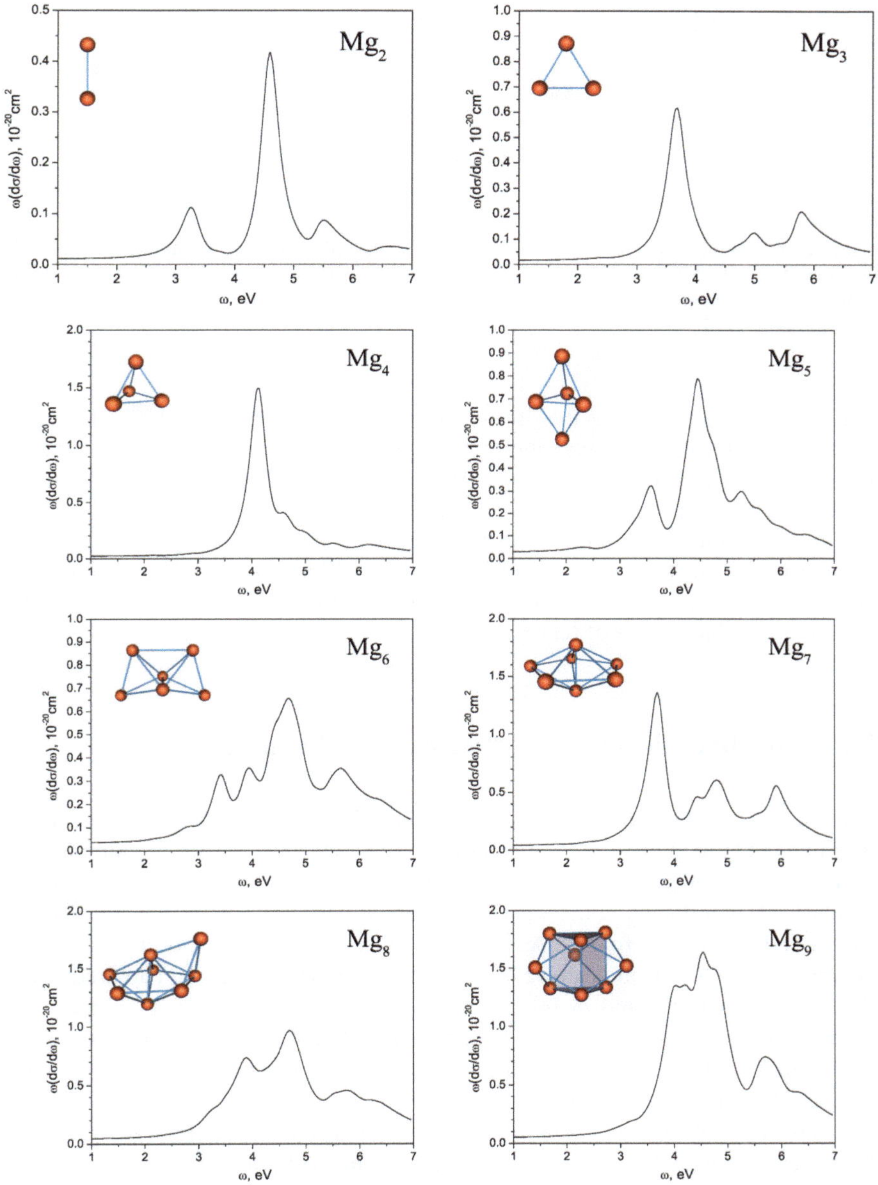

Fig. 7.14 BrS spectral dependence $\omega d\sigma/d\omega$ calculated for a 27.2 eV electron collision with small magnesium clusters. The cluster geometries, calculated in [303], are shown in the *insets* [300]

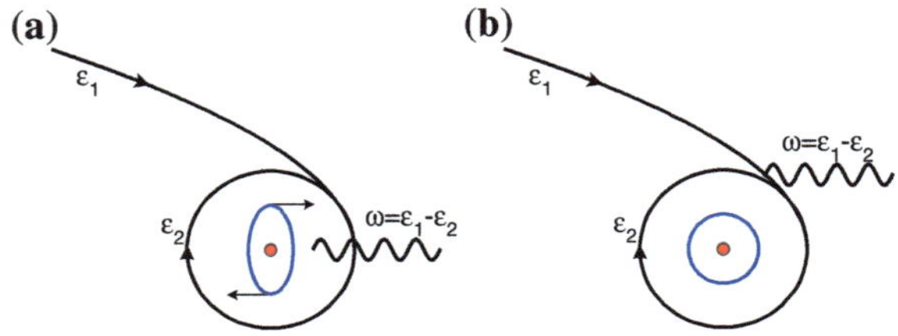

Fig. 7.15 **a** Schematic representation of the polarizational radiative electron capture process (PR). Note that the rotating dipole emits radiation in this case. **b** Schematic representation of the ordinary or direct electron capture process (RR). Note that the decelerated electron radiates in this case (BrS type emission) [115]

Figure 7.14 demonstrates the strong enhancement in the BrS spectra in the region 4–5 eV due to the resonant excitations of the plasmon oscillations of the cluster electrons. In the maxima the values of the PBrS cross section greatly exceed those of the OBrS. The PBrS spectra for Mg_2 and Mg_3 are characterized by strong resonances at $\omega \approx 4\,eV$, which can be qualitatively interpreted as the precursor of the plasmon resonance split due to the cluster deformation. For larger clusters, the plasmon resonance energy increases non-monotonously being in the range of N considered much lower than the bulk value of 6.26 eV.

7.4 Radiative Electron Capture by Metal Clusters

In this section we describe and analyze the effect of a target polarization in the process of radiative electron capture by a metallic cluster (electron-cluster recombination). In this process, which below is called Polarizational Recombination (PR), the cluster is polarized by a distant collision with a charged particle (an electron, for clarity), and the induced dipole rotates. As a result of this rotation, polarization radiation is emitted, and the electron is slowed down until eventually it can be captured [115]. This process is illustrated in Fig. 7.15a. We demonstrated that it dominates over the direct or ordinary process (we use the name Radiative Capture, RR) shown in Fig. 7.15b. In the latter case photons are emitted by the electron itself moving in the static field of the target cluster.

It is well established that atomic clusters form negative ions quite readily. Indeed, the electrical conductivity of a plasma may be altered by the presence of clusters, which may inhibit or retard its heating, since electrons, once attached, are no longer mobile. Electrons are bound to a neutral system (atom, cluster, fullerene) by the presence of a dipole field. In the case of neutral atoms, which possess no permanent dipole moment of their own, this dipole must be induced by the presence of the electron. For diatomic molecules, which possess a large intrinsic dipole, negative ions can possess several bound states, while for atoms, the corresponding negative ion usually possesses only one.

The problem of an electron bound by a dipole field was first addressed in the context of the hydrogen atom interacting with negative mesons [434]. It has been established theoretically [283, 321, 415, 431], that the minimum dipole moment to bind an electron is $0.639a_0$. Further, it has been shown [123] that a rigid dipole of moment greater than this in fact supports an infinite number of bound states. In a comprehensive theoretical study [107] the description of negative ions has been given in terms of the Dyson equation. Experimentally, the binding of an electron by the field of a molecular dipole was first observed for $LiCl^-$ [105], and there is now an extensive literature on the subject.

Most stable metallic clusters (the so-called, "magic" clusters) are spherically-symmetric, and, therefore, do not possess an intrinsic dipole moment (see, for example, [297, 298, 303, 311] and references therein). However, the dipole moment can be efficiently induced in these systems, as it is suggested by large cross sections of the photoionization of clusters in comparison to the atomic targets [94].

7.4.1 Cross Section of Radiative Capture

The amplitudes of RR and PR can be illustrated by the Feynman diagrams in Fig. 2.2. The only difference with the BrS process is, that the final electron state is now referred to the discrete spectrum.

The wavefunctions of the electrons involved in the collective motion are obtained using the spherical jellium model [128, 210, 211, 295, 297, 298]. For the projectile electrons wavefunction we use the partial-wave expansion (2.11) with the partial waves calculated in the frozen field of the cluster [211]. The wavefunction of the final state of the electron is considered in the form $\Psi_2(\mathbf{r}) = R_{n_2l_2}(r)Y_{l_2m_2}(\mathbf{n_r})$, where the radial wavefunction $R_{n_2l_2}(r)$ is a solution of the jellium model for the cluster with $N+1$ valence electrons. This solution belongs to the discrete spectrum and is characterized by the principal and orbital quantum numbers, (n_2, l_2) (the notation m_2 used above stands for the magnetic quantum number).

Similar to the derivation of the OBrS and PBrS amplitudes (see Sect. 2.2.1), the RR and PR amplitudes read[3]:

$$f_{\mathrm{ord}} = \langle 2|\mathbf{e}\cdot\mathbf{r}|\mathbf{p}_1^+\rangle \tag{7.27}$$

$$\begin{aligned} f_{\mathrm{pol}} &= -\sum_n \left[\frac{\langle 0\,|\mathbf{e}\cdot\mathbf{D}|\,n\rangle \left\langle 2n\left|\hat{V}\right|\mathbf{p}_1^{(+)}0\right\rangle}{\omega_{n0}-\omega} + \frac{\left\langle 20\left|\hat{V}\right|\mathbf{p}_1^{(+)}n\right\rangle \langle n\,|\mathbf{e}\cdot\mathbf{D}|\,0\rangle}{\omega_{n0}+\omega} \right] \\ &= -\frac{\mathrm{i}}{2\pi^2}\int \mathrm{d}\mathbf{Q}\,\frac{\mathbf{e}\cdot\mathbf{Q}}{Q^2}\,\langle 2|\exp(\mathrm{i}\mathbf{Q}\cdot\mathbf{r})|\mathbf{p}_1^+\rangle\,\alpha(\omega, Q). \end{aligned} \tag{7.28}$$

[3] To stress the similarity between the OBrS and PBrS, on one side, and RR and PR on the other side, we still use the subscripts "ord" and "pol" when referring to the RR and PR amplitudes, respectively.

Here $\hat{V} = \sum_a 1/|\mathbf{r} - \mathbf{r}_a|$ is the operator of the Coulomb interaction between of the projectile and the ath cluster electron, the sum is carried out over all delocalized electrons in the target. Other notations are as in (2.9), (2.10) and (2.16).

The *selective* differential cross section for radiative electron capture to the final $(n_2 l_2)$ state is connected with the total amplitude $f = f_{\text{ord}} + f_{\text{pol}}$ of the process by

$$\frac{d\sigma_{n_2 l_2}}{d\Omega_{\mathbf{k}}} = \frac{\omega^3}{2\pi v_1 c^3} \sum_{\lambda} \sum_{m_2=-l_2}^{l_2} \left| f_{\text{ord}} + f_{\text{pol}} \right|^2 . \tag{7.29}$$

Here the sums are carried out over the photon polarizations, λ, and over the magnetic quantum number m_2 of the captured electron.

Let us compare the contributions of the two mechanisms to the cross section. For doing this one can estimate f_{pol} assuming that the radius R_2 of the captured state is much larger than that of a cluster.[4] To this end, taking into account that due to the dipole selection rule only dipole transitions $|0\rangle|n\rangle$ contribute to the sum $\sum_n$ in (7.28), one estimates the matrix element of the Coulomb interaction $\sum_a v_a$ as follows:

$$\left\langle 2n \left| \hat{V} \right| \mathbf{p}_1^{(+)} 0 \right\rangle \approx \left\langle 2n \left| \sum_a \frac{\mathbf{r} \cdot \mathbf{r}_a}{r^3} \right| \mathbf{p}_1^{(+)} 0 \right\rangle \approx \frac{\left\langle 2 \left| \mathbf{r} \right| \mathbf{p}_1^{(+)} \right\rangle \cdot \langle n \left| \mathbf{D} \right| 0 \rangle}{R_2^3} \tag{7.30}$$

Substituting the obtained relation into the first line of (7.28), and taking into account spherical symmetry of the target, one expresses f_{pol} via f_{ord}:

$$\begin{aligned} f_{\text{pol}} &\approx -\frac{\left\langle 2 \left| \mathbf{r} \right| \mathbf{p}_1^{(+)} \right\rangle}{R_f^3} \cdot \sum_n \left[\frac{\langle 0 \left| \mathbf{e} \cdot \mathbf{D} \right| n \rangle \langle n \left| \mathbf{D} \right| 0 \rangle}{\omega_{n0} - \omega} + \frac{\langle 0 \left| \mathbf{D} \right| n \rangle \langle n \left| \mathbf{e} \cdot \mathbf{D} \right| 0 \rangle}{\omega_{n0} + \omega} \right] \\ &= -f_{\text{ord}} \frac{\alpha(\omega)}{R_2^3} \end{aligned} \tag{7.31}$$

where $\alpha(\omega)$ is the dynamic polarizability of the cluster.

To estimate the ratio of f_{pol} to f_{ord} in the vicinity of the plasmon resonance one can approximate $\alpha(\omega)$ with its imaginary part which is proportional to the photoabsorption cross section $\sigma_\gamma(\omega)$, see (4.10). Hence

$$\delta = \left| \frac{f_{\text{pol}}}{f_{\text{ord}}} \right| \sim \frac{|\alpha(\omega)|}{R_2^3} \sim \frac{\sigma_\gamma(\omega) c}{4\pi R_2^3} . \tag{7.32}$$

Substituting typical resonance values for the photoabsorption cross sections of sodium and silver cluster ions (see e.g. [94]), one concludes that the ratio δ can

[4] This assumption is fully applicable to the case of a neutral cluster which can bind an extra electron due to the long-range polarization potential.

be as high as $10^2 \dots 10^3$. Therefore, polarizational mechanism dominates the cross section by several orders of magnitude in the region of the giant resonance.

To illustrate this general statement, in Sect. 7.4.2 we present the results of calculation of the partial cross section for electron capture to the n_2s-states ($l_2 = 0$) for sodium and silver cluster ions [115].

General formulae (7.27)–(7.29) are simplified for the capture to an n_20 state. In this case, only the p-wave ($l_1 = 1$) of the incident electron in the partial-wave series (2.11) contributes to the RR and PR amplitudes. Then, the explicit expression for the partial differential cross section reads

$$\frac{d\sigma_{n_20}}{d\Omega_{\mathbf{k}}} = \frac{\omega^3}{2v_1^3c^3}\sum_{\lambda}(\mathbf{e}\cdot\mathbf{n}_{\mathbf{p}_1})^2 \times \left| \langle n_20\|r\|p_11\rangle + \frac{2}{\pi}\int_0^\infty dq\, q\, \alpha(\omega,q)\, \langle n_20\| j_1(qr)\|p_11\rangle \right|^2 . \tag{7.33}$$

Here the notation $\langle .\|.\|.\rangle$ stands for the radial matrix element.

Summing over the photon polarizations, see (2.26), and integrating over the emission angles one derives the partial cross section of the process:

$$\sigma_{n_20} = \frac{4\omega^3}{2v_1^3c^3}\left| \langle n_20\|r\|p_11\rangle + \frac{2}{\pi}\int_0^\infty dq\, q\, \alpha(\omega,q)\, \langle n_20\| j_1(qr)\|p_11\rangle \right|^2 . \tag{7.34}$$

The first term in $|\dots|^2$ is the contribution of the RR mechanism, while the second one is due to polarization of the cluster.

The resulting cross section is expressed via the generalized polarizability of the cluster. To calculate this quantity very accurately one must account for many body correlation effects. This can be done *ab initio* within the framework of the random phase approximation (see, for example, [92, 180, 211, 307] and Sect. 7.4.3 below).

Alternatively, one can apply the "G-function" method, described in Sect. 4.3, and calculate the generalized polarizability as it is suggested by (4.8): $\alpha(\omega,q) = \alpha(\omega)G(\omega,q)$. Here, the essential part of many-electron correlations is concentrated in the factor $\alpha(\omega)$. The function $G(\omega,q)$ describes the dependence of the polarizability on the transferred momentum. This function is not sensitive to many-electron correlations and shows a weak dependence on the photon energy. In the photon energy range $\omega \gg \omega_{\rm p}$, one can use the approximation $G(\omega,q) \approx F(q)/N$, where $F(q)$ and N are the form-factor of the cluster and the number of delocalized electrons, respectively. It was demonstrated in 4.3 (see also [243–245]) that the above-written approximation for $G(\omega,q)$ turns out to be a reasonable approximation over the whole ω-range rather than merely in the high-frequency regime. Therefore, to calculate $\alpha(\omega,q)$ one can use the following simple relation

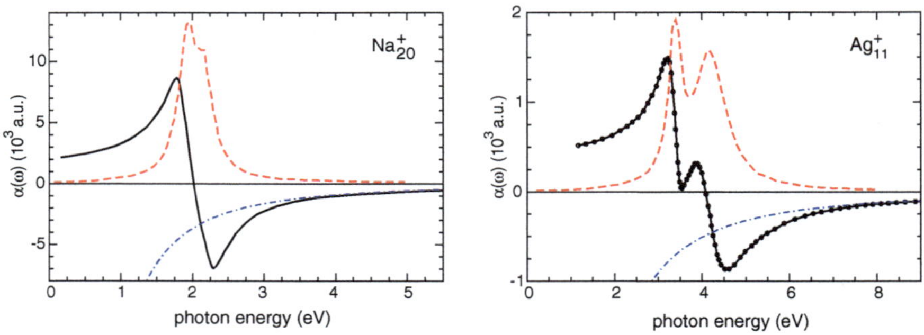

Fig. 7.16 Dynamic polarizability of the Na_{20}^+ (*left*) and Ag_{11}^+ (*right*) cluster ions. The real part is the *full curve*, the imaginary part is the *broken curve*. The *chained curve* represent the asymptotic formula $\alpha(\omega) \approx -N/\omega^2$ [115]

$$\alpha(\omega, q) \approx \alpha(\omega) F(q)/N. \tag{7.35}$$

The imaginary part of $\alpha(\omega)$ can be derived from experimental data on photoabsorption using relation (4.10). Then, the real part is calculated by applying the dispersion relation (4.11).

The approach based on (7.35) accompanied by (4.10–4.11) was used in [115] to calculate the partial cross section for electron capture for the cluster Na_{20}^+ and Ag_{11}^+.

7.4.2 Numerical Results for Na_{20}^+ and Ag_{11}^+

The results of calculations [115] are plotted in Figs. 7.16–7.19.

Figure 7.16 shows the real (full curve) and the imaginary part (broken curve) of $\alpha(\omega)$ for the Na_{20}^+ (left panel) and Ag_{11}^+ (right panel) clusters. The experimental data on the photoabsorption cross section [94] were used to obtain the values of $\mathrm{Im}\,\alpha(\omega)$. For comparison, the asymptotic dependence of the polarizability in the region of large frequencies $\alpha(\omega)\big|_{\omega\gg\omega_p} \approx \mathrm{Re}\,\alpha(\omega)\big|_{\omega\gg\omega_p} \approx -N/\omega^2$ (chained curve).

In the approximation chosen, the q-dependence of the generalized polarizability is described by the form-factor $F(q)$ of the cluster. The calculated form-factors for the Na_{20}^+ and Ag_{11}^+ clusters are plotted in Fig. 7.17. The oscillations in the q-dependence of the form-factor as well as polarizability $\alpha(\omega, q)$ of the clusters are due to the diffraction of the projectile electron at the edge of the cluster, which exists in all models of metallic clusters (see, e.g., [92]). The diffraction phenomenon also appears in variations of the cross section for fast electrons scattering as a function of scattering angle [147, 391, 421] Note that the diffraction behaviour is determined by the radius of the cluster and by the model chosen for its description. In principle, this phenomenon can be used as a test for the experimental verification of the validity of different cluster models.

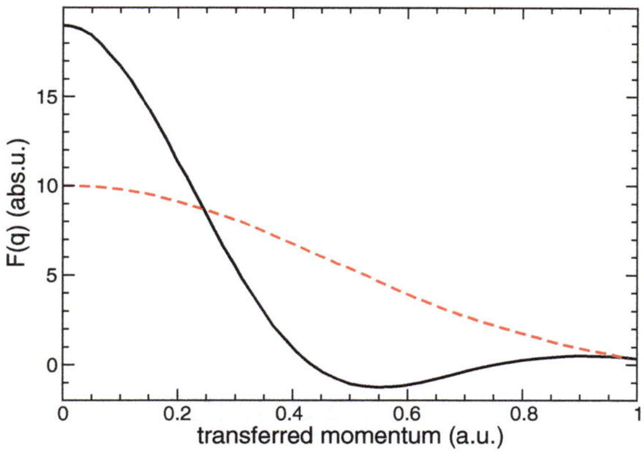

Fig. 7.17 Form-factors of the Na_{20}^+ (*full curve*) and Ag_{11}^+ (*broken curve*) cluster ions. [115]

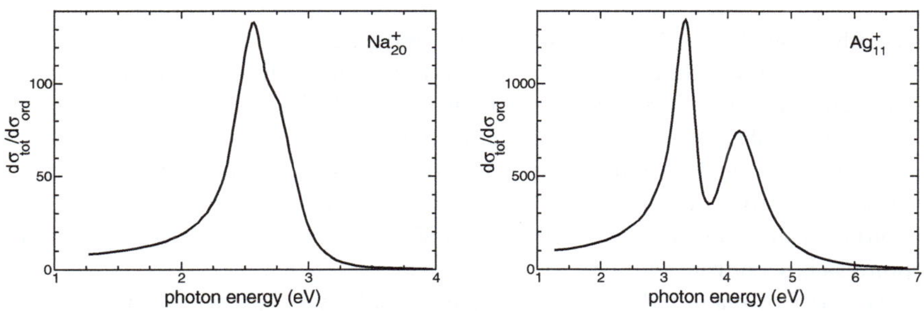

Fig. 7.18 The ratio (7.36) versus photon energy $\omega = \varepsilon_1 - \varepsilon_{n_f s}$ calculated in the vicinity of the plasmon resonance for the electron capture to an $(n_f s)$-state for the Na_{20}^+ (*left graph*) and Ag_{11}^+ (*right graph*) cluster ions [115]

Using the polarizability, one can calculate the ratio ξ of the total cross section (7.34) to the cross section of RR, represented by the term proportional to $|\langle n_2 0 \| r \| p_1 1 \rangle|^2$ in (7.34). This ratio is not very sensitive to the choice of the final state of the electron. Indeed, the major contribution to the integral in (7.34) arises from the domain $qr < 1$ there the spherical Bessel function can be approximated as $j_1(qr) \approx (qr)/3$. In this case the PR amplitude becomes proportional to the matrix element $\langle n_2 0 \| r \| p_1 1 \rangle$, so that it cancels out from the ratio, which becomes equal to

$$\xi \equiv \frac{\sigma_{\text{tot}}(\omega)}{\sigma_{\text{ord}}(\omega)} = \left| 1 + \frac{2}{3\pi} \int_0^\infty dq \; q^2 \alpha(\omega, q) \right|^2 . \tag{7.36}$$

Figure 7.18 represent the ratio (7.36) for the clusters Na_{20}^+ and Ag_{11}^+. These figures demonstrate for a specific example of the electron capture to an $n_f s$-state, that the polarizational mechanism is strongly dominant in the region of the giant resonance.

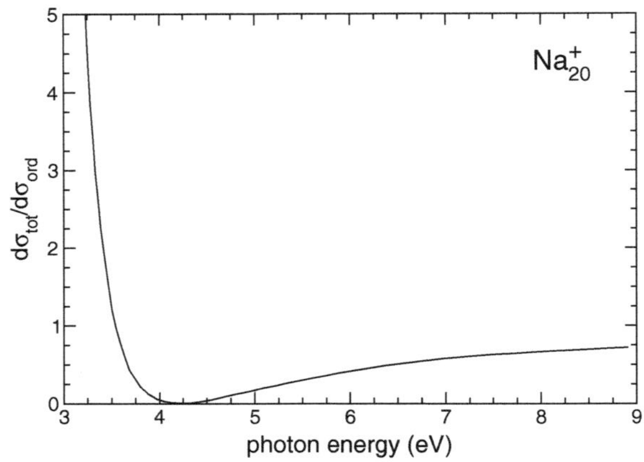

Fig. 7.19 Same as in Fig. 7.18 but for higher photon energies. Note the minimum which is due to the interference between the two mechanisms [115]

As illustrated by Fig. 7.19, towards higher frequencies, the ratio may have a minimum, arising from an interference between ordinary and polarizational mechanisms. This interference introduces an asymmetry in the shape of the giant resonance as compared with the shape of the resonance in the photoabsorption cross section. Such an asymmetry is typical for Fano profiles in atomic photoabsorption. With increasing energy of the projectile electron, $\varepsilon_1 = \omega + \varepsilon_f$, the ratio $\xi \to 1$ thus indicating, that the ordinary mechanism becomes the predominant mechanism of radiative electron capture as is seen from Fig. 7.19.

The order of magnitude of the ratio ξ is virtually independent of the size R of the cluster. The dependence of $\alpha(\omega, q)$ on the cluster size R can be estimated as R^3. The integration region over q in (7.36) is determined by $q \lesssim R^{-1}$. Therefore, the integral as well as the ratio ξ is virtually independent of the size of the cluster.

7.4.3 Non-Radiative Capture by Means of Many-Body Theory

The calculations of cross sections of the low-energy electron attachment to neutral and charged metallic clusters were carried out in [113, 200].

In the cited papers further development of the initial theoretical work on attachment [115] was made. Novel features which were introduced to the theoretical framework include the following. First, the dynamic response of the target was treated within the RPAE scheme. Thus, no experimental data on photoabsorption or any other empirical parameters were used. Second, the projectile wave functions were treated beyond the HF scheme by accounting for the polarization interaction between the electron and the cluster. This allowed one to carry out accurate quantitative analysis of the radiative capture in the domain of low incident energies energy (below the plasmon resonance). No assumption was made for a target cluster to stay in its ground stay after the collision. Instead, all bound states of the resulting negative ion were computed and accounted for as possible partial channels of the process. To this end, the cross section was calculated as the sum of all partial terms which correspond to the capture into a particular final state. As a result, the structure of the attachment spectrum becomes more complicated than within the single partial-channel scheme, Sects. 7.4.2 and 7.4.3, and can exhibit more than one resonance.

The cross section summed over all partial channels can be written as follows:

$$\sigma(p_1) = \frac{2\pi}{v_1} \sum_{l_1} \overline{\sum_{m_1}} \sum_{\nu_2 m_2} \sum_{n} \left| f_{n0,21} \right|^2 \delta\left(\varepsilon_{\nu_2} - \varepsilon_1 + \omega_{n0}\right) . \qquad (7.37)$$

Here, index n denotes the quantum numbers of the target cluster final state which can be different from the initial ground state labeled by 0; $\omega_{n0} = \varepsilon_n - \varepsilon_0$ is the excitation energy.[5] The notation ν_2 stands for a set $(n_2 l_2)$ of the quantum numbers of the captured electron, $\varepsilon_{\nu_2} < 0$ is the energy of the $(n_2 l_2)$-state. The short-hand notation ν_1 refers to the incident electron and stands for a set (p_1, l_1); the initial energy $\varepsilon_1 = p_1^2/2$. The delta-function expresses the energy conservation law and thus imposes restrictions on the sum over the final n- and ν_2-states. The bar over the summation sign indicates the averaging over magnetic quantum number m_1: $\overline{\sum_{m_1}} = (2l_1 + 1)^{-1} \sum_{m_1}$.

The general expression for the partial amplitude of the process, in which the electron is captured to the $(\nu_2 m_2)$ state and, simultaneously, the target undergoes the transition $0 \to n$, is given by:

$$f_{n0,\,21} = \int \Psi_n^*(\{\mathbf{r}_a\})\, \Psi_{\nu_2 m_2}^*(\mathbf{r})\, \hat{V}(\mathbf{r}; \{\mathbf{r}_a\})\, \Psi_0(\{\mathbf{r}_a\})\, \Psi_{\nu_1 m_1}(\mathbf{r})\, \mathrm{d}\mathbf{r}\, \mathrm{d}\{\mathbf{r}_a\} . \qquad (7.38)$$

Here $\hat{V}(\mathbf{r}; \{\mathbf{r}_a\}) = \sum_a 1/|\mathbf{r} - \mathbf{r}_a|$ is the operator of the Coulomb interaction between the projectile and target electrons, the notation $\{\mathbf{r}_a\}$ stands for the set of position vectors $\mathbf{r}_1, \ldots, \mathbf{r}_N$ of the target electrons. The initial $(j = 1)$ and final $(j = 2)$ state wavefunctions of the electron are sought in the form $\Psi_{\nu_j m_j}(\mathbf{r}) = R_{\nu_j}(r)\, Y_{l_j m_j}(\mathbf{n}_\mathbf{r})$.

The calculation of the matrix elements (7.38) was performed in [113, 200] within the PRAE framework using the frozen core HF wavefunctions obtained within the jellium model [129, 180, 210].

The negative ion wavefunctions $\Psi_{\nu_2 m_2}(\mathbf{r})$ and the energies ε_2 were calculated as solutions of Dyson's equation with the non-local polarization potential [19, 107, 175]

$$\widehat{H}^{(0)} \Psi_{\nu_2 m_2}(\mathbf{r}) + \int \Sigma_{\varepsilon_{\nu_2}}(\mathbf{r}, \mathbf{r}')\, \Psi_{\nu_2 m_2}(\mathbf{r}')\, \mathrm{d}\mathbf{r}' = \varepsilon_{\nu_2} \Psi_{\nu_2 m_2}(\mathbf{r}) . \qquad (7.39)$$

Here, $\widehat{H}$ is the static single-particle Hamiltonian of the cluster, $\Sigma_{\varepsilon_{\nu_2}}(\mathbf{r}, \mathbf{r}')$ is the energy-dependent non-local potential, which is equal to the self-energy part of

---

[5] In line with the terminology adopted for the BrS process, see Sect. 5.3 the transitions to $n \neq 0$ can be called 'inelastic' capture to distinguish from the 'elastic' channel when $n = 0$.

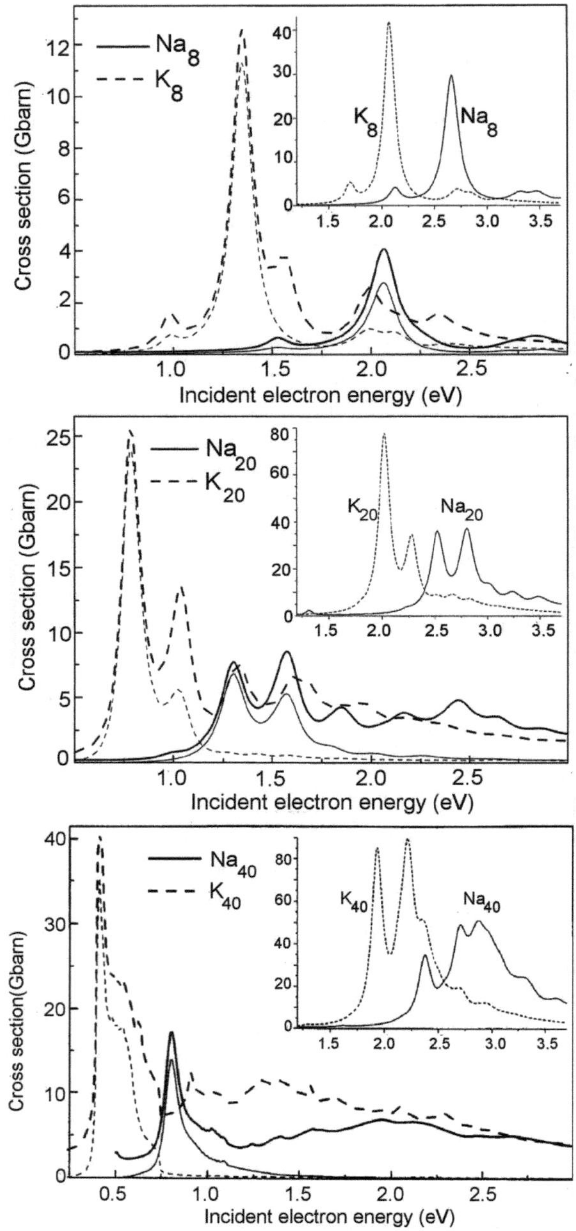

Fig. 7.20 The dependence of the total electron capture cross section on the incident electron energy for the Na_N (*full thick curves*) and K_N (*broken thick curves*) clusters, with $N = 8, 20, 40$ as indicated in the *graphs*. In each *graph*, the dominant partial contribution to the cross section is shown by *thin curves* (*broken* for the potassium clusters and *full* for the sodium ones). The *insets* present the photoabsorption cross section for the clusters. The *graphs* represent Figs. 1–3 from [200] with editorial changes

the single-electron Green function of the system 'cluster+electron'. This function includes the correlation interaction of the extra electron with the cluster as well as accounts for the exchange effect.

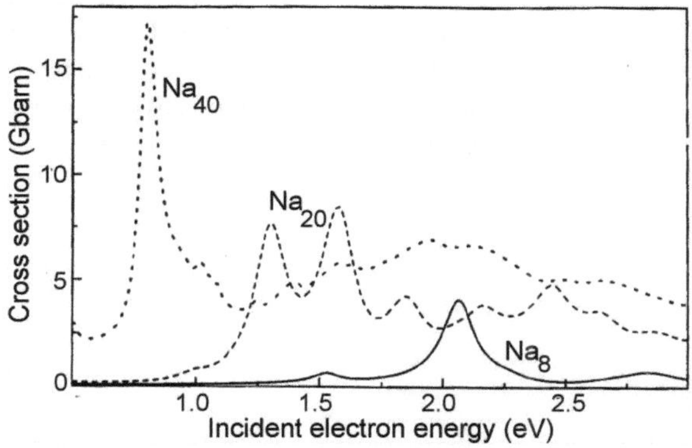

Fig. 7.21 Electron attachment cross sections for Na_8 (*full curve*), Na_{20} (*broken curve*) and Na_{40} (*dotted curve*) versus the incident electron energy. The *graph* represents Fig. 4 from [200] with editorial changes

To account for the finite lifetime of the excited states n, the widths Γ_{n0} corresponding to the cluster excitation $0 \rightarrow n$ were introduce into (7.37) according to the rule (see, for example, [81]):

$$2\pi\delta(\varepsilon_2 - \varepsilon_1 + \omega_{n0}) \rightarrow \frac{\Gamma_{n0}}{(\varepsilon_2 - \varepsilon_1 + \omega_{n0})^2 + \Gamma_{n0}^2/4}. \tag{7.40}$$

The numerical calculation of the cross section (7.37) supplemented with the substitution (7.40) was performed using the widths $\Gamma_{n0} = 0.06\omega_{n0}$ [440].

The dependence of the total electron capture cross section on the energy of the incident electron is presented in Fig. 7.20. Three graphs in the figure correspond to the collision with the sodium and potassium metal clusters containing $N = 8, 20$ and 40 atoms, respectively [200]. In these clusters all shells of delocalized electrons are filled, therefore, their dynamic dipole response is characterized by a single-plasmon resonance mode [82, 94]. The dipole and the quadrupole transitions were included when calculating the cross sections. The dipole transitions appeared to be the dominant parts (shown with thin full and broken curves), and thus determine main features of the cross section.

The calculations show that the attachment spectra are often dominated by a main peak, which corresponds to the capture in the deepest bound state of the negative ion, accompanied by secondary peaks. For the clusters with larger number of atoms, the latter tend to appear above the energy of the main peak and their number increases rapidly with N. The main peak is due to the excitation of the dipole plasmon during the electron attachment process. Its position is shifted towards lower energies compared to the dipole plasmon resonance in the photoabsorption cross section $\sigma_\gamma(\omega)$ (calculated within the RPAE) by the value of binding energy of the captured electron. The cross sections $\sigma_\gamma(\omega)$ are presented in the insets in the graphs. The difference of the plasmon resonance shapes for the two processes is due to the difference between the electric field exciting the plasmon in the photoabsorption process from that in the case of electron attachment.

The electron attachment cross sections for Na_8, Na_{20} and Na_{40} clusters are compared in Fig. 7.21. All the above-mentioned features are clearly seen. Namely, as N increases so do the peak values of the cross sections, the number of the secondary peaks, and they are shifted towards higher energies.

Chapter 8
Conclusion

In this book we presented the achievements made in the development of the theory of polarizational BrS formed in binary collisions involving structureless "elementary" charged particles, atomic particles, metallic clusters and fullerenes. Apart from what has been described above, we would like to mention that a number of algorithms and computer codes have been developed, which allow one to perform accurate quantitative analysis of the spectral and spectral-angular distributions of the total BrS formed in an arbitrary collision process involving a charged structureless particle and a multi-electron target.

The polarizational BrS problem is rather broad, because this kind of radiation can be emitted in any collision involving structured particles: nuclei, atoms, molecules or clusters. The number of various colliding pairs, different interaction forces between particles, kinematic conditions and the frequency ranges make this problem quite varied and interesting. With minor modifications introduced, the formalism discussed in the book can be made applicable to any binary collision between the particles interacting via other, not necessarily electromagnetic, fields.

The manifestation of PBrS is most pronounced in the systems the dynamic response of which to alternating external field is governed by the excitation of collective degrees of freedom. From this viewpoint, the collisions involving fullerenes, metal clusters or nanoparticles could be of the prime interest for experimental study of the phenomena since the PBrS mechanism dominates the spectrum in a broad range of photon energies.

Theory of PBrS, presented in the book, is relevant to a number of fields of modern fundamental and applied research. For example, the experiments on measuring the BrS radiation formed in collisions of ultra-relativistic heavy ions due to their mutual polarization are under discussion at the Linear Hadron Collider (LHC) facility [417]. Accurate theoretical results for the cross sections of various radiative processes, including bremsstrahlung, are needed in the variety of applications related to the radiation protection of electronic devices, living tissues, etc.

Theoretical and numerical study of the BrS spectrum radiation arisen in electron and heavy particle collisions with nano-objects can be quite relevant field of study

A. V. Korol and A. V. Solov'yov, *Polarization Bremsstrahlung*,
Springer Series on Atomic, Optical, and Plasma Physics 80,
DOI: 10.1007/978-3-642-45224-6_8,

in medical applications. One of the important medical problems, which is still to be solved, is the cancer treatment. Modern methods of treatment include, in particular, ion beam cancer therapy. This technique is based on the fact that high-energy radiation, interacting with living tissue, produces structural and chemical modifications which affect biological functions and cause death of tumor cells.

It is widely accepted that the major part of damage done by external radiation or incident heavy ions is related, directly or indirectly, to the secondary electrons produced by ionization of the medium [399]. These electrons may interact with parts of DNA molecules in the cell nuclei, generate other secondaries, such as other electrons and radicals, which can interact with DNA. Increase of a number of DNA double strand breaks and, correspondingly, the probability of the tumor cell death can be achieved by injecting many-electron nanoparticles which play a role of sensitizers.

At present, considerable interest is in using metallic nanoparticles as photosensitizers in the cancer treatment with ionizing radiation. Since the first experimental evidence [185], this activity has been expanding and a number of new experiments on the sensitization with various metallic nanoparticles have been performed recently [111, 228, 313, 344, 375]. The scientific interest in obtaining a deeper understanding of radiation damage is motivated by the development of ion-beam cancer therapy and other applications of ions interacting with biological targets [74, 264, 373]. A number of important scientific questions, especially related to DNA damage assessment on the molecular level, have not yet been resolved. Therefore, recently this field has attracted much attention from the scientific community. There are series of conferences devoted to these subjects [74, 199].

To describe properly the influence of sensitization on the mechanism of DNA double strand breaks, it is necessary to know the electronic spectrum of secondary electrons propagating through the medium as well as the photon, BrS-like, spectrum. The latter can be adequately calculated only with the account of the polarizational channel of the spectrum formed in the collision of projectile ions and secondary electrons with the nanoparticles.

Chapter 9
References

1. Abramowitz, M., Stegun, I.E.: Handbook of Mathematical Functions. Dover, New York (1964)
2. Akhiezer, A.I., Berestetsky, V.B.: Quantum Electrodymanics. Interscience Publishers, New York (1965)
3. Altman, J.C., Quarles, C.A.: Atomic-field double Bremsstrahlung. Phys. Rev. A **31**, 2744–2750 (1985)
4. Ambrose, V., Quarles, C.A., Ambrose, R.: Thin-target Bremsstrahlung at 0° from 50 keV electrons. Nucl. Instrum. Methods Phys. Res. B **124**, 457–463 (1997)
5. Ambrose, R., Kahler, D.L., Lehtihet, H.E., Quarles, C.A.: Angular dependence of thick-target Bremsstrahlung. Nucl. Instrum. Methods Phys. Res. B **56/57**, 327–329 (1991)
6. Amusia, M.Ya.: "Atomic" Bremsstrahlung spectrum. Commun. At. Mol. Phys. **11**, 123–137 (1982)
7. Amusia, M.Ya.: Atomic Bremsstrahlung. Phys. Rep. **162**, 249–335 (1988)
8. Amusia, M.Ya.: Atomic photoeffect. Plenum, New York (1990)
9. Amusia, M.Ya.: Bremsstrahlung. Energoatomizdat. Moscow (1990) (in Russian)
10. Amusia, M.Ya.: Polarization radiation or "Atomic" Bremsstrahlung. Nucl. Instrum. Methods Phys. Res. A **314**, 225–234 (1992)
11. Amusia, M.Ya.: "Atomic Bremsstrahlung": retrospectives, current status and perspectives. J. Radiat. Phys. Chem. **75**, 1232–1250 (2006)
12. Amusia, M.Ya., Avdonina, N.B., Chernysheva, L.V., Kuchiev, M.Yu.: 'Stripping' of the atom in Bremsstrahlung. J. Phys. B: At. Mol. Opt. Phys. **18**, L791–L796 (1985)
13. Amusia, M.Ya., Avdonina, N.B., Kuchiev, M.Yu., Chernysheva, L.V.: Influence of the atomic structure on the angular distribution and polarization of Bremsstrahlung. Izv. Akad. Nauk SSSR: Ser. Fiz. **50**, 1261–1266 (1986) (English Translation: Acad. Sci. USSR, Bull. (Phys.) **50**, 13 (1986))

A. V. Korol and A. V. Solov'yov, *Polarization Bremsstrahlung*,
Springer Series on Atomic, Optical, and Plasma Physics 80,
DOI: 10.1007/978-3-642-45224-6_9, © Springer-Verlag Berlin Heidelberg 2014

14. Amusia, M.Ya., Baltenkov, A.S., Gilerson, V.B.: Fast electron-atom Bremsstrahlung. Pisma v ZhTF **3**, 455–461 (1977) (English Translation: Sov. Phys.–Tech. Phys. Lett. **3**, 1105–1109 (1977))
15. Amusia, M.Ya., Baltenkov, A.S., Korol, A.V., Solov'yov, A.V.: Bremsstrahlung in collisions of atoms with slow neutrons. Zh. Eksp. Teor. Fiz. **93**, 1537–1544 (1987) (English Translation: Sov. Phys. – JETP **66**, 877–881 (1987))
16. Amusia, M.Ya., Baltenkov, A.S., Paiziev, A.A.: Electron-atom Bremsstrahlung with allowance for polarizability. Pisma v ZhETF **24**, 366–369 (1976) (English Translation: Sov. Phys. – JETP Lett. **24**, 332–334 (1976))
17. Amusia, M.Ya., Baltenkov, A.S., Zhalov, M.B., Korol, A.V., Solov'yov, A.V.: Polarizational Bremsstrahlung in scattering of particles at atoms and nuclei. In: Proceedings of 21st Winter School of Leningrad Institute of Nuclear Physics (Leningrad, January 1986), pp. 135–194 (1986) (in Russian)
18. Amusia, M.Ya., Cherepkov, N.A.: Many-electron correlations in the scattering processes. Case Stud. At. Phys. **5**, 47–179 (1975)
19. Amusia, M.Ya., Cherepkov, N.A., Chernysheva, L.V., Davidović, D.M., Radojević, V.: Slow-electron elastic scattering on argon. Phys. Rev. A **25**, 219–225 (1982)
20. Amusia, M.Ya., Chernysheva, L.V., Korol, A.V.: Bremsstrahlung spectrum on Xe for 600 eV electrons. J. Phys. B: At. Mol. Opt. Phys. **23**, 2899–2907 (1990)
21. Amusia, M.Ya., Dolmatov, V.K., Ivanov, V.K.: Photoionization of atoms with half-filled shells. Z. Eksp. Teor. Fiz. **85**, 115–123 (1983) (English translation: Sov. Phys. – JETP **58**, 67–72 (1983))
22. Amusia, M.Ya., Gribakin, G.F., Kharchenko, V.A., Korol, A.V., Solov'yov, A.V.: Light particle scattering by mesic hydrogen. Muon Cat. Fusion **2**, 143–146 (1988)
23. Amusia, M.Ya., Gulkarov, I.S., Zhalov, M.B., Potapova, T.M.: Influence of the nuclear structure on the Bremsstrahlung spectra of fast nucleons. Yadernaya Fizika **40**, 1321–1324 (1984) (in Russian)
24. Amusia, M.Ya., Ivanov, V.K., Chernysheva, L.V.: Photoionization cross section of $4d^{10}$ Xe, Cs and Ba subshells with account of electron shell rearrangement. Phys. Lett. **59A**, 191–193 (1976)
25. Amusia, M.Ya., Ivanov, V.K., Cherepkov, N.A., Chernysheva, L.V.: Interference effects in photoionization of noble gas atoms outer s-subshells. Phys. Lett. **40A**, 361–362 (1972)
26. Amusia, M.Y., Ivanov, V.K., Kupchenko, V.A.: The photoionization of atomic Eu in the vicinity of its giant resonance. Z. Phys. D **14**, 219–221 (1989)
27. Amusia, M.Ya., Korol A.V.: Electron Bremsstrahlung spectrum in the tip region. J. Phys. B: At. Mol. Opt. Phys. **22**, L571–L573 (1989)
28. Amusia, M.Ya., Korol, A.V.: Dependence of the spectral shape of BrS in charged-particle–atom collisions upon the projectile velocity. J. Phys. B: At. Mol. Opt. Phys. **24**, 3251–3265 (1991)
29. Amusia, M.Ya., Korol A.V.: Angular distribution and polarization of Bremsstrahlung radiation formed by intermediate energy particle-atom scattering. J. Phys. B: At. Mol. Opt. Phys. **25**, 2383–2392 (1992)

30. Amusia, M.Ya., Korol, A.V.: Recent developments of polarizational Bremsstrahlung. Nucl. Instrum. Methods Phys. Res. B **79**, 146–149 (1993)
31. Amusia, M.Ya., Korol, A.V.: On the continuous spectrum electromagnetic radiation in the electron-fullerene collisions. Phys. Lett. **186A**, 230–234 (1994)
32. Amusia, M.Ya., Korol, A.V., Solov'yov, A.V.: Bremsstrahlung in electron-positronium scattering. Z. Phys. D **1**, 347–349 (1986)
33. Amusia, M.Ya., Kuchiev, M.Yu., Solov'yov, A.V.: Bremsstrahlung in atom–atom collisions. Pisma v Zh. Tekhn. Fiz. **10**, 1025–1029 (1984) (English Translation: Sov. Phys. – Tech. Phys. Lett **10**, 431–432 (1984))
34. Amusia, M.Ya., Kuchiev, M.Yu., Solov'yov, A.V.: Bremsstahlung in atom-atom collisions. Zh. Eksp. Teor. Fiz. **89**, 1512–1521 (1985) (English Translation: Sov. Phys. – JETP **62**, 876–881 (1985))
35. Amusia, M.Ya., Kuchiev, M.Yu., Korol, A.V., Solovyov, A.V.: Bremsstrahlung of relativistic particles and dynamic polarizability of target atom. Zh. Eksp. Teor. Fiz. **88**, 383–389 (1985) (English Translation: Sov. Phys. – JETP **61**, 224–227 (1985))
36. Amusia, M.Ya., Korol, A.V. Solov'yov, A.V.: Coherence of radiation of target electrons in atomic collisions. Pisma v Zh. Tekhn. Fiz. **12**, 705–710 (1986) (English Translation: Sov. Phys. – Tech. Phys. Lett. textbf12, 290–292 (1986))
37. Amusia, M.Ya., Kuchiev, M.Yu., Solov'yov, A.V.: Total emission spectrum in the collision of a heavy charged particle with an atom. Pisma v Zh. Tekhn. Fiz. **11**, 1401–1404 (1985) (English Translation: Sov. Phys. – Tech. Phys. Lett. **11**, 577–579 (1885))
38. Amusia, M.Ya., Kuchiev, M.Yu., Solov'yov, A.V.: Bremstrahlung in atom-atom collisions at relativistic velocities. Zh. Tekhn. Fiz. **57**, 820–822 (1987) (English Translation: Sov. Phys. – Tech. Phys. **32**, 499–500 (1987))
39. Amusia, M.Ya., Kuchiev, M.Yu., Solov'yov, A.V.: Bremsstrahlung in collisions of atomic particles at relativistic velocities. Zh. Eksp. Teor. Fiz. **94**, 74–86 (1988) (English Translation: Sov. Phys. – JETP **67**, 41–48 (1988))
40. Amusia, M.Ya., Pratt, R.H.: Emission and absorption of radiation by structured particles. Commun. At. Mol. Phys. **28**, 247–258 (1992)
41. Amusia, M.Ya., Solov'yov, A.V.: Inelastic scattering on muonic hydrogen. J. Phys. B: At. Mol. Opt. Phys. **18**, 3663–3666 (1985)
42. Amusia, M.Ya., Solov'yov, A.V.: Continuous spectrum of electromagnetic radiation in collisions of nuclear particles. In: Sushkov, O.P. (ed.) Modern Developments in Nuclear Physics, pp. 425–437. World Scientific, Singapore (1987)
43. Amusia, M.Ya., Solov'yov, A.V.: Spectral-angular distribution of radiation in collisions of relativistic atomic particles. Zh. Eksp. Teor. Fiz. **95**, 745–756 (1990) (English Translation: Sov. Phys. – JETP **70**, 416–425 (1990))
44. Amusia, M.Ya., Solov'yov, A.V.: Polarisational Bremsstrahlung in scattering of a multicharged ion. J. Phys. B: At. Mol. Opt. Phys. **23**, 2889–2898 (1990)
45. Amusia M.Ya., Zimkina T.M., Kuchiev M.Yu.: Broad bands in the atomic emission spectra under the action of fast electrons. Zh. Tekhn. Fiz. **52**, 1424–1426 (1982) (English Translation: Sov. Phys. – Tech. Phys. **27**, 866–869 (1982))

46. Amusia M.Ya., Zon, B.A., Kretinin, I.Yu.: Polarization Bremsstrahlungin α-decay. Zh. Eksp. Teor. Fiz. **132**, 387–391 (2007) (English Translation: JETP **105**, 343–346 (2007))
47. Andersen, J.U., Bonderup, E.: Classical dielectric models of fullerenes and estimation of heat radiation. Eur. Phys. J. D **11**, 413–434 (2000)
48. Andersen, J.U., Bonderup, E.: Local field corrections for light absorption by fullerenes. Eur. Phys. J. D **11**, 435–448 (2000)
49. Astapenko, V.A.: Polarization-interference effects in the Bremsstrahlung of quasiclassical electrons on ions with a core. Zh. Eksp. Teor. Fiz. **115**, 1619–1629 (1999) (English translation: JETP **88**, 889–895 (1999))
50. Astapenko, V.A.: Theory of the polarization Bremsstrahlung from thermal electrons scattered by the debye sphere of an ion in a plasma. (English translation: Plasma Phys. Rep. **27**, 474–479 (2001))
51. Astapenko, V.A.: Bremsstrahlung of relativistic electron scattering on an atom: comparison of various channels. Phys. Lett. **361A**, 242–247 (2007)
52. Astapenko, V.A.: Bremsstrahlung of fast charged particles on clusters in a wide spectral range. Zh. Eksp. Teor. Fiz. **128**, 88–94 (2005) (English translation: JETP **101**, 73–79 (2005))
53. Astapenko, V.A.: Polarizational Bremsstrahlung radiation on a nanosphere in a dielectric. Russian Phys. J. **54**, 649–657 (2011)
54. Astapenko, V.A., Baturin, A.S., Korostylev, E.V.: Bremsstrahlung of non-relativistic electrons in metallic target with account for polarizational channel. Zh. Eksp. Teor. Fiz. **137**, 121–128 (2010) (English translation: JETP **110**, 107–113 (2010))
55. Astapenko, V.A., Baturin, A.S., Korostylev, E.V.: Bremsstrahlung spectrum of non-relativistic electrons scattered in metallic target: theory and experiment. J. Surf. Invest. X-ray Synchrotron Neutron Tech. **4**, 193–198 (2010)
56. Astapenko, V.A., Buimistrov, V.M., Krotov, Yu.A.: Bremsstrahlung accompanied by excitation and ionization of the scattering atoms. Zh. Eksp. Teor. Fiz. **93**, 825–837 (1987) (English Translation: Sov. Phys. – JETP **66**, 464–471 (1987))
57. Astapenko, V.A., Buimistrov, V.M., Krotov, Yu.A., Mikhailov, L.K., Trakhtenberg, L.I.: Dynamic Bremsstrahlung of a relativistic charged particle scattered by an atom. Zh. Eksp. Teor. Fiz. **88**, 1560–1569 (1985) (English Translation: Sov. Phys. – JETP **61**, 930–936 (1985))
58. Astapenko, V.A., Buimistrov, V.M., Krotov, Yu.A., Nasonov, N.N.: Polarization Bremsstrahlung from non-relativistic electrons penetrating a polycrystalline target. Phys. Lett. **332A**, 298–302 (2004)
59. Astapenko, V.A., Bureeva, L.A., Lisitsa, V.S.: Incoherent polarization Bremsstrahlung of a fast charged particle scattered by an atom. Laser Phys. **10**, 960–966 (2000)
60. Astapenko, V.A., Bureeva, L.A., Lisitsa, V.S.: Classical and quantum theories of PBrS in the local electron density model. Zh. Eksp. Teor. Fiz. **117** 496–510 (2000) (English translation: JETP **90**, 434–446 (2000)

61. Astapenko, V.A., Bureeva, L.A., Lisitsa, V.S.: Polarization Bremsstrahlung of a fast charged particle on a Thomas-Fermi atom. Zh. Eksp. Teor. Fiz. **117**, 906–912 (2000) (English translation: JETP **90**, 788–794 (2000))
62. Astapenko, V.A., Bureeva, L.A., Lisitsa, V.S.: Polarization radiation phenomena in plasmas with heavy ions. Phys. Scripta **T86**, 62–67 (2000)
63. Astapenko, V.A., Bureeva, L.A., Lisitsa, V.S.: Radiation loss of electrons under scattering by a Thomas Fermi atom. Zh. Eksp. Teor. Fiz. **121**, 19–30 (2002) (English translation: JETP **84**, 12–21 (2002))
64. Astapenko, V.A., Bureeva, L.A., Lisitsa, V.S.: Polarization mechanism for Bremsstrahlung and radiative recombination in a plasma with heavy ions. Fizika Plazmy, **28**, 337–345 (2002) (English translation: Plasma Physics Reports, **28**, 303–311 (2002))
65. Astapenko, V.A., Bureeva, L.A., Lisitsa, V.S.: Polarization effects in atomic transitions. Uspekhi Fiz. Nauk **172**, 155–192 (2002) (English translation: Physics – Uspekhi **45**, 149–184 (2002))
66. Astapenko, V.A., Bureeva, L.A., Lisitsa, V.S.: Statistical models in the polarization radiation theory. J. Radiat. Phys. Chem. **75**, 1287–1307 (2006)
67. Astapenko, V.A., Kukushkin, A.B.: Multiphoton static and polarization Bremsstrahlung in collisions of charged particles with multi-charged ions in a strong laser field. Zh. Eksp. Teor. Fiz. **111**, 419–439 (1997) (English translation: JETP **84**, 229–240 (1997))
68. Astapenko, V.A., Kubankin, A.S., Nasonov, N.N., Polyanskii, V.V., Pokhil, G.P., Sergienko, V.I., Khablo, V.A.: Measurement of the polarization bremsstrahlung of relativistic electrons in polycrystalline targets. Pisma v Zh. Eksp. Teor. Fiz. **84**, 341–344 (2006) (English Translation: JETP Letters, **84**, 281–284 (2006))
69. Avdonina, N.B., Amusia, M.Ya., Kuchiev, M.Yu., Chernysheva, L.V.: Bremsstrahlung of fast electrons on atoms. Zh. Tekhn. Fiz. **56**, 246–254 (1986) (English Translation: Sov. Phys. – Tech. Phys. **31**, 150–155 (1986))
70. Avdonina, N.B., Pratt, R.H.: Bremsstrahlung from ions. Nucl. Instrum. Methods Phys. Res. B **99**, 166–169 (1995)
71. Avdonina, N.B., Pratt, R.H.: Parametrization of the Coulombic Bremsstrahlung spectrum. Indian J. Phys. B **71**, 209–222 (1997)
72. Avdonina, N.B., Pratt, R.H.: Bremsstrahlung from ions. In: Duggan, J.L., Morgan, I.L. (eds.) Proceedings of 14th International Conference on Application of Accelerators in Research and Industry (AIP Conf. Proc. vol. 392, New York, AIP), pp. 249–252 (1997)
73. Avdonina, N.B., Pratt, R.H.: Bremsstrahlung spectra from atoms and ions at low relativistic energies. J. Phys. B: At. Mol. Opt. Phys. **32**, 4261–4276 (1999)
74. Baccarelli, I., Gianturco, F.A., Scifoni, E., Solovyov, A.V., Surdutovich, E. (eds.): Topical issue on molecular level assessments of radiation biodamage. Eur. Phys. J. D **60**, 1 (2010)
75. Baier, V.N., Katkov, V.M., Fadin, V.S.: Radiation by relativistic electrons. Atomizdat, Moscow, (1973) (in Russian)

76. Baier, V.N., Katkov, V.M., Strakhovenko, V.M.: Electromagnetic Processes at High Energies in Oriented Single Crystals. World Scientific, Singapore (1998)
77. Baldwin, G.C., Klaiber, G.S.: Photo-fission in heavy elements. Phys. Rev. **71**, 3–10 (1947)
78. Becker, U., Kerkhoff, H.G., Lindle, D.W., Kobrin, P.H., Ferrett, T.A., Heimann, P.A., Truesdale, C.M., Shirley, D.A.: Orbital-collapse effects in photoemission from atomic Eu. Phys. Rev. A **34**, 2858–2864 (1986)
79. Belyaev, A.K. , Ivanov, V.K., Polozkov, R.G., Tiukanov, A.S., Solov'yov, A.V., Greiner, W.: Photoabsorption of the fullerene ions C_{60}^{+} and C_{60}^{++}. Int. J. Quant. Chem. **107**, 2781–2786 (2007)
80. Belyaev, A.K., Tiukanov, A.S., Toropkin, A.I., Ivanov, V.K., Polozkov, R.G., Solov'yov, A.V.: Photoabsorption of the fullerene C_{60} and its positive ions. Phys. Scr. **80**, 048121 (2009)
81. Berestetskii, V.B., Lifshitz, E.M., Pitaevskii, L.P.: Course of Theoretical Physics, vol. 4. Quantum Electrodynamics. Pergamon Press, Oxford (1982)
82. Bertsch, G.F., Broglia, R.A.: Oscillations in Finite Quantum Systems. Cambridge University Press, Cambridge (1984)
83. Bertsch, G.F., Bulgac, A., Tomanek, D., Wang, Y.: Collective plasmon excitations in C_{60} clusters. Phys. Rev. Lett. **67**, 2690–2693 (1991)
84. Bethe, H., Heitler, W.: On the stopping of fast particles and on the creation of positive electrons. Proc. Roy. Soc. Lond. Ser. A **146**, 83–112 (1934)
85. Bethe, H.A., Salpeter, E.E.: Quantum Mechanics of One and Two Electron Atoms. Springer, Berlin (1957)
86. Bizau, J.M., Blancard, C., Cubaynes, D., Folkmann, F., Champeaux, J.P., Lemaire, J.L., Wuilleumier, F.J.: Absolute photoionization cross sections along the Xe isonuclear sequence: Xe^{3+} to Xe^{6+}. Phys. Rev. A **73**, 022718 (2006)
87. Blazhevich, S.V., Cherpunov, A.S., Grishin, V.K., Gudkov, K.A., Ishkhanov, B.S., Kamishanchenko, N.V., Kosterin, S.A., Lasutin, E.V., Nasonov, N.N., Ovchinnik, V.D., Petukhov, V.P., Pirogov, V.A., Shvedunov, V.I., Strelnitsky, V.E.: Suppression of polarization Bremsstrahlung of relativistic electrons moving through an amorphous carbon foil. Phys. Lett. **211A**, 309–312 (1996)
88. Blazhevich, S.V., Cherpunov, A.S., Grishin, V.K., Ishkhanov, B.S., Nasonov, N.N., Petukhov, V.P., Shvedunov, V.I.: Polarization Bremsstrahlung of relativistic electrons in aluminium. Phys. Lett. **254A**, 230–232 (1999)
89. Blundell, S.E., Guet, G.: Ionic structure effects on the optical response of lithium clusters. Z. Phys. D **28**, 81–85 (1993)
90. Bolotovskii, B.M.: Vavilov Cherenkov radiation: its discovery and application. Uspekhi Fiz. Nauk **179**, 1161–1173 (2009) (English translation: Physics – Uspekhi **52**, 1099–1110 (2009))
91. Borovski, I.B., Rydnik, V.I.: A model of X-ray emission generation in solids under electron bombardment. Izv. Akad. Nauk SSSR: Ser. Fiz. **31**, 1009–1015 (1967) (in Russian)
92. Brack, M.: The physics of simple metal clusters: self-consistent jellium model and semiclassical approaches. Rev. Mod. Phys. **65**, 677–732 (1993)

93. Bréchignac, C., Cahuzac, Ph., Carlier, F., Leygnier, J.: Collective excitation in closed-shell potassium cluster ions. Chem. Phys. Lett. **164**, 433–437 (1989)
94. Bréchignac, C., Connerade, J.P.: Giant resonances in free atoms and in clusters. J. Phys. B: At. Mol. Opt. Phys. **27**, 3795–3828 (1994) (Erratum: ibid. **28**, 131 (1995))
95. Briggs, J.S.: Collision broadening and molecular orbital X-rays. J. Phys. B: At. Mol. Phys. **7**, 47–54 (1974)
96. Briggs, J.S., Detmann, K.: Theory of continuum emission during ion-atom collisions. J. Phys. B: At. Mol. Phys. **10**, 1113–1132 (1977)
97. Brown, G.V., Beiersdorfer, P., Chen, H., Scofield, J.H., Boyce, K.R., Kelley, R.L., Kilbourne, C.A., Porter, F.S., Gu, M.F., Kahn, S.M., Szymkowiak, A.E.: Energy-dependent excitation cross section measurements of the diagnostic lines of Fe XVII. Phys. Rev. Lett. **96**, 253201 (2006)
98. Buimistrov, V.M., Krotov, Yu.A., Trakhtenberg, L.I.: Photon emission in collisions of a proton or positron with an atom. Zh. Eksp. Teor. Fiz. **79**, 808–816 (1980) (English Translation: Sov. Phys. – JETP **52**, 411–415 (1980))
99. Buimistrov, V.M., Trakhtenberg, L.I.: Bremsstrahlung cross section in scattering of an electron by a hydrogen atom. Zh. Eksp. Teor. Fiz. **69**, 108–114 (1975) (English Translation: Sov. Phys. – JETP **42**, 54–57 (1975))
100. Buimistrov, V.M., Trakhtenberg, L.I.: The role of atomic electrons in Bremsstrahlung. Zh. Eksp. Teor. Fiz. **23**, 850–853 (1977) (English Translation: Sov. Phys. – JETP **46**, 447–448 (1986))
101. Buimistrov, V.M., Trakhtenberg, L.I.: On the disparity between differential cross sections of electron and positron Bremsstrahlung. Phys. Lett. **69A**, 261–262 (1978)
102. Bureeva, L., Lisitsa, V.: Polarization recombination as a new channel for recombination of free electrons on complex ions. J. Phys. B: At. Mol. Opt. Phys. **31**, 1477–1485 (1998)
103. Bureeva, L., Lisitsa, V.: Polarization recombination, a new recombination channel. Phys. Scr. T**80**, 163–164 (1999)
104. Campbell, A.: K X-ray yields from elements of low atomic number. Proc. Roy. Soc. Lond. A **274** 319–342 (1963)
105. Carlsten, J.L., Peterson, J.R., Linneberger, W.C.: Binding of an electron by the field of a molecular dipole–$LiCl^-$. Chem. Phys. Lett. **37**, 5–8 (1976)
106. Casting, R., Descamps, J.: Sur les bases physiques de l'analyse ponctuelle par spectrographie X. J. Phys. Radium **16**, 304–317 (1955)
107. Chernysheva, L.V., Gribakin, G.F., Ivanov, V.K., Kuchiev, M.Yu.: Many-body calculation of negative ions using the Dyson equation. J. Phys. B: At. Mol. Opt. Phys. **21**, L419–425 (1988)
108. Chernysheva, L.V., Yakhontov, V.L.: Two-program package to calculate the ground and excited state wave functions in the Hartree-Fock-Dirac approximation. Comput. Phys. Commun. **119**, 232–249 (1999)
109. Chen, Z., Msezane, A.: Effect of C_{60} giant resonance on the photoabsorption of encaged atoms. Phys. Rev. A **86**, 063405 (2012)

110. Chen, Z., Msezane, A.: Photoabsorption spectrum of the Xe@C_{60} endohedral fullerene. Eur. Phys. J. D **66**, 184 (2012)
111. Chen, W., Zhang, J.: Using nanoparticles to enable simultaneous radiation and photodynamic therapies for cancer treatment. J. Nanosci. Nanotechnol. **6**, 1159–1166 (2006)
112. Connerade, J.P., Esteva, J.M., Karnatak, R.C.: Giant Resonances in Atoms, Molecules and Solids. Vol. 151 of NATO Advanced Study Institute, Series B: Physics. Plenum, New York (1986)
113. Connerade, J.-P., Gerchikov, L.G., Ipatov, A.N., Solov'yov, A.V.: Polarization effects in electron attachment to metallic clusters. J. Phys. B: At. Mol. Opt. Phys. **32**, 877–894 (1999)
114. Connerade, J.P., Solov'yov, A.V.: Giant resonances in photon emission spectra of metal clusters. J. Phys. B: At. Mol. Opt. Phys. **29**, 3529–3547 (1996)
115. Connerade, J.P., Solov'yov, A.V.: Radiative electron capture by metallic clusters. J. Phys. B: At. Mol. Opt. Phys. **29**, 365–375 (1996)
116. Connerade, J.-P., Solov'yov, A.V.: Formalism for multiphoton plasmon excitation in jellium clusters. Phys. Rev. A **66**, 013207 (2002)
117. Connerade, J.-P., Solov'yov, A.V. (eds.): Latest Advances in Atomic Clusters Collision: Fission, Fusion, Electron, Ion and Photon Impact, pp. 1–383. Imperial College Press, London (2004)
118. Connerade, J.-P., Solov'yov, A.V.: Dynamical screening of a confined atom by a fullerene. J. Phys. B: At. Mol. Opt. Phys. **38**, 807–813 (2005)
119. Connerade, J.-P., Solov'yov, A.V. (eds.): Latest Advances in Atomic Clusters Collision: Structure and Dynamics from the Nuclear to the Biological Scale, pp. 1–442. Imperial College Press, London (2008)
120. Cooper, J.W.: Interaction of maxima in the absorption of soft X rays. Phys. Rev. Lett. **13**, 762–764 (1964)
121. Cooper, J., Zare, R.N.: Angular distribution of photoelectrons. J. Chem. Phys. **48**, 942–943 (1968) (Erratum: *ibid.* **49**, 4252 (1968))
122. Cosslet, V., Thomas, R.: Multiple scattering of 5–30 keV electrons in evaporated metal films: I. Total transmission and angular distribution. Br. J. Appl. Phys. **15**, 883–908 (1964)
123. Crawford, O.H.: Bound states of a charged particle in a dipole field. Proc. Phys. Soc. **91**, 279–284 (1967)
124. Dubois, A., Maquet, A.: Bremsstrahlung radiation emitted in fast-electron–H-atom collisions. Phys. Rev. A **40**, 4288–4297 (1989)
125. Dubois, A., Maquet, A., Jetzke, S.: Electron–H-atom collisions in the presence of a laser field: one-photon free-free transitions. Phys. Rev. A **34**, 1888–1895 (1986)
126. Dupuoy, G., Perrier, F., Verdier, G., Arnal, F.: Transmission d'electrons monociétiques à travers des feuilles métalliques minces. Compt. Rend. **258**, 3655 (1964)
127. Eisenberg, J.M., Greiner, W.: Nuclear Theory. North Holland, Amsterdam (1970)

128. Ekardt, W. Work function of small metal particles: self-consistent spherical jellium-background model. Phys. Rev. B **29** 1558–1564 (1984)
129. Ekardt, W.: Dynamical polarizability of small metal particles: self-consistent spherical jellium background model. Phys. Rev. Lett. **52**, 1925–1928 (1984)
130. Ekardt, W. (ed.): Metal Clusters. Wiley, New York (1999)
131. Elwert G and Haug E.: Calculation of Bremsstrahlung cross sections with Sommerfeld-Maue eigenfunctions. Phys. Rev. **183**, 90–105 (1969)
132. Fano, U., Cooper, J.W.: Spectral distribution of atomic oscillator strengths. Rev. Mod. Phys. **40**, 441–507 (1968) (Addendum: *ibid.* **41**, 724–725 (1969))
133. Feng, I.J., Pratt, R.H., Tseng, H.K.: Positron Bremsstrahlung. Phys. Rev. A **24**, 1358–1363 (1981)
134. Fermi, E., Amaldi, E., Anderson, H.L., Persico, E., Rasetti, F., Smith, C.S., Wattenberg, A., Segrè, E. (eds.): The Collected Papers of Enrico Fermi: Volume I. University of Chicago Press, Chicago (1962)
135. Florescu, A., Obolensky, O.I., Pratt, R.H.: Structures in low-energy classical Bremsstrahlung. J. Phys. B: At. Mol. Opt. Phys. **35**, 2911–2925 (2002)
136. Foresman, J.B., Frisch, A.: Exploring Chemistry with Electronic Structure Methods. Gaussian Inc., Pittsburgh (1996)
137. Fock, V.A.: Fundamentals of Quantum Mechanics. MiR Publishers, Moscow (1978) (ISBN: 0828551979)
138. Fowler, P.W., Manolopoulos, D.E.: An Atlas of Fullerenes. Dover Publications, New York (2007)
139. Frank, W., Gipper, P., Kann, K.-H., Sodan, H., Tretyakov, Yu.P.: Quasimolecular K X-ray excitation by bombarding La targets with La and Xe ions. Phys. Lett. **59B**, 41–44 (1975)
140. Freeman, A.J., Watson, R.E.: Theoretical investigation of some magnetic and spectroscopic properties of rare-earth ions. Phys. Rev. **127**, 2058–2064 (1962)
141. Frisch, M.J., et al.: Gaussian 98 (Revision A.9). Gaussian Inc., Pittsburgh (1998)
142. Frommhold, L.: Electron-atom Bremsstrahlung and the sonoluminescence of rare gas bubbles. Phys. Rev. E **58**, 1899–1905 (1998)
143. Fritsch, W., Wille, U.: On the theory of molecular-orbital X-rays. J. Phys. B: At. Mol. Phys. **12**, L335–L340 (1979)
144. Fujimoto, F., Komaki, K., Ishido, K.: Surface plasma oscillations in aluminum fine particles. J. Phys. Soc. Japan **23**, 1186–1186 (1967)
145. Fujimoto, F., Komaki, K.: Plasma oscillations excited by a fast electron in a metallic particle. J. Phys. Soc. Japan **25**, 1679–1687 (1968)
146. Gerchikov, L.G., Connerade, J.-P., Solov'yov, A.V., Greiner, W.: Scattering of electrons on metal clusters and fullerenes. J. Phys. B: At. Mol. Opt. Phys. **30**, 4133–4161 (1997)
147. Gerchikov, L.G., Efimov, P.V., Mikoushkin, V.M., Solov'yov, A.V.: Diffraction of fast electrons on the fullerene C60 molecule. Phys. Rev. Lett. **81**, 2707–2710 (1998)

148. Gerchikov, L.G., Ipatov, A.N., Solov'yov, A.V.: Many-body treatment of the photon emission process in electron-cluster collisions. J. Phys. B: At. Mol. Opt. Phys. **31**, 2331–2341 (1998)
149. Gerchikov, L.G., Ipatov, A.N., Polozkov, R.G., Solov'yov, A.V.: Surface- and volume-plasmon excitations in electron inelastic scattering on metal clusters. Phys. Rev. A **62**, 043201 (2000)
150. Gerchikov, L.G., Ipatov, A.N., Solov'yov, A.V.: Many-body treatment of electron inelastic scattering on metal clusters. J. Phys. B: At. Mol. Opt. Phys. **30**, 5939–5959 (1997)
151. Gerchikov, L.G., Ipatov, A.N., Solov'yov, A.V., Greiner, W.: Excitation of multipole plasmon resonances in clusters by fast electron impact. J. Phys. B: At. Mol. Opt. Phys. **31**, 3065–3077 (1998)
152. Gerchikov, L.G., Ipatov, A.N., Solov'yov, A.V., Greiner, W.: Non-adiabatic electron-ion coupling in dynamical jellium model for metal clusters. J. Phys. B: At. Mol. Opt. Phys. **33**, 4905–4926 (2000).
153. Gerchikov, L.G., Korol, A.V., Lyalin, A.V., Solov'yov, A.V.: Giant resonances in photon emission spectra of atoms, clusters and solids. AIP Conf. Proc. **389**, 447–464 (1997)
154. Gerchikov, L.G., Solov'yov, A.V.: Photon emission in electron-cluster collisions in the vicinity of a plasmon resonance. Z. Phys. D **42**, 279–287 (1997)
155. Gerchikov, L.G., Solov'yov, A.V., Greiner, W.: Dynamical jellium model for metallic clusters. Int. J. Mod. Phys. E **8**, 289–298 (1999)
156. Ginzburg, V.L., Tsytovich, V.N.: Transition Radiation and Transition Scattering. Adam Hilger, Bristol (1990)
157. Gnatchenko, E.V., Tkachenko, A.A., Verkhovtseva E.T.: Bremsstrahlung in the scattering of 0.3-2 keV electrons on Ar atoms. Optika i Spektroskopiya **92**, 17–20 (2002) (English translation: Opt. Spectrosc. **92**, 13–16 (2002))
158. Gnatchenko, E.V., Tkachenko, A.A., Verkhovtseva E.T.: Specific features of ultrasoft X-ray Bremsstrahlung in scattering of intermediate energy electrons by argon atoms. Surf. Rev. Lett. **9**, 651–654 (2002)
159. Gnatchenko, E.V., Tkachenko, A.A., Verkhovtseva, E.T., Zon, B.A.: Ultrasoft X-ray bremsstrahlung isochromatic spectra from 300–2000 eV electrons on Ar and Kr. Phys. Rev. Lett. **95**, 023002 (2005)
160. Gnatchenko, E.V., Tkachenko, A.A., Nechay, A.N.: Absolute differential Bremsstrahlung cross section for the scattering of 0.6 keV electrons by xenon atoms. Pis'ma v ZhETF **86**, 344–348 (2007) (English Translation: JETP Letters **86**, 292–292 (2007))
161. Gnatchenko, E.V., Tkachenko, A.A., Nechay, A.N.: Absolute differential Bremsstrahlung cross section for 0.42-keV electrons scattered by Ar, Kr and xenon atoms. Phys. Rev. A **80**, 022707 (2009)
162. Gnatchenko, E.V., Nechay, A.N., Samovarov, V.N., Tkachenko, A.A.: Polarization Bremsstrahlung from xenon atoms and clusters: a cooperative effect contribution. Phys. Rev. A **82**, 012702 (2010)

163. Gnatchenko, E.V., Nechay, A.N., Samovarov, V.N., Tkachenko, A.A.: Polarization Bremsstrahlung spectrum of xenon clusters: detection of the contribution of collective interactions. Low Temp. Phys. **36**, 196 (2010)
164. Goldhaber, M., Teller, E.: On nuclear dipole vibrations. Phys. Rev. **74**, 1046–1049 (1948)
165. Golovinski, P.A.: Stimulated Bremsstrahlung effect in the scattering of electrons by atoms subject to the polarization of a target. Zh. Eksp. Teor. Fiz. **94**, 87–94 (1988) (English Translation: Sov. Phys. – JETP **68**, 1346–1350 (1988))
166. Golovinski, P.A., Zon, B.A.: Bremsstrahlung in collisions of electrons with negative ions. Zh. Tekhn. Fiz. **50**, 1847–1851 (1980) (English Translation: Sov. Phys. – Tech. Phys. **25**, 1076–1077 (1980))
167. González, A.D., Miraglia, J.E., Garibotti, C.R.: Continuum emission during ion-atom collisions at high projectile energies. Phys. Rev. A **34**, 2834–2842 (1986)
168. Gonzalez, A.D., Pacher, M.C., Miraglia, J.E.: Retardation effects in atomic Bremsstrahlung. Phys. Rev. A **37**, 4974–4977 (1988)
169. Gorokhov, D.A., Suris, R.A., Cheianov, V.V.: Electron-energy-loss spectroscopy of the C_{60} molecule. Phys. Lett. **223A**, 116–122 (1996)
170. Gostishchev, N.A., Kubankin, A.S., Nasonov, N.N., Polyanskii, V.V., Sergienko, V.I., Khablo, V.A.: Angular dependence of the coherent peak position in the polarization Bremsstrahlung spectrum of relativistic electrons in polycrystalline targets. Pis'ma v ZhTF **34**, 78–82 (2008) (English Translation: Tech. Phys. Lett. **34**, 763–764 (2008))
171. Gradshteyn, I.S., Ryzhik, I.M.: Table of Integrals, Series and Products. Academic Press, New York (1965)
172. Green, M.: The angular distribution of characteristic x radiation and its origin within a solid target. Proc. Phys. Soc. **83**, 435–452 (1964)
173. Greiner, W., Reinhardt, J.: Heavy ion atomic physics (theory). In: Bromley, D.A. (ed.) Heavy Ion Science, vol. 5, pp. 3–318. Plenum Press, New York (1985)
174. Greiner, W., Reinhardt, J.: Quantum Electrodynamics. Springer, Berlin (1992)
175. Gribakin, G.F., Gul'tsev, B.V., Ivanov, V.K., Kuchiev, M.Yu.: Interaction of an alkaline-earth atom with an electron: scattering, negative ion and photodetachment. J. Phys. B: At. Mol. Opt. Phys. **23**, 4505–4519 (1990)
176. Griffits, D.J.: Introduction to Electrodynamics. Prentice-Hall Inc., Upper Saddle River (1999)
177. Grishin, V.K.: Polarization radiation from relativistic electrons in matter containing fullerenes. Nucl. Instrum. Methods Phys. Res. B **227**, 82–86 (2005)
178. Gros, M., Müller, B., Greiner, W.: Radiative transitions induced in heavy-ion collisions. J. Phys. B: At. Mol. Phys. **9**, 1849–1856 (1976)
179. Guet, C., Hobza, P., Spiegelman, F., David, F. (eds.): Atomic Clusters and Nanoparticles, NATO Advanced Study Institute, les Houches Session LXXIII, les Houches, 2000. EDP Sciences and Springer, Berlin (2001)
180. Guet, C., Johnson, W.R.: Dipole excitations of closed-shell alkali-metal clusters. Phys. Rev. B **45**, 11283–11287 (1992)

181. ter Haar, D.: Elements of Hamiltonian Mechanics. Pergamon Press, Oxford (1991)
182. Haberland, H. (ed.): Clusters of Atoms and Molecules, Theory, Experiment and Clusters of Atoms. Springer Series in Chemical Physics, vol. 52. Springer, Berlin (1994)
183. Haensel, R., Keitel, G., Schreiber, P., Kunz, C.: Optical absorption of solid krypton and xenon in the far ultraviolet. Phys. Rev. **188**, 1375–1380 (1969)
184. Hahn, Y.: Electron-ion recombination processes—an overview. Rep. Progr. Phys. **60**, 691–759 (1997)
185. Hainfeld, J.J., Slatkin, D.N., Smilowitz, H.M.: The use of gold nanoparticles to enhance radiotherapy in mice. Phys. Med. Biol. **49**, N309–N315 (2004)
186. Hammer, D., Frommhold, L.: Polarization Bremsstrahlung spectra of electron–rare-gas atom collisions at temperatures from 5 to 40 kK. Phys. Rev. A **64**, 024705 (2001) (Erratum: ibid. **64**, 059901(E) (2001))
187. Hammer, D., Frommhold, L.: Light emission of sonoluminescent bubbles containing a rare gas and water vapor. Phys. Rev. E **65**, 046309 (2002)
188. Hammer, D., Frommhold, L.: Electron-ion Bremsstrahlung spectra calculations for sonoluminescence. Phys. Rev. E **66**, 056303 (2002)
189. Haug, E., Nakel, E.: The Elementary Process of Bremsstrahlung. World Scientific, Singapore (2004)
190. Haygood, R., Williams, S., Quarles, C.A.: Bremsstrahlung from gas atom targets: study of background processes. J. Radiat. Phys. Chem. **75**, 1688–1692 (2006)
191. de Heer, W.A.: The physics of simple metal clusters: experimental aspects and simple models. Rev. Mod. Phys. **65**, 611– 676 (1993)
192. Heitler, W.: The Quantum Theory of Radiation. Dover, Toronto (1984)
193. Henke, B.L., Gullikson, E.M., Davis, J.C.: X-ray interactions: photoabsorption, scattering, transmission, and reflection at E=50-30000 eV, Z=1-92. At. Data Nucl. Data Tables **54**, 181–342 (1993) (also at http://henke.lbl.gov/optical_constants/)
194. Herlert, A., Krückeberg, S., Schweikhard, L., Vogel, M., Walther, C.: First observation of doubly charged negative gold cluster ions. Phys. Scr. **T80**, 200–202 (1999)
195. Hertel, I.V., Steger, H., de Vries, J., Weisser, B., Mezel, C., Kamke, B., Kamke, W.: Giant plasmon excitation in free C60 and C70 molecules studied by photoionization. Phys. Rev. Lett. **68**, 784 (1992)
196. Hippler, R.: Two-photon Bremsstrahlung of free atoms. Phys. Rev. Lett. **66**, 2197–2199 (1991)
197. Hubbard, D.F., Rose, M.E.: Nuclear structure effects in Bremsstrahlung. Nucl. Phys. **84**, 337–379 (1966)
198. Hubbel, J.H., Seltzer, S.M.: Tables of X-ray mass attenuation coefficients and mass energy-absorption coefficients from 1 keV to 20 MeV for elements Z = 1 to 92 and 48 additional substances of dosimetric interest. NISTIR 5632 (web version http://www.nist.gov/pml/data/xraycoef/index.cfm)

199. Huber, B.A., Malot, C., Domaracka, A., Solovyov, A.V. (eds.): 1st Nano-IBCT conference 2011 radiation damage of biomolecular systems: nanoscale insights into ion beam cancer therapy. J. Phys.: Conf. Ser. **373** (2012)
200. Ipatov, A.N., Connerade, J.-P., Gerchikov, L.G., Solov'yov, A.V.: Electron attachment to metallic clusters. J. Phys. B: At. Mol. Opt. Phys. **31**, L27–L33 (1998)
201. Ishii, K.: High energy limit of atomic Bremsstrahlung. Nucl. Instrum. Methods Phys. Res. B **99**, 163–165 (1995)
202. Ishii, K., Yamazaki, H., Matsuyama, S., Galster, W., Satoh, T., Budnar, M.: Contribution of atomic Bremsstrahlung in PIXE spectra and screening effect in atomic Bremsstrahlung. X-Ray Spectrom. **34**, 363–365 (2005)
203. Ishii, K.: Continuous X-rays produced in light-ion atom collisions. J. Radiat. Phys. Chem. **75**, 1135–1163 (2006)
204. Ishii, K., Morita, S.: Continuum X-rays produced by light-ion–atom collision. Phys. Rev. A **30**, 2278–2286 (1984)
205. Ishii, K., Morita, S.: Atomic Bremsstrahlung produced by heavy-ion bombardments. Phys. Rev. A **31**, 1168–1170 (1985)
206. Ishii, K., Morita, S.: Scaling law for a continuum of X-rays produced by light-ion–atom collisions. Nucl. Instrum. Methods Phys. Res. B **22**, 68–71 (1986)
207. Ishii, K., Morita, S.: Theoretical estimation of PIXE detection limits. Nucl. Instrum. Methods Phys. Res. B **34**, 209–216 (1988)
208. Ishii, K., Morita, S., Tawara, W.: Bremsstrahlung induced by proton and ^{3}He–ion bombardments in the 1-4-Mev/amu energy range. Phys. Rev. **13**, 131–138 (1976)
209. Ivanov, V.K.: Collective properties in negative ion photodetachment. In: Connerade, J.-P.I (ed.) Correlations in Clusters and Related Systems, pp. 73–91. World Scientific, Singapore (1996)
210. Ivanov, V.K., Ipatov, A.N., Kharchenko, V.A., Zhizhin, M.L.: Electronic structure and adiabatic oscillations of metal clusters. Pisma JETP **58**, 649–655 (1993) (English translation: JETP Letters, **58**, 629–635 (1993))
211. Ivanov, V.K., Ipatov, A.N., Kharchenko, V.A., Zhizhin, M.L.: Size dependence of electronic structure and adiabatic type of collective vibration in small metal clusters. Phys. Rev. A **50**, 1459–1464 (1994)
212. Ivanov, V.K., Kashenock, G.Yu., Polozkov, R.G., Solov'yov, A.V.: Photoionization cross sections of the fullerenes C20 and C60 calculated in a simple spherical model. J. Phys. B: At. Mol. Opt. Phys. **34**, L669–L678 (2001)
213. Ivanov, V.K., Kashenock, G.Yu., Polozkov, R.G., Solov'yov, A.V.: Method of calculation of fullerenes photoionization cross sections based on local density and random phase approximations. Zh. Eksp. Teor. Fiz. **123**, 744–756 (2003) (English translation: JETP **96**, 658–668 (2003))
214. Jackson, J.D.: Classical Electrodynamics. Wiley, Hoboken (1999)
215. Jakubassa-Amundsen, D.H.: Bremsstrahlung during electron capture to continuum. J. Radiat. Phys. Chem. **75**, 1319–1329 (2006)

216. Jänkälä, K., Mikkelä, M.-H., Huttula, M.: Valence photoionization of free, neutral, and size-varied alkali metal clusters. J. Phys. B: At. Mol. Opt. Phys. **44**, 105101 (2011)
217. Jänkälä, K., Tchaplyguine, M., Mikkelä, M.-H., O. Björneholm, O., Huttula, M: Photon energy dependent valence band response of metallic nanoparticles. Phys. Rev. Lett. **107**, 183401 (2011)
218. Jellinek, J. (ed.): Theory of Atomic and Molecular Clusters: with a Glimpse at Experiments. Springer Series in Cluster Physics. Springer, Berlin (1999)
219. Kafle, B.P., Katayanagi, H., Prodhan, M., Yagi, H., Huang, C., Mitsuke, K.: Absolute total photoionization cross section of C60 in the range of 25 120 eV: revisited. J. Phys. Soc. Japan **77**, 014302 (2008)
220. Kahlerm D.L., Liu, J., Quarles, C.A.: Double Bremsstrahlung. Phys. Rev. Lett. **68**, 1690–1693 (1992)
221. Kamyshanchenko, N., Nasonov, N., Pokhil, G.: Suppression of Density Effect in Polarization Bremsstrahlung for Relativistic Charged Particles Crossing a Thin Target. Nucl. Instrum. Methods Phys. Res. B **173**, 195–202 (2001)
222. Kaplan, I.G.: Giant resonances in atoms, atomic clusters, fullerenes, condensed matter, and Nuclei. Zeit. f. Physik D **40**, 375–380 (1997)
223. Keller, S., Dreizler, R.M.: Relativistic independent particle approximation study of triply differential cross sections for electron-atom Bremsstrahlung. J. Phys. B: At. Mol. Opt. Phys. **30**, 3257–3266 (1997)
224. Keller, J.M. Coplan, M.A.: Electron energy loss spectroscopy of C60. Chem. Phys. Lett. **193**, 89–92 (1992)
225. Kelly, H.P.: Many body calculations of Photoionization cross sections. Phys. Scr. **T17**, 109–119 (1987)
226. Kidun, O., Fominykh, N., Berakdar, J.: Multiple ionization of the Fullerene by a single photon. Comp. Mater. Sci. **35**, 354–358 (2006)
227. Kim, L., Pratt, R.H.: Numerical calculation of classical Bremsstrahlung. Phys. Rev. A **36**, 45–58 (1987)
228. Kim, J-K., Seo, S-J., Kim, K-H., Kim, T-J., Chung, M-H., Kim, K-R., Yang, T-K.: Therapeutic application of metallic nanoparticles combined with particle-induced X-ray emission effect. Nanotechnology **21**, 425102 (2010)
229. Kirsch, J., Betz, W., Reinhardt, J., Müller, B., Greiner, W., Soff, G.: The impact parameter dependence of the Pb + Pb K MO spectrum. Z. Phys. A **292**, 227–234 (1979)
230. Kissel, L., Quarles, C.A., Pratt, R.H.: Shape functions for atomic-field Bremsstrahlung from electrons of kinetic energy 1–500 keV on selected neutral atoms $1 \leq Z \leq 92$. At. Data Nucl. Data Tables **28**, 381–460 (1983)
231. Kittel, C.: Introduction to Solid State Physics. Wiley, USA (2005)
232. Knight, W.D., Clemenger, K., de Heer, W.A., Saunders, W.A., Chou, M.Y., Cohen, M.L.: Electronic shell structure and abundances of sodium clusters. Phys. Rev. Lett. **52**, 2141–2143 (1984)
233. Koch, H.W., Motz, J.W.: Bremsstrahlung cross-section formulas and related data. Rev. Mod. Phys. **116**, 920–955 (1959)

234. Kogan, V.I., Kukushkin, A.B., Lisitsa, V.S.: Kramers electrodynamics and electron-atomic radiative-collisional processes. Phys. Rep. **213**, 1–116 (1992)
235. Köhn, C., Ebert, U.: Angular distribution of Bremsstrahlung photons and of positrons for calculations of terrestrial gamma-ray flashes and positron beams. Atmos. Res. **135–136**, 432–465 (2014)
236. Korol, A.V.: Stripping of the Atom in the Bremsstrahlung process for entermediate-energy electron-atom collisions. J. Phys. B: At. Mol. Opt. Phys. **25**, L341–L344 (1992)
237. Korol, A.V.: The soft-photon Bremsstrahlung in an electron and a positron collision with a Hydrogen atom. J. Phys. B: At. Mol. Opt. Phys. **27**, 4765–4777 (1994)
238. Korol, A.V., Currell, F.J., Gribakin, G.F.: The role of target polarization in electron–ion recombination. J. Phys. B: At. Mol. Opt. Phys. **37**, 2411–2428 (2004)
239. Korol, A.V., Gribakin, G.F., Currell, F.J.: Effect of target polarization in electron–ion recombination. Phys. Rev. Lett. **97**, 223201 (2006)
240. Korol, A.V., Kuchiev, M.Yu., Solov'yov, A.V.: Bremsstrahlung in collision of a slow charged particle with an exited Hydrogen. J. Phys. B: At. Mol. Opt. Phys. **25**, 3379–3392 (1992)
241. Korol, A.V., Lyalin, A.G., Obolensky, O.I., Solovyov, A.V.: Manifestation of the Bethe Ridge in the Polarizational Bremsstrahlung Process. J. Phys. B: At. Mol. Opt. Phys. **33**, L179–L186 (2000)
242. Korol, A.V., Lyalin, A.G., Solov'yov, A.V., Shulakov, A.S.: Semiempirical Method for Polarizational Bremsstrahlung Calculation. J. Electron. Spectr. Relat. Phenom. **79**, 323–326 (1996)
243. Korol, A.V., Lyalin, A.G., Solov'yov, A.V., Shulakov, A.S.: Electron BrS Spectra on La near the 4d threshold. J. Phys. B: At. Mol. Opt. Phys. **29**, L611–L617 (1996)
244. Korol, A.V., Lyalin, A.G., Solov'yov, A.V.: New method for the polarizational Bremsstrahlung calculation. J. Phys. B: At. Mol. Opt. Phys. **28**, 4947–4962 (1995)
245. Korol, A.V., Lyalin, A.G., Solov'yov, A.V., Shulakov, A.S.: Theoretical treatment of Bremsstrahlung spectra in the vicinity of gaint atomic resonances: application to Ba. J. Phys. B: At. Mol. Opt. Phys. **28**, L155–L160 (1995)
246. Korol, A.V., Lyalin, A.G., Solov'yov, A.V.: Theoretical treatment of Bremsstrahlung process in the Vicinity of giant atomic resonances. Phys. Rev. A **53**, 2230–2238 (1996)
247. Korol, A.V., Lyalin, A.G., Solov'yov, A.V.: Bremsstrahlung in collisions of electrons with atoms and clusters. In: Campbell, D.M., Kleinpoppen, H. (eds.) Selected Topics on Electron Physics, pp. 263–278. Plenum, New York (1996)
248. Korol, A.V., Lyalin, A.G., Solov'yov, A.V., Shulakov, A.S.: Bremsstrahlung in intermediate-energy electron scattering by a La Atom. Zh. Eksp. Teor. Fiz. **109**, 1174–1187 (1996) (English Translation: Sov. Phys. – JETP **82**, 631–638 (1985))

249. Korol, A.V., Lyalin, A.G., Solov'yov, A.V.: Total Bremsstrahlung spectra of 1–25 keV electrons on Ne and Ar. J. Phys. B: At. Mol. Opt. Phys. **30**, L115–L121 (1997)
250. Korol, A.V., Lyalin, A.G., Solov'yov, A.V.: Bremmstrahlung cross section for electron–atom with polarization mechanism taken into account in a broad spectral range. Optika i Spektroskopiya **86**, 552–558 (1999) (English Translation: Opt. Spectrosc. **86**, 486-489 (1999))
251. Korol, A.V., Lyalin, A.G., Solovyov, A.V.: Bremsstrahlung calculations in the 'Stripping' approximation. In: XX1 ICPEAC, Book of Abstracts, p. 240. Sendai, Japan (1999)
252. Korol, A.V., Lyalin, A.G., Solovyov, A.V., Avdonina, N.B., Pratt, R.H.: On the stripping approximation in the Bremsstrahlung process. J. Phys. B: At. Mol. Opt. Phys. **35**, 1197–1210 (2002)
253. Korol, A.V., Lyalin, A.G., Obolensky,O.I., Solovyov, A.V.: Investigation of the role of the polarizational mechanism of emission by atoms in a wide range of photon frequencies. Zh. Eksp. Teor. Fiz. **114**, 458–473 (1998) (English Translation: JETP **87**, 251–259 (1998))
254. Korol, A.V., Lyalin, A.G., Obolenski, O.I., Solov'yov, A.V., Solovjev, I.A.: Polarizational Bremsstrahlung on atoms and ions: relativistic and non-relativistic cases. In: Duggan, J.L., Morgan, I.L. (eds.) AIP Conference Proceedings. Vol. 576, pp. 64–67. AIP, New York (2001)
255. Korol, A.V., Lyalin, A.G., Obolensky, O.I., Solov'yov, A.V., Solovjev, I.A.: Relativistic description of the polarizational mechanism of the elastic Bremsstrahlung process. Zh. Eksp. Teor. Fiz. **121**, 819–837 (2002) (English Translation: JETP **94**, 704–719 (2002))
256. Korol, A.V., Obolensky, O.I., Solov'yov, A.V.: Scaling behaviour for the inner-shell polarizational Bremsstrahlung cross section. J. Phys. B: At. Mol. Opt. Phys. **31**, 5347–5353 (1998)
257. Korol, A.V., Obolensky, O.I., Solov'yov, A.V.: Features of the Bremsstrahlung accompanying collisions with a Hydrogen atom in an excited state. Zh. Tekh. Fiz. **69**, 713 (1999) (English Translation: Techn. Phys. **44**, 1135-1140 (1999))
258. Korol, A.V., Obolensky, O.I., Solov'yov, A.V., Solovjev, I.A.: The full relativistic description of the Bremsstrahlung process in a charged particle–atom collision. J. Phys. B: At. Mol. Opt. Phys. **34**, 1589–1617 (2001)
259. Korol, A.V., Obolensky, O.I., Solov'yov, A.V., Solovjev, I.A.: Relativistic calculation of polarizational Bremsstrahlung spectra. Surf. Rev. Lett. **9**, 1191–1195 (2002)
260. Korol, A.V., Solov'yov, A.V.: Polarizational Bremsstrahlung of electrons in collisions with atoms and clusters. J. Phys. B: At. Mol. Opt. Phys. **30**, 1105–1150 (1997).
261. Korol, A.V., Solov'yov, A.V.: Polarizational Bremsstrahlung in non-relativistic collisions. J. Radiat. Phys. Chem. **75**, 1266–1286 (2006)
262. Korol, A.V., Solov'yov, A.V.: Relativistic effects in polarizational Bremsstrahlung. J. Radiat. Phys. Chem. **75**, 1251–1265 (2006)

263. Korol, A.V., Solov'yov, A.V.: Comment on "photoexcitation of a volume Plasmon in C60 Ions". Phys. Rev. Lett. **98**, 179601 (2007)
264. Korol, A.V., Solov'yov, A.V. (eds.): Topical issue on atomic cluster collisions. Eur. Phys. J. D **67**, (2013)
265. Kracke, G., Alber, G., Briggs, J.S., Maquet, A.: Two-Photon Bremsstrahlung in electron collisions with noble gas atoms. J. Phys. B: At. Mol. Opt. Phys. **26**, L561–L566 (1993)
266. Kracke, G., Alber, G., Briggs, J.S., Maquet, A., Véniard, V.: Two-photon free-free transitions in laser-assisted Electron-Hydrogen scattering. J. Phys. B: At. Mol. Opt. Phys. **27**, 3241–3256 (1994)
267. Kreibig, U., Zacharias, P.: Surface plasma resonances in small spherical silver and gold particles. Z. Phys. **231**, 128–143 (1970)
268. Kreibig, U., Vollmer, U.: Optical Properties of Metal Clusters. Springer, Berlin (1995)
269. Kubo, R.: Electronic properties of metallic fine particles. I. J. Phys. Soc. Jpn. **17**, 975–986 (1962)
270. Kuchiev, M.Yu., Schäfer, A., Solov'yov, A.V., Greiner, W.: New interference phenomenon in Quasimolecular radiation spectrum. J. Phys. B: At. Mol. Opt. Phys. **25**, 1853–1864 (1992)
271. Kuchiev, M.Yu., Solov'yov, A.V.: Fresnel peculiarity of the response function of a quantum Adiabatic system. Phys. Lett. **166A** 168–172 (1992)
272. Kumakhov, M.A., Komarov, F.F.: Radiation from Charged Particles in Solids. AIP, New York (1989)
273. Kurkina, L.I.: Low-energy electron Bremsstrahlung from metal clusters. Phys. Sol. State **46**, 557–562 (2004)
274. Kurkina, L.I.: Resonance behaviour of the Bremsstrahlung process in collisions of the low-energy electrons with Alkali-Metal clusters. J. Phys. B: At. Mol. Opt. Phys. **37**, 2649–2660 (2004)
275. Kurkina, L.I.: Resonances in low-energy electron cluster Bremsstrahlung. Comp. Mater. Sci. **35**, 342–346 (2006)
276. Kutzner, M., Zikri Altun, Kelly, H.P.: Photoionization of atomic Barium including effects of relaxation and polarization. Phys. Rev. A **41** 3612–3623 (1990)
277. Lambin, Ph., Lucas, A.A., Vigneron, J.-P.: Polarization waves and van der Waals cohesion of C60 Fullerite. Phys. Rev. B **46** 1794–1803 (1992)
278. Landau, L.D., Lifshitz, E.M.: Course of theoretical physics. In: The Classical Theory of Fields. vol. 2, Pergamon Press, Oxford (1971)
279. Landau, L.D., Lifshitz, E.M.: Course of theoretical physics. In: Quantum Mechanics. Non-relativistic Theory. vol. 3, Elsevier, Oxford (2004)
280. Lifshitz, E.M, Pitaevskii, L.P.: Course of theoretical physics. Physical Kinetics, vol. 10, Pergamon Press, Oxford (1981)
281. Lee, C.M., Kissel, L., Pratt, R.H., Tseng, H.K.: Electron Bremsstrahlung spectrum, 1–500 keV. Phys. Rev. A **13**, 1714–1727 (1976) (Erratum: Phys. Rev. A **24**, 28662867 (1981))
282. Lee, C.M., Pratt, R.H., Tseng, H.K.: Bremsstrahlung spectrum from atomic ions. Phys. Rev. A **16** 2169–2173 (1977)

283. Levy-Leblanc J.M.: Electron capture by polar molecules. Phys. Rev. **153**, 1–4 (1967)
284. Lichten, W.: Molecular-orbital x rays. Phys. Rev. A **9**, 1458–1461 (1974).
285. Chamberlain, M.B., Burr, A.F., Liefeld, R.J.: Excitation-Eenergy-Dependent features in the continuum spectrum of cerium near the Mα and Mβ X-Ray emission lines. Phys. Rev. A **9**, 663–667 (1974)
286. Liefeld, R.J., Burr, A.F., Chamberlain, M.B.: Electron-Excitation Energy-Dependent features in the M-Series x-Ray spectrum of Lanthanum. Phys. Rev. A **9**, 316–322 (1974)
287. Lindgren, I., Morrison, J.: Atomic Many-Body Theory. Springer, Berlin (1986)
288. Lisitsa V.S.: Stark Broadening of Hydrogen Lines in Plasmas. Uspekhi Fiz. Nauk **122**, 449–495 (1977) (English translation: Sov. Phys. – Uspekhi **20**, 603–630 (1977))
289. Lo, S., Korol, A.V., Solov'yov, A.V.: Dynamical screening of an atom confined within a finite-width Fullerene. J. Phys. B: At. Mol. Opt. Phys. **40**, 3973–3981 (2007)
290. Lo, S., Korol, A.V., Solov'yov, A.V.: Dynamical screening of an Endohedral atom. Phys. Rev. A **79**, 063201 (2009)
291. Ludziejewski, T., Stöhlker, T., Keller, S., Beyer, H., Bosch, F., Brinzanescu, O., Dunford, R.W., Franzke, B., Kozhuharov, C., Liesen, D., Livingston, A.E., Menzel, G., Meier, J., Mokler, P.H., Reich, H., Rymuza, P., Stachura, Z., Steck, M., Stenner, L., Swiat, P., Warczak, A.: Electron Bremsstrahlung in collisions of 223 MeV/u He-Like Uranium Ions with Gaseous targets. J. Phys. B: At. Mol. Opt. Phys. **31**, 2601–2609 (1998)
292. Lundqvist, S., March, N.H.: Theory of the Inhomogeneous Electron Gas. Plenum Press, New York (1983)
293. Lushnikov, A.A., Simonov, A.J.: Surface Plasmons in small metal particles. Z. Physik **270**, 17–24 (1974)
294. Lushnikov, A.A., Simonov, A.J.: Excitation of surface Plasmons in metal particles by fast electrons and X Rays. Z. Physik B **21**, 357–352 (1972)
295. Lyalin, A.G., Matveentzev, A., Solov'yov, I.A., Solov'yov, A.V., Greiner, W.: On the applicability of deformed Jellium model to the description of metal clusters. Eur. Phys. J. D **24**, 15–18 (2003)
296. Lyalin, A.G., Obolensky, O.I., Solov'yov, I.A., Solov'yov, A.V., Greiner, W.: Molecular dynamics simulations of cluster fission and fusion processes. Phys. Scr. **T110**, 319–324 (2004)
297. Lyalin, A.G., Semenov S K, Solov'yov, A.V., Cherepkov, N.A., Greiner, W.: Hartree-Fock deformed Jellium model for metallic clusters. J. Phys. B: At. Mol. Opt. Phys. **33**, 3653–3664 (2000)
298. Lyalin, A.G., Semenov, S.K., Solov'yov, A.V., Cherepkov, N.A., Connerade, J.-P., Greiner, W.: Hartree-Fock deformed Jellium calculations for metallic clusters. J. Chin. Chem. Soc. **48**, 419–426 (2001)
299. Lyalin, A.G., Shylakov, A.S.: Aplication of the atomic BrS to describe the resonance emission in K-Spectrum of Boron in Hexagonal BN crystal. Fizika Tverdogo Tela **36**, 3185–3193 (1994) (in Russian)

300. Lyalin, A.G., Solov'yov, A.V.: Polarizational Bremsstrahlung from atomic clusters. J. Radiat. Phys. Chem. **75**, 1358–1379 (2006)
301. Lyalin, A.G., Solov'yov, A.V., Greiner, W.: Comparative study of metal-cluster fission in Hartree-Fock and local-density approximations. Phys. Rev. A **65**, 043202 (2002)
302. Lyalin, A.G., Solov'yov, A.V., Greiner, W., Semenov, S.K.: Hartree-Fock approach for metal-cluster fission. Phys. Rev. A **65**, 023201 (2002)
303. Lyalin, A.G., Solov'yov, I.A., Solov'yov, A.V., Greiner, W.: Evolution of the electronic and ionic structure of Mg clusters with increase in cluster size. Phys. Rev. A **67**, 063203 (2003)
304. Lysaght, M., Kilbane, D., Murphy, N., Cummings, A., Dunne, P., O'Sullivan, G.: Opacity of Neutral and low ion stages of Sn at the wavelength 13.5 nm used in extreme-ultraviolet lithography. Phys. Rev. A **72**, 014502 (2005)
305. Macek, J.H., Briggs, J.S.: Collision broadening and molecular orbital X-rays. II. J. Phys. B: At. Mol. Opt. Phys. **7**, 1312–1322 (1974)
306. Madjet, M.E., Chakraborty, H.S., Rost, J.M., Manson, S.T.: Photoionization of C_{60}: a model study. J. Phys. B **41**, 105101 (2008)
307. Madjet, M., Guet, C., Johnson, W.R.: Comparative study of exchange-correlation effects on the electronic and optical properties of Alkali-Metal clusters. Phys. Rev. A **51**, 1327 (1995)
308. Manfredi, G., Hervieux, P.-A., Haas, F.: Nonlinear dynamics of electron-positron clusters. New J. Phys. **14**, 075012 (2012)
309. Prudnikov, A.P., Brychkov, Yu.A., Marichev, O.I.: Tables in integrals. In: More Special Functions, vol. 3, Gordon and Breach, New York (1989)
310. *Mathematica*®, Worfram Research Corp. http://www.wolfram.com
311. Matveentzev, A., Lyalin, A., Solov'yov, I.A., Solov'yov, A.V., Greiner, W.: On the applicability of Jellium model to the description of Alkali clusters. Int. J. Mod. Phys. E **12** 81–107 (2003)
312. G. Mazzitelli, G., Mattioli, M.: Ionization balance for optically thin plasmas: rate coefficients for Cu, Zn, Ga, and Ge Ions. At. Data Nucl. Data Tables **82**, 313–356 (2002)
313. McMahon, S.J., Hyland, W.B., Muir, M.F., Coulter, J.A., Jain, S., Butterworth, K.T., Schettino, G., Dickson, G.R., Hounsell, A.R., O'Sullivan, J.M., Prise, K.M., Hirst, D.G., Currel, F.J.: Biological consequences of nanoscale energy deposition near irradiated heavy atom nanoparticles. Sci. Rep. **1**, 18 (2011) (Corrigendum: *ibid.* **3**, 1725 (2013))
314. Meiwes-Broer, K. -H. (ed.): Metal clusters at surfaces: structure, quantum properties, physical chemistry. In: Springer Series in Cluster Physics. Springer, Berlin (2000)
315. Meron, M., Johnson, B.M., Jones, K.W., Schuch, R., Schmidt-Böcking, H., Tserruya, I.: Investigations of molecular orbitals with Accel-Decel beams. Nucl. Instrum. Methods B **10/11**, 64–68 (1985)
316. Meyerhof, W.E., Saylor, T.K., Lazarus, S.M., Little, W.A., Triplett, B.B., Chase, L.F., Jr.: Observation of Z = 70 Quasiatomic K X Rays from 30- and 60-MeV

35Br + 35Br collisions. Phys. Rev. Lett. **30**, 1279–1282 (1973) (Erratum: *ibid.* **32**, 502–502 (1974))
317. Meyerhof, W.E., Clark, D.L., Stoller, Ch., Morenzoni, E., Wölfli, W., Folkmann, F., Vincent, P., Mokler, P.H., Armbruste, P.: Molecular-orbital X-rays from 1 GeV ^{208}Pb + ^{208}Pb and ^{209}Bi + ^{209}Bi collisions. Phys. Lett. **70A**, 303–306 (1979)
318. Migdal, A.B.: Quadrupole and dipole γ-Radiation of nuclei. J. Phys. (USSR) **8**, 331 (1944)
319. Migdal, A.B.: Bremsstrahlung and pair production in condensed madia at high-energies. Phys. Rev. **103**, 1811–1820 (1956)
320. Migdal, A.B.: Qualitative Methods In Quantum Theory. Westview Press, USA (2000)
321. Mittleman, M.H., Myerscough, V.P.: Minimum moment required to bind a charged particle to an extended dipole. Phys. Lett. **23**, 545–546 (1966)
322. Mrówczyński, St.: Positronium Bremsstrahlung. Phys. Lett. **113A**, 17–20 (1985)
323. Mrówczyński, St.: Interaction of elementary atoms with matter. Phys. Rev. A **33**, 1549–1555 (1986)
324. Nahar, S.N., Pradhan, A.K.: Self-consistent R-Matrix approach to Photoionization and unified ElectronIon recombination. J. Radiat. Phys. Chem. **70**, 323–344 (2004)
325. Näher, U., Bjørnholm, S., Frauendorf, S., Garcias, F., Guet, C.: Fission of metal clusters. Phys. Rep. **285**, 245–320 (1997)
326. Nakamura, K. et al. (Particle Data Group): Review of particle physics. J. Phys. G: Nucl. Part. Phys. **37**, 075021 (2010)
327. Nakel, W.: The elementary process of Bremsstrahlung. Phys. Rep. **243**, 317 (1994)
328. Nasonov, N.N.: Collective effects in the polarization Bremsstrahlung of relativistic electrons in condensed media. Nucl. Instrum. Methods Phys. Res. B 1998 **145**, 19–24 (1998)
329. Nasonov, N., Zhukova, P., Sergienko, V.: Polarization Bremsstrahlung in a backward direction for medium structure diagnostics. J. Phys.: Conf. Ser. **236** 012017 (2010)
330. Newton, R.G.: Scattering theory of waves and particles. Springer, New York (1982)
331. Nikulin, V.K., Trzhaskovskaya, M.B.: Comment on "Energy-Dependent Excitation cross section measurements of the diagnostic lines of Fe XVII". Phys. Rev. Lett. **108**, 139301 (2012)
332. Nitta, H.: Coherent polarization radiation from crystals irradiated with relativistic charged particles. J. Radiat. Phys. Chem. **75**, 1397–1408 (2006)
333. Nordgren, J., Guo, J.: Instrumentation for soft X-ray emission spectroscopy. J. Electr. Spectr. Rel. Phenom. **110-111**, 1–13 (2000)
334. Obolensky, O.I., Pratt, R.H.: Review of structures in low-energy bremsstrahlung: Classical and quantum descriptions. J. Radiat. Phys. Chem. **75**, 1308–1318 (2006)

335. Obolensky, O.I., Pratt, R.H.: Comparison of theory and experiment for polarizational Bremsstrahlung. J. Radiat. Phys. Chem. **75**, 2239–2246 (2006)
336. Ohtsuki, Yoshi-Hiko: Charged beam interactions with solids. Taylor & Fransis, New York (1989)
337. Orban, I., Glans, P., Altun, Z., Lindroth, E., Källberg, A., Schuch, R.: Determination of the recombination rate coefficients for Na-like SiIV forming Mg-like SiIII. A&A **459**, 291–296 (2006)
338. Östling, D., Apell, S.P., Rosén, A.: Theory for collective resonances of the C60 molecule. Europhys. Lett. **21**, 539–544 (1993)
339. Ozawa, K., Chang, J.H., Yamamoto, Y., Morita, S., Ishii, K.: Atomic Bremsstrahlung produced by light-ion–atom collisions below 1 MeV/amu. Phys. Rev. A **33**, 3018–3022 (1986)
340. Palazzi, G.D.: High-energy Bremsstrahlung and electron pair production in thin crystals. Rev. Mod. Phys. **40**, 611–631 (1968)
341. Percival, C., Seaton, M.J.: The polarization of atomic line radiation excited by electron impact. Phil. Trans. R. Soc. Lond. A. **251**, 113–138 (1958)
342. Philibert, J., Weinryb, E.: Étude de lémission électronique secondaire dans la microsonde de Caslaing. Compt. Rend. **255**, 2757–2759 (1962)
343. Pindzola, M.S., Kelly, H.P.: Free-Free radiative absorption coefficient for the negative argon ion. Phys. Rev. A **14**, 204–210 (1976)
344. Porcel, E., Liehn, S., Remita, H., Usami, N., Kobayashi, K., Furusawa, Y., Le Sech, C., Lacombe, S.: Platinum nanoparticles: a promising material for future cancer therapy? Nanotechnology **21**, 085103 (2010)
345. Portillo, S., C.A. Quarles, C.A.: Absolute doubly differential Electron-Bremsstrahlung cross sections from free atoms. AIP Conf. Proc. **680**, 115 (2003)
346. Portillo, S., Quarles, C.A.: Absolute doubly differential cross sections for electron Bremsstrahlung from rare gas atoms at 28 and 50 keV. Phys. Rev. Lett. **91**, 173201 (2003)
347. Pratt, R.H.: Electron Bremsstrahlung spectrum. Comments At. Mol. Phys. **10**, 121–131 (1981)
348. Pratt, R.H.: Coupling of the radiation field to the electron-atom collision system. In: Lutz, H.O., Briggs, J.S., Kleinpoppen, H. (eds.), Fundamental Processes in Energetic Atomic Collisions, pp. 145–182. Plenum, New York (1984)
349. Pratt, R.H., Feng, I.J.: The electron Bremsstrahlung spectrum from neutral atoms and ions. In: Barnett, C.F., Harrison, M.F.A. (eds.) Applied Atomic Collision Physics. vol. 2, pp. 307–321. Academic Press, Orlando (1984)
350. Pratt, R.H., Feng, I.J.: Electron-Atom Bremsstrahlung. In: Craseman, B. (ed.), Atomic Inner-Shell Physics, pp. 533–580. Plenum Press, New York (1985)
351. Pratt, R.H., Shaffer, C.D., Avdonina, N.B., Tong, X.M., Florescu, V.: New developments in the theory of Bremsstrahlung. Nucl. Instrum. Methods Phys. Res. B **99**, 156–159 (1995)
352. Pratt, R.H., Tseng, H.K., Lee, C.M., Kissel, L., MacCallum, C., Riley, M.: Bremsstrahlung energy spectra from electrons of kinetic energy 1 keV $\leq T_1 \leq$

2000 keV Incident on Neutral atoms $2 \leq Z \leq 92$. At. Data Nucl. Data Tables **20**, 175–209 (1977) (Erratum: 1981. ibid. 26, 477-481 (1981))
353. Press, W.H., Flannery, B.P., Teukolsky, S.A., Vetterling, W.T.: Numerical Recipies. The Art of Scientific Computing. University Press, Cambridge (1988)
354. Quarles, C.A.: Bremsstrahlung: an Experimentalist's personal perspective on the postmodern. Era. Radiat. Phys. Chem. **59**, 159–169 (2000)
355. Quarles, C.A., Portillo, S.: Looking for polarization Bremsstrahlung in the midst of inner-shell ionization: analysis of the X-Ray spectrum in electron interactions with thin-film targets. In: Duggan, J.L., Morgan, I.L. (eds.) AIP Conference Proceedings vol. 475, pp. 174–177. AIP, New York (1999)
356. Quarles, C.A., Portillo, S.: Review of Bremsstrahlung experiments with free gas atom targets. Nucl. Instrum. Methods Phys. Res. B **241**, 14–18 (2005)
357. Quarles, C.A., Portillo, S.: Review of absolute doubly differential cross-section experiments and cross-section ratios for electron Bremsstrahlung from rare gas atom and thin-film targets. J. Radiat. Phys. Chem. **75**, 1187–1200 (2006)
358. Quarles, C.A., Pratt, R.H. (eds.): Bremsstrahlung: theory and experiment. J. Radiat. Phys. Chem. **75**, 1113–1430 (2006)
359. Rapoport, L.P., Zon, B.A., Manakov, N.L.: Theory of Multiphoton Processes in Atoms. Atomizdat, Moscow (1978) (in Russian)
360. Ratzig, A.A., Smirnov, B.M.: Reference Data on Atoms, Molecules and Ions. Springer, Berlin (1985)
361. A. Reinköster, Korica, S., Prumper, G., Viefhaus, J., Godehusen, K., Schwarzkopf, O., Mast, M., Becker, U.: The Photoionization and Fragmentation of C_{60} in the energy range 26–130 eV. J. Phys. B: At. Mol. Opt. Phys. **37**, 2135–2144 (2004)
362. Requena, S., Gonzales, D., Williams, S.: Angular dependence of Bremsstrahlung produced by 17.5-keV Electrons incident on thick Ag. Phys. Rev. A **83**, 022712 (2011)
363. Requena, S., Williams, S., Quarles, C.A: A comparison of the Bremsstrahlung yields from 53 keV electrons on gold targets produced by PENELOPE and experiment. Nucl. Instrum. Methods Phys. Res. B **268**, 3561–3563 (2010)
364. Richter, M., Meyer, M., Pahler, M., Prescher, T., Raven, E.V., Sonntag, B., Wetzel, H.E.: Experimental study of atomic 4d giant resonances by photoabsorption and photoelectron spectroscopy: Ba, La, and Ce. Phys. Rev. A **39**, 5666–5675 (1989)
365. Romanikhin, V.P.: The Brensstrahlung of electrons on atoms calculated in a broad range of photon energies. Zh. Eksp. Teor. Fiz. **121**, 286–290 (2002) (English translation: JETP **94**, 239–243 (2002))
366. Rotkin, V.V., Suris, R.A.: Spherical quantum well model of the C60 molecule. Mol. Mater. **4** 87–94 (1994)
367. Rotkin, V.V., Suris, R.A.: Calculation of electronic structure of Fullerene within spherical quantum well model. Fiz. Tverdogo Tela **36**, 3569–3581 (1994) (English translation: Solid State Phys. **36**, 1899–1905 (1994))

368. Rubio, A., Alonso, J.A., Blase, X., Balbas, L.C., Louie, S.G.: Ab Initio Photoabsorption spectra and structures of small semiconductor and metal clusters. Phys. Rev. Lett. **77**, 247–250 (1996) (Erratum: ibid. **77**, 5442 (1996))
369. Saris, F.W., Van der Weg, W.F., Tawara, T., Laubert, R.: Radiative transitions between Quasimolecular levels during energetic Atom-Atom collisions. Phys. Rev. Lett. **28**, 717–720 (1972)
370. Sattler, K.D.: Handbook of Nanophysics: Clusters and Fullerenes. CRC Press, Boca Raton (2010)
371. Sauter, F.: Über den atomaren Photoeffekt bei großer Härte der anregenden Strahlung. Ann. Phys. **401**, 217–248 (1931)
372. Sauter, F. Zur unrelativistischen Theorie des kontinuierlichen Röntgenspektrums. Ann. Phys. **410**, 486–496 (1933)
373. Schardt, D., Elsässer, T., Schulz-Ertner, D.: Heavy-Ion Tumor therapy: physical and radiobiological benefits. Rev. Mod. Phys. **82**, 383–425 (2010)
374. Sauter, F.: Über die Bremsstrahlung schneller Elektronen. Ann. Phys. **412**, 404–412 (1934)
375. Sech, C.L., Kobayashi, K., Usami, N., Furusawa, Y., Porcel, E., Lacombe, S.: Comment on "Therapeutic application of metallic nanoparticles combined with particle-induced X-ray emission effect". Nanotechnology **23**, 078001 (2012)
376. Seltzer, S.M., Berger, M.J.: Tables of Energy Losses and Ranges of Electrons and Positrons. NASA Special Publication No. 3012. NASA, Greenbelt, MD, 1964
377. Schuch, R., Schmidt-Böcking, H., Tserruya, I., Johnson, B.M., Jones, K.W., Meron, M.: X-ray spectroscopy of Cl-Ar quasi-molecular orbitals from $1s\sigma$–$2p\pi$ transitions. Z. Phys. A **320**, 185–189 (1985)
378. Scully, S.W.J., Emmons, E.D., Gharaibeh, M.F., Phaneuf, R.A., Kilcoyne, A.L.D., Schlachter, A.S., Schippers, S., Müller, A., Chakraborty, H.S.,. Madjet, M.E., Rost, J.M.: Photoexcitation of a volume plasmon in C_{60} ions. Phys. Rev. Lett. **94**, 065503 (2005)
379. Scully, S.W.J., et al.: Reply. Phys. Rev. Lett. **98**, 179602 (2007)
380. Shaffer, C. David, Pratt, R.H.: Comparison of relativistic partial-wave calculations of triply differential electron-atom bremsstrahlung with simpler theories. Phys. Rev. A **56**, 3653–3658 (1997)
381. Shaffer, C. David, Tong, Xiao-Min, Pratt, R.H.: Triply differential cross section and polarization correlations in electron bremsstrahlung emission. Phys. Rev. A **53**, 4158–4163 (1996)
382. Shulakov, A.S., Zimkina, T.M., Brajko, A.P., Radchenko, V.I., Fomichev, V.A.: Short-Wave Limit of the Bremsstrahlung spectra for 60–140 eV electrons in Lanthanum. Fiz. Tverd. Tela. **25**, 789–795 (1983)
383. Selby, K., Vollmer, M., Masui, J., Kresin, V., de Heer, W.A., Knight, W.D.: Surface plasma resonances in free metal clusters. Phys. Rev. B **40**, 5417–5427 (1989)
384. Selby. K., Kresin, V., Vollmer, M., Masui, J., de Heer, W.A., Scheidemann, A., Knight, W.D.: Photoabsorption spectra of sodium clusters. Phys. Rev. B **43**, 4565–4572 (1991)

385. Sobelman, I.I.: Atomic spectra and radiative transitions. Springer, Berlin (1979)
386. Sohval, A.R., Delvaille, J.P., Kalata, K., Schnopper, H.W.: Knock-on bremsstrahlung in heavy-ion collisions with thick targets. J. Phys. B: At. Mol. Phys. **8**, L426–L428 (1975)
387. Solov'yov, A.V.: Multipole series for polarizational BrS cross section. In: Dalgarno, A., Freund R.S., Lubell, M.S., Lucatorto, T.B. (eds.), 16th ICPEACS, Book of Abstracts (1989)
388. Solov'yov, A.V.: Polarizational Bremsstrahlung in collisions of slow atomic particles. Z. Phys. D **24**, 5–13 (1992)
389. Solov'yov, A.V.: 5 EPS Conf. on At. Mol. Phys., Abstracts of Contributed Papers. p. 711. Edinburgh, UK (1995)
390. Solov'yov, A.V.: Nuclear Physics Parallels in Atomic Cluster Physics. pp. 351–362 in Ref. [179] (2001)
391. Solov'yov, A.V.: Plasmon excitations in metal clusters and fullerenes. Int. J. Modern Phys. B **19**, 4143–4184 (2005)
392. Solov'yov, A.V., Ivanov, V.K., Polozkov, R.G.: Electron-positron quantum droplets. Eur. Phys. J D **40**, 313–316 (2006)
393. Solov'yov, I.A., Solov'yov, A.V., Greiner, W.: Structure and properties of small sodium clusters. Phys. Rev. A **65**, 053203 (2002)
394. Solov'yov, I.A., Solov'yov, A.V., Greiner, W.: Optical response of small magnesium clusters. J. Phys. B: At. Mol. Opt. Phys. **37**, L137–L145 (2004)
395. Sommerfeld, A.: Atomic Structure and Spectral Lines. Methuen & Co, London (1934)
396. Sommerfeld, A.: Über die Beugung und Bremsung der Elektronen. Annalen der Physik **403**, 257–330 (1931)
397. Sternheimer, R.M.: Electronic polarizabilities of ions from the Hartree-Fock wave functions. Phys. Rev. **96**, 951–968 (1954)
398. Storm, E.: Calculated Bremsstrahlung spectra from Thick Tungsten targets. Phys. Rev. A **5**, 2328–2338 (1972)
399. Surdutovich, E., Solovyov, A.V.: Multiscale physics of Ion-Beam cancer therapy. J. Phys. Conf. Ser. **373**, 012001 (2012)
400. Ter-Mikaelian, M.: Influence of the medium on electromagnetic processes at high energies. Erevan, USSR (1969) (in Russian) (English edition. Ter-Mikaelian, M.: High-Energy Electromagnetic Processes in Condensed Media. Wiley Interscience, New York (1972))
401. Tkachenko, A.A., Gnatchenko, E.V., Verkhovtseva, E.T.: Peculiarities in Bremsstrahlung in low-energy electron scattering from Xe. Optika i Spektroskopiya, **78**, 208–211 (1995) (English translation: Opt. Spectrosc. **78**, 183–185 (1995))
402. Trzhaskovskaya, M.B., Nikulin, V.K., Clark, R.E.H.: Radiative recombination and photoionization cross sections for heavy element impurities in plasmas. At. Data Nucl. Data Tables **94**, 71–139 (2008)
403. Trzhaskovskaya, M.B., Nikulin, V.K., Clark, R.E.H.: Radiative recombination and photoionization cross sections for heavy element impurities in plasmas: II.

Ions of Si, Cl, Ar, Ti, Cr, Kr, and Xe. At. Data Nucl. Data Tables **95**, 987–1050 (2009)

404. Tsai, Y.-S.: Pair production and Bremsstrahlung of charged leptons. Rev. Mod. Phys. **46** (1974) 815–851 (Erratum: ibid. **49** (1977) 421–423)
405. Tseng, H.K.: Nonrelativistic calculation of electron Bremsstrahlung angular distribution of recoiled electrons. Phys. Rev. A **39**, 1012–1015 (1989)
406. Tseng, H.K.: Nonrelativistic calculation of an elementary process of electron Bremsstrahlung from Atoms. Phys. Rev. A **40**, 6826–6830 (1989)
407. Tseng, H.K.: Relativistic calculation of an elementary process of electron Bremsstrahlung from atoms. J. Phys. B: At. Mol. Opt. Phys. **30**, L317–L321 (1997) (Corrigendum: ibid. **33**, 1471 (2000))
408. Tseng, H.K.: Relativistic calculation of the unpolarized triple-differential cross section and the polarization correlation of the electron Bremsstrahlung from atoms. J. Phys. B: At. Mol. Opt. Phys. **35**, 1129–1142 (2002)
409. Tseng, H.K., Pratt, R.H.: Comments on the calculation of relativistic Bremsstrahlung cross sections. Phys. Rev. A **1**, 528–531 (1970)
410. Tseng, H.K., Pratt, R.H.: Exact screened calculations of atomic-field Bremsstrahlung. Phys. Rev. A **3**, 100–115 (1971)
411. Tseng, H.K., Pratt, R.H.: Electron Bremsstrahlung energy spectra above 2 MeV. Phys. Rev. A **19**, 1525–1528 (1979)
412. Tseng, H.K., Pratt, R.H., Lee, C.M.: Electron Bremsstrahlung angular distributions in the 1–500 keV energy range. Phys. Rev. A **19** 187–195 (1979)
413. Tserruya, I., Schuch, R., Schmidt-Böcking, H., Barrette, J., Da-Hai, Wang, Johnson, B.M., Meron, M., Jones, K.W.: Interference effects in the Quasimolecular K X-Ray production probability for 10-MeV Cl^{16+}–Ar collisions. Phys. Rev. Lett. **50**, 30–33 (1983)
414. Tsytovich, V.N., Ojringel, I.M. (eds.): Polarizational Bremsstrahlung of Particles and Atoms. Nauka, Moscow (1987) (English translation by Plenum Press, New York (1993))
415. Turner, J.F., Fox, K.: Minimum dipole moment required to bind an electron to a finite dipole. Phys. Lett. **23**, 547–549 (1966)
416. Überall, H.: High-energy interference effect of Bremsstrahlung and pair production in crystals. Phys. Rev. **103**, 1055–1067 (1965)
417. Uggerhøj, U.I.: Private communication (2012)
418. Varfolomeev, A.A.: Coherent Non-elastic Scattering of Neutrino from electrons of a medium. Yad. Fizika **28**, 1034–1039 (1978) (in Russian)
419. Varfolomeev, A.A.: Coherent interaction of Neutrino with Dense medium. Yad. Fizika **31**, 1268–1275 (1980) (in Russian)
420. Varshalovich, D.A., Moskalev, A.N., Khersonskii, V.K.: Quantum Theory of Angular Momentum. World Scientific, New York (1988)
421. Verkhovtsev, A.V., Korol, A.V., Solov'yov, A.V., Bolognesi, P., Ruocco, A., Avaldi, L.: Interplay of the volume and surface plasmons in the electron energy loss spectra of C60. J. Phys. B: At. Mol. Opt. Phys. **45**, 141002 (2012)
422. Verkhovtsev, A.V., Korol, A.V., Solov'yov, A.V.: Formalism for collective excitations in fullerenes. Eur. Phys. J. D **66**, 253 (2012)

423. Verkhovtsev, A.V., Korol, A.V., Solov'yov, A.V.: Quantum and classical features of the photoionization spectrum of C_{60}. Phys. Rev. A, **88**, 043201 (2013)
424. Verkhovtsev, A.V., Polozkov, R.G., Ivanov, V.K., Korol, A.V., Solov'yov, A.V.: Hybridization-Related correction to the Jellium model for fullerenes. J. Phys. B: At. Mol. Opt. Phys. **45**, 215101 (2012)
425. Verkhovtseva, E.T., Gnatchenko, E.V.: X-Ray Bremsstrahlung in the scattering of intermediate energy electrons on atoms of inert gases. Low Temp. Phys. **28**, 270–278 (2012)
426. Verkhovtseva, E.T., Gnatchenko, E.V., Pogrebnyak, P.S.: Investigation of the connection between "Giant" resonances and "Atomic" Bremsstrahlung. J. Phys. B: At. Mol. Opt. Phys. **16**, L613–L616 (1983)
427. Verkhovtseva, E.T., Gnatchenko, E.V., Pogrebnyak, P.S., Tkachenko, A.A.: Electron-produced ultrasoft X-ray spectrum of Xe. J. Phys. B: At. Mol. Opt. Phys. **19**, 2089–2118 (1986)
428. Verkhovtseva, E.T., Gnatchenko, E.V., Zon, B.A., Nekipelov, A.A., Tkachenko, A.A.: Bremsstrahlung in electron scattering by Xenon. Zh. Eksp. Teor. Fiz. **98**, 797–807 (1990) (English Translation: JETP **71**, 443–448 (1990))
429. Verkhovtseva, E.T., Gnatchenko, E.V., Tkachenko, A.A., Zon, B.A.: Resonance structures in polarization Bremsstrahlung from electron–atom collisions. J. Radiat. Phys. Chem. **75**, 1115–1134 (2006)
430. Verweyen, A., Guthöhrlein, G.R, Gerhard E., Brühl, S., Rienecker, C., Sonntag, B: Atomic Bremsstrahlung of Xe, Ba and Eu in the energy range of 4d giant resonances. In: Abstracts of the 17th International Conference on X-Ray and Inner-Shell Processes, p. 170. Hamburg (1996)
431. Wallis, R.F., Herman, R., Milnes, H.W.: Energy levels of an electron in the field of a finite dipole. J. Mol. Spec. **4**, 51–74 (1960)
432. Wendin, G., Nuroh, K.: Bremsstrahlung Resonances and appearance-potential spectroscopy near the 3d thresholds in metallic Ba, La, and Ce. Phys. Rev. Lett. **39**, 48–51 (1977)
433. West, J.B., Woodruff, P.R., Codling, K., Houlgate, R.G.: The 4d, 5s and 5p partial photoionization cross sections of Xenon above the 4d threshold. J. Phys. B: At. Mol. Phys. **9**, 407–410 (1976)
434. Wightman, A.S.: Moderation of negative mesons in hydrogen I: moderation from high energies to capture by an H_2 molecule. Phys. Rev. **77**, 521–528 (1950)
435. Williams, S.: Polarizational Bremsstrahlung: a review. Chap. 3 in: Joon I. Jang (ed.) New Developments in Photon and Materials Research, Nova Science Publishers (2013)
436. Williams, S., Quarles, C.A. Absolute Bremsstrahlung yields at 135° from 53-keV Electrons on gold film targets. Phys. Rev. A **78**, 062704 (2008)
437. Wolf, H.W., Bruhn, R., Radler, K., Sonntag, B.: Atomic Character of the 4d-Absorption of Ce Metal: An experimental proof. Phys. Lett. **59A**, 67–69 (1976)
438. Yakhontov, V.L., Amusia, M.Ya.: Radiative double-electron capture in collisions of fast heavy ions with solid carbon targets. Phys. Rev. A **55**, 1952–1961 (1997)

439. Yakhontov, V.L.: Relativistic linear response wave functions and dynamic scattering tensor for the $ns_{1/2}$ states in Hydrogenlike atoms. Phys. Rev. Lett. **91**, 093001 (2003)
440. Yannouleas, C., Broglia, R.A.: Collective and single-particle aspects in the optical response of metal Microclusters. Phys. Rev. A **44**, 5793–5802 (1991)
441. Yatsyshin, P.I., Polozkov, R.G., Ivanov, V.K., Solov'yov, A.V.: Structure of electron-positron clusters in the Hartree-Fock approximation. Phys. Scripta **80**, 093001 (2009)
442. Yerokhin, V.A., Surzhykov, A.: Off-Resonant Dielectronic recombination in a collision of an electron with a heavy hydrogen-like atom. Phys. Rev. A **81**, 062703 (2010)
443. Yerokhin, V.A., Surzhykov, A.: Electron-Atom Bremsstrahlung: double-differential cross section and polarization correlations. Phys. Rev. A **82**, 062702 (2010)
444. Zapryagaev, S.A., Manakov, N.L., Palchikov, V.G.: The Theory of One- and Two-Electron Multicharged Ions. Energoatomizdat, Moscow (1985) (in Russian)
445. Zhdanov, V.P.: Static approximation in calculating the Bremsstrahlung of electrons on atoms and ions. Zh. Eksp. Teor. Fiz. **72**, 112–121 (1977) (English translation: Sov. Phys. – JETP, **46**, 57–61 (1977))
446. Zhdanov, V.P.: 1978. Bremsstrahlung spectra for 0.4–5 keV electrons colliding with neutral and low-ionized atoms. Fizika Plazmi **4**, 128–133 (1978) (in Russian)
447. Zimkina, T.M., Fomichev, V.A., Gribovskii, S.: Fizika Tverdogo Tela **15**, 2685–2688 (1973)
448. Zimkina, T.M., Fomichev, V.A., Gribovskii, S., Zhukova, I.I.: Peculiar features in the photoabsorption spectra of rare-earth metals from the Lanthanum group. Fizika Tverdogo Tela **9**, 1447–1449 (1967) (English Translation: Sov. Phys. – Solid State **9**, 1128 (1967))
449. Zimkina, T.M., Gribovskii, S.: The features of photoionization absorption in the ultrasoft X-ray region. J. Physique Coll. C-4, (Suppl. 10) **32**, 282–289 (1971)
450. Zimkina, T.M., Lyakhovskaya, I.I., Shulakov, A.S., Alaverdov, V.I., Rasuvaeva, M.V.: X-ray absorption and emission spectra of La and Th in compounds with O and F in the energy region from 60 to 140 eV. Fiz. Tverd. Tela **25** 26–32 (1983) (in Russian)
451. Zimkina, T.M., Shulakov, A.S., Brajko, A.P.: Intense resonance X-ray emission bands of La and Th in the region of 4d- and 5d- Thresholds of ionization. Fiz. Tverd. Tela **23** 2006–2009 (1981) (in Russian)
452. Zimkina, T.M., Shulakov, A.S., Brajko, A.P., Stepanov, A.P., Fomichev, V.A.: Structure of 4d-emission spectra of rare-earth elements from La to Yb. Fiz. Tverd. Tela. **25**, 1981–1986 (1984) (English translation: Sov. Phys. – Solid State **26**, 1201 (1984))
453. Zimkina, T.M., Shulakov, A.S., Brajko, A.P., Stepanov, A.P., Fomichev, V.A.: Atomic effects in emission spectroscopy of solid state shell levels. Izv. Akad. Nauk SSSR: Ser. Fiz. **48**, 1263–1272 (1984) (in Russian)

454. Zon, B.A.: Bremsstrahlung in collisions between electrons and atoms. Zh. Eksp. Teor. Fiz. **73**, 128–133 (1977) (English Translation: Sov. Phys. – JETP **46**, 65–68 (1976))
455. Zon, B.A.: Absorption of optical radiation in weakly ionized gases. Zh. Eksp. Teor. Fiz. **77**, 44–51 (1977) (English Translation: Sov. Phys. – JETP **50**, 23–26 (1979))
456. Zon, B.A.: Bremsstrahlung in intermediate-energy electron scattering by noble gas atom. Zh. Exp. Teor. Phys. **107**, 1176–1179 (1995) (English translation: JETP **80**, 655–656 (1995))

Index

A. V. Korol and A. V. Solov'yov, *Polarization Bremsstrahlung*,
Springer Series on Atomic, Optical, and Plasma Physics 80,
DOI: 10.1007/978-3-642-45224-6, © Springer-Verlag Berlin Heidelberg 2014

MIX
Papier aus verantwortungsvollen Quellen
Paper from responsible sources
FSC® C105338

If you have any concerns about our products,
you can contact us on
ProductSafety@springernature.com

In case Publisher is established outside the EU,
the EU authorized representative is:
Springer Nature Customer Service Center GmbH
Europaplatz 3, 69115 Heidelberg, Germany

Printed by Libri Plureos GmbH
in Hamburg, Germany